Pharmaceutical Applications of Cell and Tissue Culture to Drug Transport

NATO ASI Series

Advanced Science Institutes Series

A series presenting the results of activities sponsored by the NATO Science Committee, which aims at the dissemination of advanced scientific and technological knowledge, with a view to strengthening links between scientific communities.

The series is published by an international board of publishers in conjunction with the NATO Scientific Affairs Division

A	**Life Sciences**	Plenum Publishing Corporation
B	**Physics**	New York and London
C	**Mathematical and Physical Sciences**	Kluwer Academic Publishers
D	**Behavioral and Social Sciences**	Dordrecht, Boston, and London
E	**Applied Sciences**	
F	**Computer and Systems Sciences**	Springer-Verlag
G	**Ecological Sciences**	Berlin, Heidelberg, New York, London,
H	**Cell Biology**	Paris, Tokyo, Hong Kong, and Barcelona
I	**Global Environmental Change**	

Recent Volumes in this Series

Volume 218—Pharmaceutical Applications of Cell and Tissue Culture to Drug Transport
edited by Glynn Wilson, S. S. Davis, L. Illum, and Alain Zweibaum

Volume 219—Atherosclerotic Plaques: Advances in Imaging for Sequential Quantitative Evaluation
edited by Robert W. Wissler

Volume 220—The Superfamily of *ras*-Related Genes
edited by Demetrios A. Spandidos

Volume 221—New Trends in Pharmacokinetics
edited by Aldo Rescigno and Ajit K. Thakur

Volume 222—The Changing Visual System: Maturation and Aging in the Central Nervous System
edited by P. Bagnoli and W. Hodos

Volume 223—Mechanisms in Fibre Carcinogenesis
edited by Robert C. Brown, John A. Hoskins, and Neil F. Johnson

Volume 224—Drug Epidemiology and Post-Marketing Surveillance
edited by Brian L. Strom and Giampaolo P. Velo

Series A: Life Sciences

Pharmaceutical Applications of Cell and Tissue Culture to Drug Transport

Edited by

Glynn Wilson

SmithKline Beecham Pharmaceuticals
King of Prussia, Pennsylvania

S. S. Davis and L. Illum

University of Nottingham
Nottingham, United Kingdom

and

Alain Zweibaum

INSERM
Paris, France

Plenum Press
New York and London
Published in cooperation with NATO Scientific Affairs Division

Proceedings of a NATO Advanced Research Workshop on
Pharmaceutical Applications of Cell and Tissue Culture,
held September 4-9, 1989,
in Bandol, France

Library of Congress Cataloging-in-Publication Data

NATO Advanced Research Workshop on Pharmaceutical Applications of Cell and Tissue Culture (1989 : Bandol, France)
Pharmaceutical applications of cell and tissue culture to drug transport / edited by Glynn Wilson ... [et al.].
p. cm. -- (NATO ASI series. Series A, Life sciences ; v. 218)
"Proceedings of a NATO Advanced Research Workshop on Pharmaceutical Applications of Cell and Tissue Culture, held September 4-9, 1989, in Bandol, France"--T.p. verso.
"Published in cooperation with NATO Scientific Affairs Division."
Includes bibliographical references and index.
ISBN 0-306-44082-2
1. Pharmacology--Research--Methodology--Congresses. 2. Tissue culture--Congresses. 3. Cell culture--Congresses. 4. Human cell culture--Congresses. I. Wilson, Glynn. II. North Atlantic Treaty Organization. Scientific Affairs Division. III. Title. IV. Series.
[DNLM: 1. Biological Transport--drug effects--congresses. 2. Cells, Cultured--physiology--congresses. 3. Receptors, Drug--congresses. 4. Tissue Culture--congresses. QV 38 N278p 1989]
RM301.25.N38 1989
615'.7--dc20
DNLM/DLC
for Library of Congress 91-39349
CIP

ISBN 978-1-4757-0288-0 ISBN 978-1-4757-0286-6 (eBook)
DOI 10.1007/978-1-4757-0286-6

Softcover reprint of the hardcover 1st edition 1991
A Division of Plenum Publishing Corporation
233 Spring Street, New York, N.Y. 10013

SPONSORSHIP

Anotecs Separations Ltd., Banbury (GB)

Bayer AG Leverkusen (D)

Beecham Pharmaceuticals, London, (GB)

Ciba-Geigy, Horsham (GB)

Delagrange, Chilly-Mazarin (F)

Eli Lilly, Indianapolis (USA)

Glaxo Group Research, Ware (GB)

Lipha, Lyon (F)

Merck Sharp & Dohme, Hoddesdon (GB)

Millipore S.A. St. Quentin Yvelines (F)

Pfizer Central Research, Sandwich (GB)

Reckitt & Colman, Hull (GB)

Sandoz AG, Basle (S)

PREFACE

In recent years there have been rapid advances in the growth and differentiation of mammalian cells in culture. This has led to increasing use of such in vitro systems in a wide variety of studies on fundamental aspects of cell structure and function, including normal growth and metabolism, mechanisms of differentiation and oncogenesis, mechanisms of protein and membrane synthesis and cell polarity. Recent advances in our ability to grow cells, including human cells, on permeable supports, to generate confluent cellular barriers with the morphological polarity corresponding to their in vivo counterparts has greatly facilitated such studies. In particular these new techniques have led to an increasing interest in the use of cell and tissue culture systems as a means for examining the transport of drugs across epithelial and endothelial barriers. An obvious question is whether these new in vitro methodologies will provide convenient systems that can substitute for and replace animal models. Various research groups both in academia and in the pharmaceutical industry have been investigating these types of methodologies in order to develop convenient well characterized systems that can be used to examine basic aspects of transcellular transport and to evaluate the permeability of drug molecules and delivery systems. Of particular note is use of confluent cell layers to study the transport of large molecules such as peptides and proteins produced through recombinant DNA technology.

While there is much interest and increasing activity in this area many fundamental questions remain to be answered. Of prime importance is how well these systems resemble (morphologically and biochemically) normal cellular barriers, and to what extent they can be used to predict absorption and metabolism in vivo.

Although there continue to be rapid advances in our understanding of the cellular and molecular biology of epithelial and endothelial cells, our general understanding of the transport of a variety of molecules across cellular barriers is still very poor. Therefore, the development and characterization of cell and tissue culture methodologies, specifically to study drug transport, must be viewed against the background need to acquire fundamental information on transport processes and mechanisms. Thus, in many cases, the development and use of these new experimental systems for transport studies has proceeded without a clear understanding of many of their fundamental biological properties. This has meant that definitive characterization of such systems, to determine the relevance of their functional properties in relation to those of cells and tissues in vivo has not been possible in many cases.

To address these issues a three-day meeting (Advanced Research Workshop), sponsored by the NATO Science Foundation and various pharmaceutical and biotechnology companies, was held in Bandol, France during September 4-9, 1989, and this book comprises the various contributions presented. The purpose of the meeting was to bring together researchers from academic and industrial laboratories. To our knowledge this was the first time that these groups had been brought together for a meeting dedicated to this topic. A particular goal was to examine the realities of exploiting these systems for evaluating and predicting transport phenomena against the state of our knowledge on the structural and functional properties of such systems.

The book considers first the growth and characterization of cell and tissue culture systems, including the requirement for morphological polarity, differentiation, and functioning of specific transport pathways. The major biological barriers, i.e., the nasal, buccal and intestinal epithelium and the vascular endothelium, that separate the site of drug administration from site of drug action are examined in turn. In addition the importance of macrophage and hematopoietic cell systems is discussed. In each case the presenters have focused on the characterization of the respective in vitro systems and their use for transport or other functional studies. The effects of substrata or three-dimensional matrices together with cellular immortalization procedures is considered in several later chapters. The penultimate chapter contrasts the potential utility of new cell culture systems with established in vivo methodologies. A final chapter summarizes the views expressed at the panel discussion.

We wish to express our appreciation to the NATO Science Foundation for the financial support of the Workshop as part of the Double Jump Program and to the various pharmaceutical and biotechnology companies who also provided assistance. Thanks are also due to Joanne Miller for assistance in compiling this volume.

March 1, 1990

G. Wilson
S.S. Davis
L. Illum
A. Zweibaum

CONTENTS

PHARMACEUTICAL APPLICATIONS OF CELL CULTURE: AN OVERVIEW

Ronald T. Borchardt, Ismael J. Hidalgo, Kathleen M. Hillgren and Ming Hu

Department of Pharmaceutical Chemistry
The University of Kansas
Lawrence, KS 66045 USA

INTRODUCTION

A major challenge confronting the pharmaceutical scientist in the future will be the selective and efficient delivery of the next generation of drugs. Many of these drugs will be discovered by synthetic chemists through rational drug design or by molecular biologists through recombinant DNA technology. In rational drug design, drug candidates are developed with molecular characteristics that permit optimal interaction with the specific macromolecules (*e.g.*, receptor, enzyme) which mediates the desirable therapeutic effect.[1] However, rational drug design does not necessarily mean rational drug delivery, which strives to incorporate into a molecule the molecular properties necessary for the optimal delivery between the point of administration and the final target site in the body.[2]

DNA technology has for the first time made it possible to produce large quantities of highly pure proteins for pharmaceutical applications.[3] Many of these proteins are endogenous molecules (*e.g.*, insulin) which are very potent and very specific in producing their pharmacological effects. However, delivery of these proteins by routes other than the parenteral route is very problematic because of the inherent physicochemical properties of these molecules (*e.g.*, size, charge, hydrophilic characteristics) and their propensity to undergo metabolism at epithelial barriers (*e.g.*, intestinal mucosa).[4]

In an effort to develop novel strategies for delivery of drug candidates arising from rational drug design and recombinant DNA technology, some pharmaceutical scientists have begun to employ the technique of cell culture to study drug transport and metabolism at specific biological barriers.[5] These cell culture techniques afford the opportunity (a) to rapidly assess the potential permeability and metabolism of a drug; (b) to elucidate the mechanism(s) of drug transport and the pathway(s) of drug degradation (or activation); (c) to rapidly evaluate strategies for achieving drug targeting, enhancing drug transport and minimizing drug metabolism; (d) to employ human rather than animal tissues; and (e) to minimize time-consuming, expensive and sometimes controversial animal studies.

As described in detail in this book, pharmaceutical scientists have established and validated and are using cell culture systems that mimic many of the *in vivo* biological barriers that prevent efficient delivery of drugs from their point of administration to their final target site. These biological barriers, which

Pharmaceutical Applications of Cell and Tissue Culture to Drug Transport
Edited by G. Wilson *et al.*, Plenum Press, New York, 1991

are physical and/or metabolic, can be divided into the following general categories: (a) epithelial barriers; (b) endothelial barriers; and (c) cellular barriers. Included in the category of epithelial barriers would be the cells that constitute the intestinal, rectal, buccal, sublingual and nasal mucosa, as well as cells that constitute the epidermis of the skin. The endothelial barriers would include the cells that constitute organs such as the liver, lung and kidney, as well as the cells that constitute the reticular endothelial system (*e.g.*, macrophages) and the vasculature (*e.g.*, brain capillary endothelial cells). Cellular barriers would include the plasma membrane and various subcellular compartments (*e.g.*, lysosomes) which are not the intended target of the drug.

The objective of this chapter is to discuss the factors that are important in the selection and validation of the cell culture system which would best mimic these biological barriers. In addition, the factors important in selecting (a) the porous membrane that is used as a matrix onto which the cells are grown, and (b) the diffusion apparatus which is used to conduct the transport studies will be discussed. The influence of the culturing conditions and/or the media used to conduct the transport experiments will also be considered. To illustrate these points, the discussion will be focused on an adenocarcinoma cell line (Caco-2) which is being developed in several laboratories as an *in vitro* model of a polarized intestinal epithelium.

SELECTION AND VALIDATION OF A CELL CULTURE MODEL THAT MIMICS A BIOLOGICAL BARRIER

In order to successfully mimic a biological barrier with an *in vitro* cell culture system, the selection of the cell line becomes particularly important. For example, the transport and metabolic properties of cultured cells can vary depending on (a) whether the cells are primary cultures, passaged lines or transformed lines; (b) the number of times the cells have been passaged; (c) the phenotypic stability of the cell line; (d) the heterogeneity of the cell line; and (e) the inherent ability of the cell line to undergo differentiation. After selecting an appropriate cell culture system, the system's barrier properties, as well as its morphological and biochemical characteristics, need to be determined to insure that it exhibits the properties of the biological barrier. As an example, we describe below the characteristics of the Caco-2 system, which suggest that it could be an *in vitro* model of a polarized intestinal epithelium.

The epithelium of the small intestine, which is the major site of drug absorption, consists of a monolayer of cells having considerable cellular heterogeneity with respect to both morphology and function.[6] The intestinal epithelial cells of primary interest, from the standpoint of drug absorption and metabolism, are the villus cells, which are the fully differentiated cells located at the upper two-thirds of the villi. An *in vitro* cell culture system consisting of a monolayer of viable, polarized and fully differentiated villus cells, similar to that found in the small intestine, would be a valuable tool in the study of drug and nutrient transport and metabolism. Attempts to culture intestinal epithelial cells (*e.g.*, crypt cells) or to establish cell lines derived from enterocytes have not been successful.[7-9]

Recently, alternative approaches have been considered, which include the utilization of some transformed cell lines. Several human colon carcinoma cell lines (*e.g.*, Caco-2, HT-29, SW116, LS174T, SW-480)[10] have been reported to undergo varying degrees of enterocytic differentiation in culture. The most extensively studied cell lines have been the HT-29 and Caco-2. These cell lines, established by Jorgen Fogh[11,12] have received a great deal of attention in recent years because of their ability to express the morphological features of mature enterocytes or goblet cells.[13-15] They have been widely utilized as *in vitro* tools for studying intestinal epithelial differentiation and function.[16-20]

When HT-29 cells are cultured in the presence of glucose, they grow as a multilayer of unpolarized, undifferentiated cells and do not express any characteristics of enterocytes.[13,17] However, when the glucose in the medium is replaced with galactose, they express moderate enterocytic differentiation.[13] Caco-2 cells, on the other hand, undergo spontaneous enterocytic differentiation in culture,[14,18] which starts as soon as the cells achieve monolayer density (*i.e.*, 7 days) and is completed within 20 days. That Caco-2 cells form numerous domes spontaneously after reaching confluence is consistent with their ability to undertake transepithelial ionic transport.[13,17] The ability of Caco-2 cells to achieve a higher degree of enterocytic differentiation than that expressed by the HT-29 cell line and their spontaneous dome formation make this cell line a more relevant *in vitro* model for the investigation of intestinal differentiation and transport processes associated with intestinal cells.[15]

Recently our laboratory reported that Caco-2 cells could be grown onto collagen-coated polycarbonate membranes having a 3.0 µm pore size.[21] Caco-2 cells reached monolayer density 6-7 days after seeding onto collagen-coated polycarbonate membranes. Close examination of cell morphology between days 3 and 15 showed that as the length of time in culture and cell number increased, changes in cell dimensions also occurred. For example, from day 3 to day 16, cell height increased by 489% and cell width decreased by 42%. Electron microscopic examination revealed that by day 3 in culture, adjacent cells had formed occluding junctional complexes (Figure 1A); however, intercellular spaces were prominent. By day 6 in culture the lateral membranes of neighboring cells were strongly interdigitated and contained numerous desmosomes. Occluding junctional complexes were present only at the apical surface of the now functionally polarized cell monolayer. Concomitant with these observations was the progressive formation of a brush border with microvilli becoming more numerous and more organized. After 16 days in culture, Caco-2 cell monolayers consisted of cells ≈30 µm in height that possessed a morphology similar to that described for the simple, columnar epithelium of the small intestine (Figure 1B). Thus, the cells appeared to undergo differentiation from "crypt type cells" to "villus type cells."

Because epithelial cells have demonstrated an asymmetric distribution of alkaline phosphatase on their luminal surface, the specific activity of this enzyme in brush border membranes and ultrastructural distribution throughout time in culture as a measure of the degree of junctional polarity was determined.[21] Alkaline phosphatase activity increased two- to threefold from day 5 to day 20 in culture. Alkaline phosphatase cytochemistry showed that the majority of the enzyme activity was located at the apical membrane at day 6 in culture.

The Caco-2 cell monolayers grown onto polycarbonate membranes for 10 days exhibited a formidable barrier function, as judged by the leakage (<0.25%/hr) of lucifer yellow CH (mol. wt. 453), polyethylene glycol (mol. wt. 4,000), inulin (mol. wt. 5,000) and dextran (mol. wt. 70,000). Moreover, the lack of apical to basolateral or basolateral to apical diffusion of horseradish peroxidase (mol. wt. 40,000), despite its ability to penetrate the intercellular space after basolateral administration, was in agreement with similar observations by Phillips et al.[22] in rat *in vivo*.

In support of the morphologic and transport studies, monolayer integrity also was evaluated by measuring transepithelial electrical resistance (TEER). TEER values increased from 96.6 ± 22 at day 3 to 173.5 ± 10.9 Ωcm^2 at day 6 and remained constant through 30 days.[21] From these studies we concluded that Caco-2 cells grown onto collagen-coated polycarbonate membranes form a tight monolayer of polarized epithelial cells and thus represent a potential transport model system for the small intestinal epithelium.

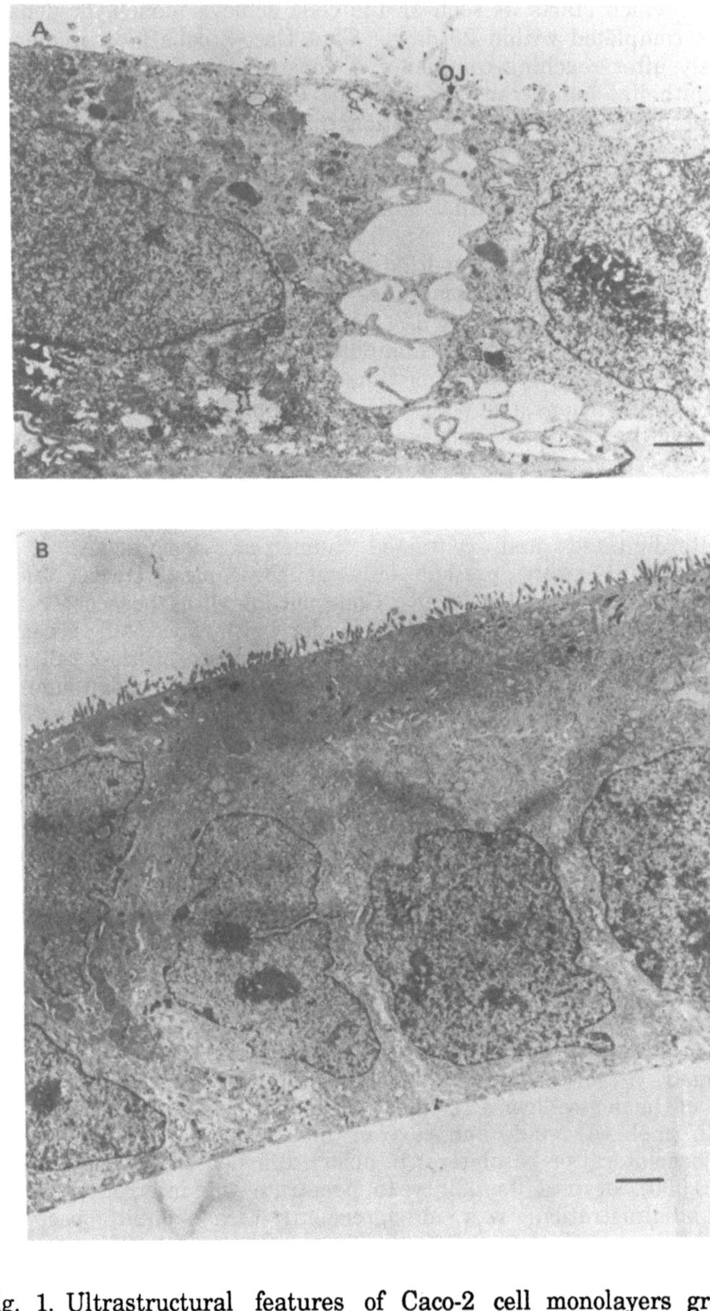

Fig. 1. Ultrastructural features of Caco-2 cell monolayers grown into microporous membranes. Panel A, at 3 days of culture, cells are nearly confluent as defined by the appearance of occluding functions; however, the lateral membranes of adjacent cells remain associated loosely. Panel B, at 6 days of culture, the monolayer is comprised of cells that have become columnar in shape with a well-defined brush border.

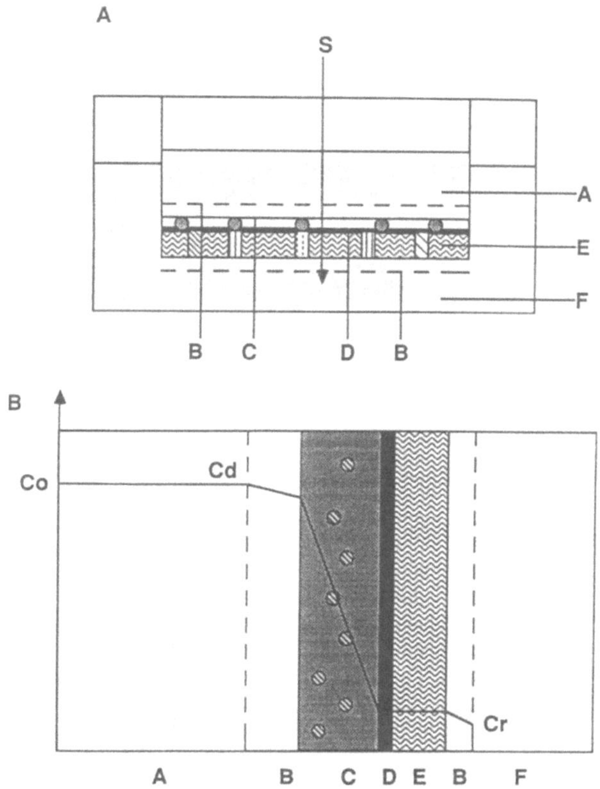

Fig. 2. Potential barriers to solute transport in a cell culture system grown onto a microporous membrane. Panel A, monolayer cell culture system grown onto a microporous membrane; Panel B, concentration profile for solute (**S**) with the largest concentration drop produced by the cell monolayer. A, apical side; B, aqueous boundary layer; C, cell monolayer; D, supporting matrix; E, microporous membrane; F, basolateral side; Co, original concentration of solute; Cd, donor side concentration; Cr, receiver side concentration.

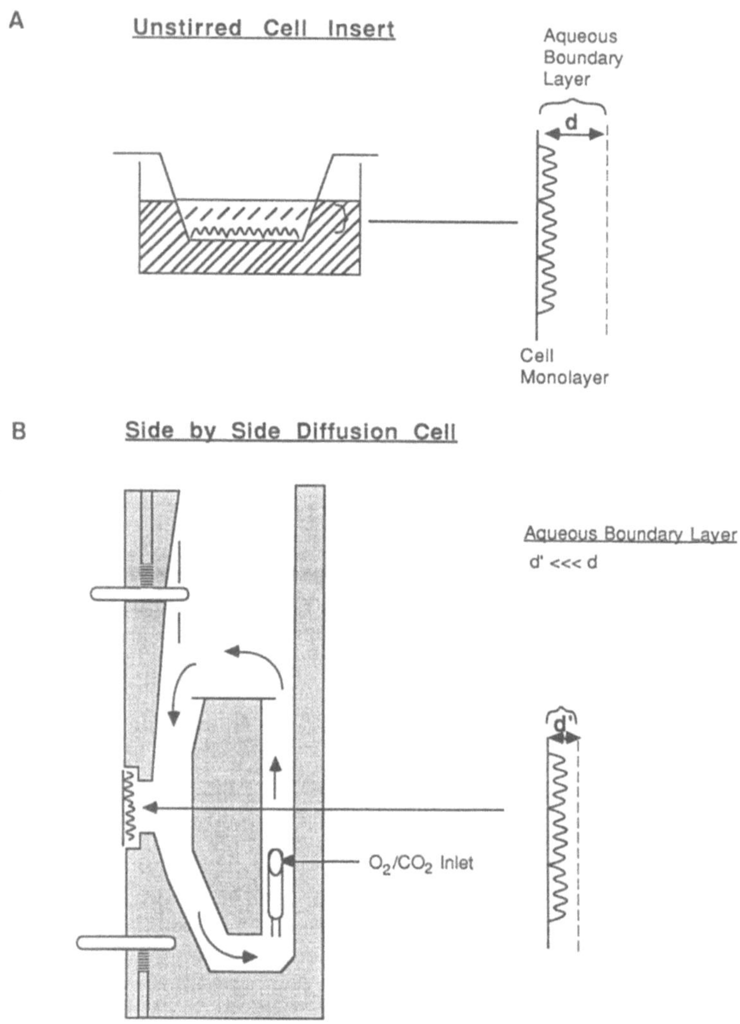

Fig. 3. Diagrammatic depiction of the effect of stirring on the aqueous boundary layer over a cell culture monolayer. Panel A, maximum aqueous boundary layer in a stagnant cell-insert device. Panel B, reduced aqueous boundary layer in a well-stirred side-by-side diffusion cell.

SELECTION AND VALIDATION OF THE MICROPOROUS MEMBRANE AND THE DIFFUSION APPARATUS

The development of a cell culture system for conducting transcellular transport studies requires a microporous membrane which, by itself and after treatment with an appropriate matrix material (*e.g.*, collagen), will support cell attachment and cell growth. Ideally, this microporous membrane should also be (a) sufficiently translucent so that the development of the cell monolayer can be verified by microscopic techniques; (b) readily permeable to hydrophilic and hydrophobic solutes; and (c) readily permeable to both low and high molecular weight solutes.

Many microporous membranes for cell culture (*e.g.*, polycarbonate, nitrocellulose) are now commercially available in different surface areas and pore sizes.[23] Careful selection of the microporous membrane, including the physicochemical properties of the membrane, its pore size and surface area, and the nature and thickness of the supporting matrix (*e.g.*, collagen) is critical in order to avoid generating artifactual data in transport experiments. As illustrated in Figure 2, the ideal diffusional characteristics of a cell culture model system are afforded when the major diffusional barrier is provided by the cell monolayer and not the microporous membrane or the supporting matrix (Fig. 2B). In conducting transport studies with cells cultured on microporous membranes, it is essential that control experiments be conducted using the microporous membrane alone and the microporous membrane coated with the supporting matrix (*e.g.*, collagen). The results of these experiments will indicate whether the solute is freely permeable through the microporous membrane and the supporting matrix and that the diffusional barrier is provided by the cell monolayer.

Another critical factor, particularly for lipophilic molecules, is the selection of the diffusion apparatus, which can influence the thickness of the aqueous boundary layer adjacent to the surface of the cell monolayer.[24] Many investigators have chosen to use the cell-inserts (polycarbonate membrane, 3 μm pore size, 4.71 cm^2 surface area) illustrated in Figure 2A. Unfortunately, these cell-insert systems are generally unstirred, thus creating a significant aqueous boundary layer over the cell monolayer (Fig. 3A). For very lipophilic compounds (*e.g.*, testosterone) this aqueous boundary layer could be the diffusional barrier to flux of the solutes across the cell culture model system, thus yielding artifactual data not representative of the *in vivo* situation.

Ideally, the diffusion apparatus should have mixing (stirring) capabilities which would allow for minimization of the thickness of the aqueous boundary layer over the cell monolayer. In most apparatus used for transcellular transport studies, mixing is performed by mechanical stirring.[25] However, a problem frequently encountered with this type of diffusion apparatus is the poor reproducibility of transport data resulting from unpredictable edge damage inflicted on the monolayers during the process of mounting the membrane in the diffusion apparatus and/or mechanical damage done to the monolayer during the transport experiment itself. For example, von Bonsdorff et al.[25] reported that approximately 80% of their Madin-Darby canine kidney (MDCK) cell monolayers grown onto nitrocellulose filters had to be discarded due to high leakage.

Recently our laboratory[24] developed a novel side-by-side diffusion apparatus which can be used for cell monolayers grown in miniaturized cell-inserts (polycarbonate membrane, 0.4 μm pore size, 0.33 cm^2 surface area) (Figure 4). In this apparatus the mixing, which is provided by gas lift (O_2/CO_2) and controlled at the appropriate flow rate by needle valves, should reduce significantly the thickness of the aqueous boundary layer (Figure 3B). To validate this diffusion apparatus, we have conducted experiments with the Caco-2 cell culture system, studying the effect of mixing on the morphology of the cell monolayer and on the permeability of hydrophilic (*e.g.*, [^{14}C]-mannitol) and lipophilic (*e.g.*, [^{3}H]-testosterone) solutes.[24]

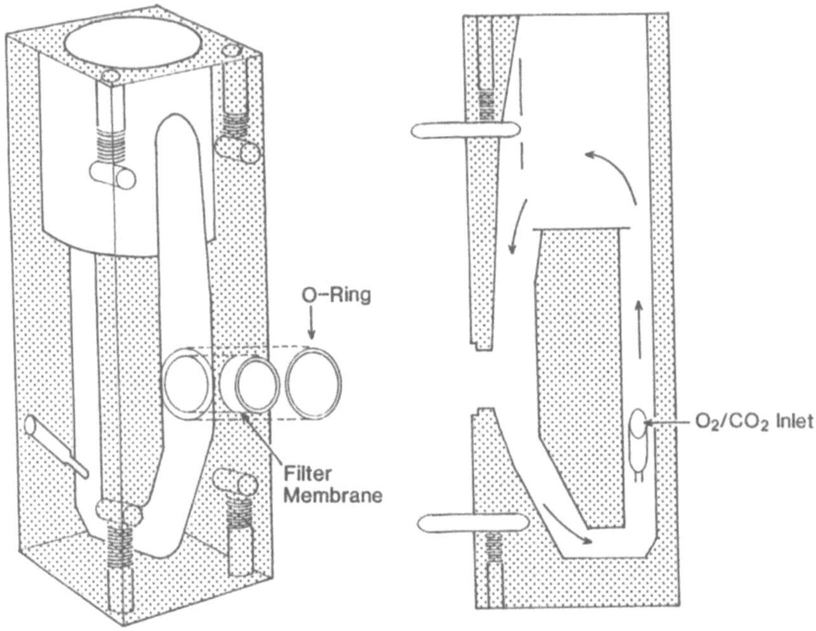

Fig. 4. Diagram of the side-by-side diffusion apparatus developed for cell culture systems. The inserts (polycarbonate membrane, 0.4 µm pore size, 0.33 cm^2) are mounted using O-rings and the media is stirred with O_2/CO_2 gas lift.

To determine whether the flow rate used to achieve mixing would result in cell damage, cell monolayers were examined by light microscopy and transmission electron microscopy. Examination of Caco-2 cell monolayers that were not exposed to flow (control) as well as monolayers that were subjected to the highest flow rate (40 mL/min) for 2 hr did not reveal any discernible gaps or holes in the monolayer, the shape and the density of the microvilli and the appearance of the tight junctions and desmosomes were unchanged, and the cells remained attached to the microporous membrane.[24] Using scanning electron microscopy (Figure 5), we were able to show that the characteristic morphological features of the surface of the cell monolayer (*e.g.*, cavities, floral-like pattern) were unaffected by the flow rate.

Further confirmation that the flow did not affect the integrity of the cell monolayer was obtained by determining the permeability of [^{14}C]-mannitol, a marker of paracellular flux. Even at the highest flow rates studied (40 mL/min), the permeability of [^{14}C]-mannitol was unchanged, suggesting that the flow did not damage the cell monolayer. In contrast to the permeability of [^{14}C]-mannitol, [^{3}H]-testosterone permeability was dramatically increased by increasing the flow rate (*e.g.*, flow rate (mL/min), permeability (cm/sec): 0, 1.04×10^{-6}; 15, 2.83×10^{-6}; 40, 3.38×10^{-6}). These data indicate that in unstirred monolayers a significant aqueous boundary layer exists which can influence the permeability of lipophilic solutes. This aqueous boundary layer can be reduced substantially by adjusting the flow rate of the O_2/CO_2 in the diffusion apparatus shown in Figure 4.

Therefore, in developing a cell culture model system, one needs to carefully select not only the cell line, but also an appropriate microporous membrane, a suitable supporting matrix, and a diffusion apparatus that will give accurate and reliable transport data.

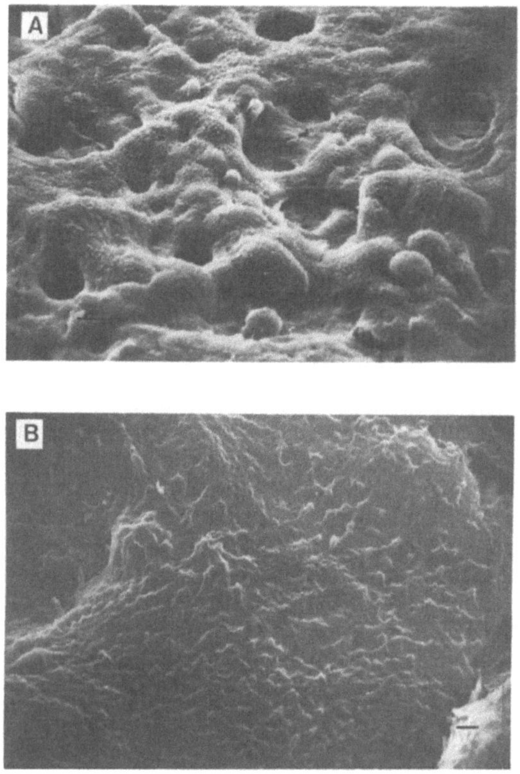

Fig. 5. Scanning electron micrograph (SEM) showing the surface of the Caco-2 cell monolayer after exposure to flow. Caco-2 cell monolayers were exposed to flow (40 mL/min) for 120 min in the apparatus shown in Fig. 4. SEMs show the characteristic morphological appearance of the surface of the Caco-2 cells (Panel A, cavities; Panel B, floral-like pattern).

INFLUENCE OF CULTURING CONDITIONS AND THE COMPOSITION OF TRANSPORT MEDIA ON THE PERMEABILITY CHARACTERISTICS OF THE CELL MONOLAYER

Once a cell line has been selected and validated as having the characteristics of the biological barrier of interest, it can then be used for transport and/or metabolism studies. However, it is important to consider that the properties of the cell monolayer may vary depending on the culturing conditions. The variables include (a) the cell seeding density; (b) whether the cells have reached confluency; (c) the stage of cellular differentiation; and (d) the presence or absence of essential nutrients, growth factors or associated cells that might influence the properties of the cell monolayer. The composition of the transport media (*e.g.*, concentration of the solute, temperature, pH, presence or absence of proteins that might bind the solute, presence or absence of competing solutes) can also influence the properties of the cell monolayer.

To illustrate how these factors can influence the permeability of a specific solute, we have described below the transport characteristics of phenylalanine (PHE) by the large neutral amino acid (LNAA) transport system in Caco-2 cells.[26,27] The apparent transport of PHE was dependent on the days in culture. For example, the apical-to-basolateral (AP → BL) transport of 0.1 mM PHE decreased from 1987.5 pmol mg protein^{-1} hr^{-1} at day 8 to 514 pmol mg protein^{-1} hr^{-1} at day 25, while the transport in the opposite direction (BL → AP) decreased from 359.3 pmol mg protein^{-1} hr^{-1} to 53.9 pmol mg protein^{-1} hr^{-1}. In addition to showing the dependency of PHE transport on days in culture, these data also illustrate the polarity of the transport mechanism where at day 25 the AP → BL transport of PHE is 12.5X faster than that in the BL → AP direction. The transport of PHE is also dependent on the number of times that the Caco-2 cells have been passaged. For example, the AP → BL transport of 2.5 mM PHE was 36 nmol mg protein^{-1} hr^{-1} at passage 26 versus 21.6 nmol mg protein^{-1} hr^{-1} at passage 62. The transport of PHE was also temperature dependent (E_a = 12.0 Kcal/mol) and concentration dependent (K_m = 0.45 mM, V_{max} =55.6 pmol mg^{-1} min^{-1}). The amount of PHE transported across a monolayer of Caco-2 cells is also dependent on the time that has elapsed since the cells were last fed (post-feeding time) (Figure 6). As shown in Figure 6, the AP → BL transport of PHE can also be influenced by the components in the transport media, including: (a) the presence or absence of glucose in the AP and/or BL side; (b) the pH gradient across the cell monolayer; and (c) the concentration of unlabeled PHE on the AP side. In addition to unlabeled PHE on the AP side inhibiting the AP → BL transport of labeled PHE, this transporter is also sensitive to inhibition by other amino acids (*e.g.*, leucine, lysine, histidine, tryptophan and tyrosine) and amino acid-type drugs (*e.g.*, L-DOPA, α-methyl DOPA) (Hidalgo and Borchardt, unpublished data).

Since the transport of PHE in Caco-2 cells is carrier-mediated, the system is particularly sensitive to the stage of differentiation of the cells (*e.g.*, passage number and days in culture) and the properties of the transport media (*e.g.*, pH, presence or absence of glucose, amino acids, etc.). However, even with solutes that are transported by passive mechanisms, one needs to be concerned with variables such as whether the cells have reached confluency, and the components of transport media.

Thus, in designing transport experiments using a cell culture model system grown onto a microporous membrane, attempts should be made to mimic as closely as possible the environment which exists at the *in vivo* biological barrier of interest.

PHARMACEUTICAL APPLICATIONS OF CELL CULTURE

For many years cell culture systems have been employed by pharmaceutical scientists to assess the pharmacological and toxicological effects of drugs, to elucidate the mechanisms by which drugs produce their biological effects, and to identify potential molecular targets for rational drug design. However, only recently have pharmaceutical scientists begun to employ cell culture systems to assess the potential permeability and metabolism of drugs, to elucidate the mechanisms of drug transport and drug metabolism, and to evaluate strategies for achieving drug targeting, enhancing drug transport and minimizing drug metabolism at specific biological barriers.

The cell culture systems grown onto microporous membranes have significant potential to aid in elucidating the various pathways by which drugs penetrate biological barriers. These pathways, illustrated in Figure 7, include: (a) passive diffusion; (b) active or facilitated diffusion (carrier-mediated); (c) paracellular diffusion; (d) vesicular-mediated transcytosis (fluid phase, adsorptive, receptor-mediated); (e) altered paracellular diffusion (glucose- or adjuvant-induced); and (f) passive diffusion with incorporation into lipoprotein particles.

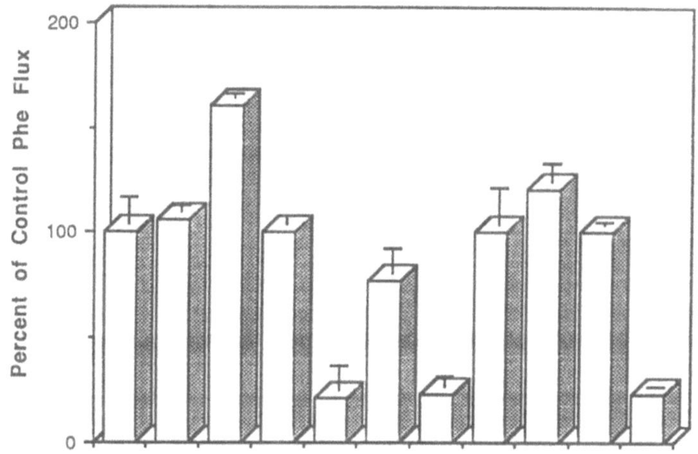

P.F.T (hr)	3.5	24	49	24	24	24	24	24	24	24	24
Glucose AP	+	+	+	+	-	+	-	+	+	+	+
Glucose BL	+	+	+	+	+	-	-	+	+	+	+
pH AP	7.4	7.4	7.4	7.4	7.4	7.4	7.4	7.4	5.5	7.4	7.4
pH BL	7.4	7.4	7.4	7.4	7.4	7.4	7.4	7.4	7.4	7.4	7.4
Unlabeled Phe (mM)	0.01	0.01	0.01	0.01	0.01	0.01	0.01	0.01	0.01	0.01	10

Fig. 6. Effects of pH, glucose, amino acids and post-feeding time on transport of phenylalanine across monolayers of Caco-2 cells. Experiments were performed in cell-inserts. Transcellular transport of C-14 labeled phenylalanine was monitored at the basolateral side of the cell inserts. The glucose concentration was 25 mM when present. Arabinose (25 mM) was used as osmotic agent when glucose was not present. P.F.T., post-feeding time, is the time between last feeding and the actual experiment. The osmolarity of the experimental media is approximately 300 mOsm/kg.

As an example, we have illustrated below how pharmaceutical scientists are using human adenocarcinoma cells (*e.g.*, Caco-2, HT-29) as model systems of the polarized intestinal epithelium. Initially Caco-2 cells, as well as HT-29 cells, were developed as an *in vitro* model for studying intestinal epithelial differentiation and function.[16-20] However, more recently the Caco-2 cell line has been used to elucidate the mechanism of epithelial transcellular transport of nutrients, including bile acids[28] and large neutral amino acids.[26,27] The characteristics of these carrier-mediated transport systems (Figure 7, Pathway B) are described in detail by Hidalgo and Borchardt elsewhere in this book.

Recently the Caco-2 cell culture system was utilized to study the binding and transport of proteins including epidermal growth factor (EGF)[29] and cobalamin (Vitamin B12) (as described elsewhere in this book by Hassan et al.) via receptor-mediated processes (Figure 7, Pathway D). The Caco-2 cell system has also been shown to be a useful model to study the production and secretion of lipoprotein particles in a polarized epithelial cell system.[30,31] The results of these studies suggest that the Caco-2 system might be useful for the study of the initial step (incorporation into lipoprotein particles) by which highly lipophilic drugs enter the lymphatic system (Figure 7, Pathway F).

Pharmaceutical scientists have begun to use the Caco-2 cell culture system to study drug transport and drug metabolism. Artursson et al. (as described elsewhere in this book) have used the Caco-2 system to study the relative contribution of the paracellular (Figure 7, Pathway C) and transcellular (Figure 7, Pathway A) pathways of the absorption of beta-adrenergic antagonists. Hu and Borchardt (unpublished data) used this cell culture system to study the mechanism of transcellular transport of the antihypertensive agent, α-methyl DOPA. This amino acid derivative was shown to be transported by the large neutral amino acid transport system (Figure 7, Pathway B) in the Caco-2 cells. Dantzig et al.[32] has used the HT-29 cell line to study the uptake of the antibiotic cefalexin by a dipeptide transport system (Figure 7, Pathway B). In addition to being used as a model for studying drug transport, the Caco-2 system may be useful for the study of intestinal metabolism. For example, Caco-2 cells have been shown to have the enzymes alkaline phosphatase, aminopeptidase N and sucrase-isomaltase.[14] The activity of sucrase-isomaltase in Caco-2 cells was nearly identical to that found in the ileum of adult humans.[16] A recent study from our laboratory has shown the presence of phenolsulfotransferase in Caco-2 cells (Baranczyk-Kuzma et al., unpublished data), an enzyme important in Phase II drug metabolism.

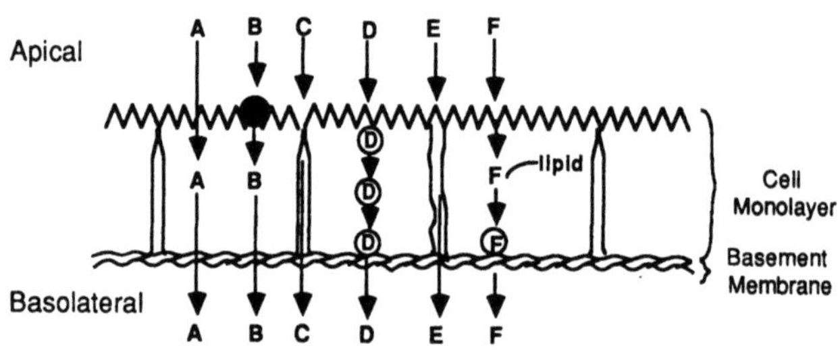

Fig. 7. Possible pathways for drug transport in a polarized epithelial cell monolayer. Pathways: A, passive diffusion; B, active or facilitated diffusion (carrier-mediated); C, paracellular diffusion; D, vesicular-mediated transcytosis (fluid phase, adsorptive, receptor-mediated); E, "altered" paracellular diffusion; F, passive diffusion with incorporation into lipoprotein particles.

CONCLUSIONS

The use of cell culturing techniques to study the transport and metabolic characteristics of specific biological barriers to drug delivery is in its infancy. However, as pharmaceutical scientists develop more sophisticated systems, establish the crucial *in vitro-in vivo* correlations to validate the cell culture models, and learn to appreciate the potential advantages of these techniques, it is likely that these systems will find widespread utility in the pharmaceutical industry as tools in drug discovery. In addition, these techniques are likely to be very useful to pharmaceutical scientists interested in elucidating mechanisms of drug transport and in devising novel strategies to enhance drug permeability.

ACKNOWLEDGEMENTS

The authors' research on cell culture models for drug transport and metabolism has been supported by grants from The Upjohn Company, American Heart Association and Merck Sharp & Dohme-Inter$_X$.

REFERENCES

1. G. R. Marshall, Computer-aided drug design, Ann. Rev. Pharmacol. Toxicol. 27:193-213 (1987).
2. E. Tomlinson, Theory and practice of site-specific drug delivery, Adv. Drug Del. Rev. 1:87-198 (1987).
3. D. Blohm, C. Bollschweiler, and H. Hillen, Protein pharmaceuticals, Angew. Chem. Int. Ed. Eng., 27:207-225 (1988).
4. V. H. L. Lee., ed., "Peptide and Protein Drug Delivery: Fundamentals and Technology," Marcel Dekker, New York (1990).
5. G. Wilson, S. S. Davis, and L. Illum, eds., "Pharmaceutical Applications of Cell and Tissue Culture," Plenum, New York (1990).
6. J. Madara and J. S. Trier, Functional morphology of the mucosa of the small intestine, in: "Physiology of the Gastrointestinal Tract," L. R. Johnson, ed., Vol. 2, pp. 1209-1249, Raven, New York (1987).
7. L. Lichtenberger, L. R. Miller, D. N. Erwin, and L. R. Johnson, Effect of pentagastrin on adult rat duodenal cells in culture, Gastroenterology 65:242-251 (1973).
8. F. Raul, M. Kedinger, P. Simon, J. Grenier, and K. Haffen, Behaviour of isolated rat intestinal cells maintained in suspension or monolayer cultures, Biol. Cell 33:163-168 (1978).
9. A. Quaroni, J. Wands, R. L. Trelstad, and K. J. Isselbacher, Epithelioid cell cultures from rat small intestine. Characterization by morphologic and immunologic criteria, J. Cell Biol. 80:248-265 (1979).
10. Y. S. Chung, I. S. Song, R. H. Erickson, M. H. Sleisenger, and Y. S. Kim, Effect of growth and sodium butyrate on brush border membrane-associated hydrolases in human colorectal cancer cell lines, Cancer Res. 45:2976-2982 (1985).
11. J. Fogh and G. Trempe, New human tumor cell lines, in: "Human Tumor Cells in vitro," J. Fogh, ed., pp. 115-141, Plenum, New York (1975).
12. J. Fogh, J. M. Fogh, and T. Orfeo, One hundred and twenty-seven cultured human tumor cell lines producing tumors in nude mice, J. Natl. Cancer Inst. 59:221-225 (1977).
13. M. Pinto, M.-D. Appay, P. Simon-Assmann, G. Chevalier, N. Dracopoli, J. Fogh, and A. Zweibaum, Enterocytic differentiation of cultured human colon cancer cells by replacement of glucose by galactose in the medium, Biol. Cell 44:193-196 (1982).
14. M. Pinto, S. Robine-Leon, M.-D. Appay, M. Kedinger, N. Triadou, E. Dussaulx, B. Lacroix, P. Simon-Assmann, K. Haffen, J. Fogh, and A. Zweibaum, Enterocyte-like differentiation and polarization of the human colon carcinoma cell line Caco-2 in culture, Biol. Cell 47:323-330 (1983).
15. M. Rousset, The human colon carcinoma cell lines HT-29 and Caco-2: two in vitro models for the study of intestinal differentiation, Biochimie 68:1035-1040 (1986).
16. A. Zweibaum, N. Triadou, M. Kedinger, C. Augeron, S. Robine-Leon, M. Pinto, M. Rousset, and K. Haffen, Sucrase-isomaltase: A marker of foetal and malignant epithelial cells of the human colon, Int. J. Cancer 32:407-412 (1983).
17. A. Zweibaum, M. Pinto, G. Chevalier, E. Dussaulx, N. Triadou, B. Lacroix, K. Haffen, J.-L. Brun, and M. Rousset, Enterocytic differentiation of a subpopulation of the human colon tumor cell line HT-29 selected for growth in sugar-free medium and its inhibition by glucose, J. Cell Physiol. 122:21-29 (1985).
18. M. Rousset, M. Laburthe, M. Pinto, G. Chevalier, C. Rouyer-Fesard, E. Dussaulx, G. Trugnan, N. Boige, J.-L. Brun, and A. Zweibaum, Enterocytic differentiation and glucose utilization in the human colon tumor cell line Caco-2: modulation by forskolin, J. Cell Physiol. 123:377-385 (1985).

19. E. Grasset, M. Pinto, E. Dussaulx, A. Zweibaum, and J.-F. Desjeux, Epithelial properties of the human colonic carcinoma cell line Caco-2: Electrical parameters, Am. J. Physiol. 247:C260-C267 (1984).
20. A. Le Bivic, M. Hirn, and H. Reggio, HT-29 cells are an in vitro model for the generation of cell polarity in epithelia during embryonic differentiation, Proc. Natl. Acad. Sci. USA 85:136-140 (1988).
21. I. J. Hidalgo, T. J. Raub, and R. T. Borchardt, Characterization of the human colon carcinoma cell line (Caco-2) as a model system for intestinal epithelial permeability, Gastroenterology 96:736-749 (1989).
22. T. E. Phillips, T. L. Phillips, and M. Neutra, Macromolecules can pass through occluding junctions of rat ileal epithelium during cholinergic stimulation, Cell Tissue Res. 247:547-554 (1987).
23. M. V. Shah, K. L. Audus, and R. T. Borchardt, The application of bovine brain microvessel endothelial-cell monolayers grown onto polycarbonate membranes in vitro to estimate the potential permeability of solutes through the blood brain barrier, Pharm. Res. 6:624-627 (1984).
24. I. J. Hidalgo, K. M. Hillgren, G. M. Grass, and R. T. Borchardt, Characterization of the aqueous boundary layer in Caco-2 cells using a novel diffusion cell, Pharm. Res. 6:000 (Abstr.) (1989).
25. C.-H. Von Bonsdorff, S. D. Fuller, and K. Simons, Apical and basolateral endocytosis in Madin-Darby canine kidney (MDCK) cells grown on nitrocellulose filters, EMBO J. 4:2781-2792 (1985).
26. I. J. Hidalgo and R. T. Borchardt, Amino acid transport in a novel model system of the intestinal epithelium (Caco-2 cells), Pharm. Res. 5:S110 1988) (Abstr.).
27. M. Hu and R. T. Borchardt, Effect of pH and glucose on L-phenylalanine transport across an intestinal epithelial cell model system (Caco-2), Pharm. Res. 6:000 (Abstr.) (1989).
28. I. J. Hidalgo and R. T. Borchardt, Transport of taurocholic acid in an intestinal epithelial model system (Caco-2 cell), Pharm. Res. 5:S110 (Abstr.) (1989).
29. I. J. Hidalgo, A. Kato, and R. T. Borchardt, Binding of epidermal growth factor by human colon carcinoma cell (Caco-2) monolayers, Biochem. Biophys. Res. Commun. 160:317-324 (1989).
30. T. E. Hughes, W. V. Sasak, J. M. Ordovas, T. M. Forte, S. Lamon-Fava, and E. J. Schaefer, A novel cell line (Caco-2) for the study of intestinal lipoprotein synthesis, J. Biol. Chem. 262:3762-3767 (1987).
31. M. G. Traber, H. J. Kayden, and M. J. Rindler, Polarized secretion of newly synthesized lipoproteins by the Caco-2 human intestinal cell line, J. Lipid Res. 28:1350-1363 (1987).
32. A. Dantzig and L. Bergin, Carrier-mediated uptake of cephalexin in human intestinal cells, Biochem. Biophys. Res. Commun. 155:1082-1087 (1988).

GROWTH AND CHARACTERIZATION OF CELL AND TISSUE CULTURES FOR THE STUDY OF DRUG TRANSPORT

Glynn Wilson

Department of Drug Delivery, SmithKline
Beecham Pharmaceuticals, 709 Swedeland
Road, King of Prussia, PA 19406, USA

INTRODUCTION

Recent advances in cell and tissue culture methodologies, particularly the growth of differentiated human cells, are providing new and potentially powerful tools for cellular and molecular biology studies on the processes involved in differentiation and in the intracellular sorting of molecules and membranes. These techniques are also beginning to be used for investigating the transport of drug molecules and delivery systems into specific cells and across specific biological barriers. A particular need is to have in vitro systems that can be employed to devise new strategies for the absorption and delivery of new drug classes e.g. peptides and proteins. Cell and tissue culture systems provide the potential for rapidly evaluating the permeability and metabolism of a drug, for defining the mechanisms of transport of drugs and delivery systems, and for testing novel strategies for enhancing drug transport and drug targeting. In addition they may provide the opportunity to use human rather than animal tissues. Cell and tissue culture systems, that display many of the morphological and functional properties of in vivo cell layers, have been established. These include epithelial barriers that form the intestinal, rectal, buccal, sublingual and nasal mucosae; cells that form the epidermis of the skin, and vascular endothelial barriers e.g. brain capillary endothelial cells. This review will discuss factors that influence the growth and characterization of in vitro systems as it pertains to their use in studies on the transport of drug molecules. Intestinal epithelial systems will be used to illustrate specific points.

GROWTH AND CHARACTERIZATION OF CELL CULTURE SYSTEMS

The evolution of filter-chamber cultures based on the design of mini-Marbrook chambers (1) has provided a major impetus to the development of cell culture systems for studying the transport of drugs and drug delivery systems across biological barriers. In these systems cell monolayers can be grown on microporous filters contained within a filter chamber (fig.1).

Separation of the culture fluids on the apical and basolateral sides of the cell layer allows transport studies to be performed in either direction i.e. from the apical (mucosal) to the basolateral (serosal) side and vice versa. A number of variations on the basic theme (shown in fig.1), which differ in the composition of the microporous membrane, the design of the filter chamber, and whether the system is stagnant or stirred, have been used (2). The growth of cell monolayers in these systems has provided novel techniques for exploring the fundamental aspects of the transport of drugs and delivery systems across defined cellular barriers, and for discriminating

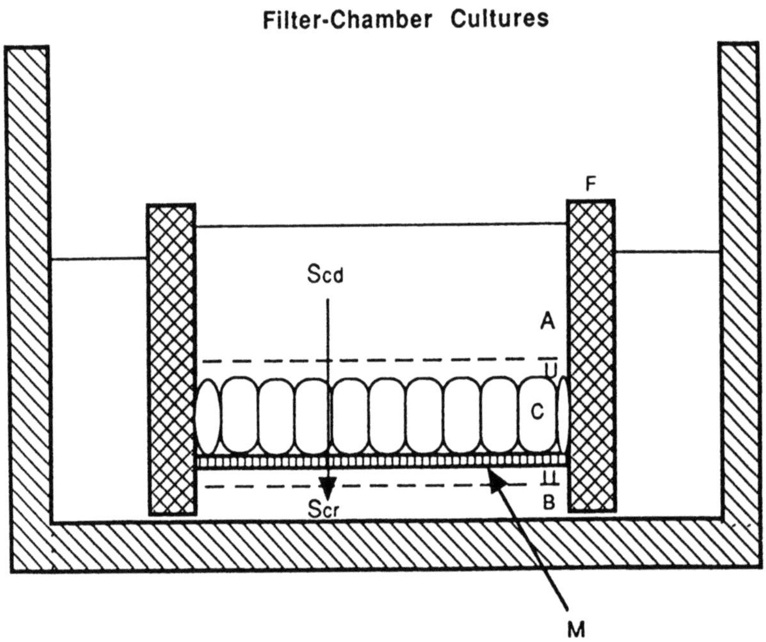

FIG. 1. Monolayers of cells grown on microporous membranes in chambers. A, apical culture medium; B, basolateral culture medium; C, cell monolayer; M, microporous membrane; U, unstirred water layer; Scd, solute concentration donor side; Scr, solute concentration receiver side; F, filter chamber.

between different routes and mechanisms shown. The establishment of in vitro cell culture systems that mimic normal biological barriers in vivo requires careful selection of the source of cells that will produce a relevant system when grown under defined culture conditions (see discussion by Fogh in this volume), and rigorous characterization of the morphological and functional properties of the cell layers obtained. The general requirements for cell and tissue culture systems are shown in Table 1. It is clear from many previous studies that the transport and metabolic properties of cultured cells can be greatly influenced by a number of variables pertaining to the cells themselves or the environment of the cell culture system (Table 1).

For example the source of cells can be primary cultures, passaged lines, or transformed lines. The number of times the cells have been passaged, the phenotypic stability of the cell line, the heterogeneity of the cell line, and the ability of the cell line to undergo differentiation can also have profound effects on the properties of the final culture system. In addition the properties of cultured cells can also depend on a number of components in the cell culture system, and the presence of other cell types.

TABLE I. Factors Affecting the Growth of Cellular Barriers in Culture*

CELL RELATED	CULTURE SYSTEM RELATED
Primary culture or stable cell line	Media components
Normal or transformed	Seeding density
Differentiation potential	Microporous membrane properties
Passage number	Hydrodynamic forces
Cell heterogeneity	Extracellular matrix
Viability	Feeding regimen
Phenotypic stability	Cost

* cf references 1,3,9,10,11,29

Establishment of the physiological relevance of in vitro systems for transport studies requires rigorous characterization of the cell or tissue layer. A number of criteria (Table 2) have been used to compare the morphological, biochemical and transport properties of in vitro biological barriers with those of the equivalent barrier in vivo. Most of these properties reflect the attainment of cell polarity and the presence of tight junctions between adjacent cells (3-6). While many of the parameters shown in Table 2 are routinely used to characterize cellular barriers, an important consideration in characterizing systems for transport studies is to have well defined in vivo criteria for comparison. A number of the criteria used, e.g. aspects of the biosynthetic and endocytic pathways of proteins and carrier-mediated transport pathways, have not yet been well characterized in vivo, and are currently the focus of basic cell and molecular biology research. Thus, the properties of these pathways cannot yet be used as rigorous criteria for characterizing cellular barriers in vitro. It is also important to select parameters most relevant to specific aims. For example, optimal transcellular transport of cobalamin across human Caco-2 cells occurs at a later time during the cell culture than either the development of maximal electrical resistance across the monolayers, or the optimal expression of brush border hydrolases (7). Therefore, these properties, in isolation, are not appropriate in characterizing the Caco-2 system for measuring cobalamin transport.

TABLE II. Parameters for the Characterization of In Vitro Systems

Parameter	Measurement
Morphology	Polarity, ultrastructure
Tight-junctions	Presence, permeability
Ion transport	Isc, PD
Metabolic properties	Glycolysis, proteolysis
Cell surface markers	Properties & polarity of enzymes, receptors, lipids
Biosynthesis	Structure and fidelity of sorting of secretory, cytoplasmic and membrane proteins, lipoproteins and lipids
Endocytosis	Efficiency and fidelity of intracellular routing of different vesicular pathways
Carrier-mediated transport	Rate, polarity and specificity of transport
Transepithelial solute flux	Dimensions of paracellular pathway
Solute diffusion	Dimensions of unstirred water layers
Permeability	Diffusion of low molecular weight compounds and macromolecules

TRANSPORT AND PERMEABILITY PROPERTIES OF HUMAN CACO-2 CELLS

The majority of in vitro methodologies (8) that have been used to study the transport of drugs across the intestinal epithelium do not have the morphological and functional properties of normal adult human epithelial layers. In addition they lack the viability and versatility required for quantitive measurements of transepithelial drug transport and for the examination of transport mechanisms. The use of isolated human intestinal epithelial cells has been slow to progress because these cells are difficult to culture and have limited viability (9). Recently attention has turned to human adenocarcinoma cell lines that reproducibly display a number of properties characteristic of differentiated intestinal cells (10-13). The HT-29 and Caco-2 cell lines (3) have been widely used to study intestinal epithelial differentiation and function because of their ability to express morphological and biochemical features of adult differentiated enterocytes and goblet cells (14). As the Caco-2 cell line displays the most highly differentiated properties under standard culture conditions (13) it appears to be the most relevant in vitro system for investigating transepithelial transport processes and as such has become the focus of attention for such studies.

A number of laboratories (7,13,15) have demonstrated that Caco-2 cells can be routinely grown as confluent monolayers on microporous filters. The monolayers develop an enterocytic morphology typical of villus cells and a polarity of a number of brush-border enzymes. Depending upon the exact experimental conditions used full expression of these properties is achieved between approximately 15-20 days in culture (7,15). Establishment of the barrier function of the monolayers can be demonstrated by lack of passage of a number of permeability markers (7,15). The inability of horseradish peroxidase (mol.wt. 40,000) to cross the tight junctions developed between adjacent Caco-2 cells demonstrates that the barrier properties of the in vitro system to this macromolecular probe are similar to those of the small intestine in vivo (16). The small amount of transcellular transport of horseradish peroxidase reflects its transport through cells in endocytic vesicles similar to that described in vivo (15). The integrity of the monolayers has also been routinely demonstrated by measuring the transepithelial electrical resistance (17). A range of values between 150-400 ohms cm^2 have been reported indicating that in different laboratories, under different culture conditions, Caco-2 monolayers can display the electrical properties of either small intestinal or colonic enterocytes (7,15,18,19).

Caco-2 monolayers are being widely used to study the transepithelial transport pathways. A number of specific transport systems that absorb nutrients and macromolecules from the small intestine are functional in the Caco-2 systems (Table 3). While detailed kinetic and molecular comparisons between these systems in vitro and in vivo are not yet available the systems involved in the carrier-mediated transport of bile acids (7,20) and large neutral amino acids (21) show many of the basic properties (i.e. specificity, saturability, competitiveness, unidirectionality) found in the small intestine in vivo. The development of these systems is also a function of the time in culture. The functioning of these carrier-mediated systems and the electrical properties of the cell monolayer can be reproducibly achieved and maintained for several weeks in culture (7,15,20,21).

TABLE III

Transport Route	Polarity	Reference
1. Carrier-mediated		
a) Bile acids	Apical	(7,20)
b) Amino acids	Apical	(21)
2. Endocytic		
a) IF-Cbl	Apical	(7,22,23)
b) EGF	Basolateral	(27)
c) Transferrin	Basolateral	(28)
3. Passive diffusion		
a) Beta blockers	None	(7)
b) Peptides		(25)
4. Permeability to macromolecules		(7,15)

A number of receptor-mediated endocytic systems that are involved in the transport of proteins and protein bound ligands in vivo are also functional and show the expected polarity in filter grown Caco-2 cells (Table 4). The transcellular transport of cobalamin (vitamin B12) mediated via a specific receptor that binds and internalizes intrinsic-factor-cobalamin and secretes transcobalamin II (7,22,23) is of particular interest following reports that cobalamin-drug conjugates cross the intestinal epithelium in vivo (24). Caco-2 is the only human cell line reported to transcellularly transport cobalamin via the intrinsic factor receptor. In the absence of many details on the mechanism of these endocytic pathways Caco-2 cells have become an important in vitro system for studying the basic cellular and molecular biology of these processes. A number of studies on the use of Caco-2 cell cultures to study drug transport have been reported. Studies on a series of beta-blockers and peptides (7,25) have indicated that the Caco-2 system may be useful for predicting the in vivo absorbability of a range of drugs, and for distinguishing the relative contributions of the paracellular and transcellular pathways, although many more examples are required to test the extent of such correlations between transport across the human cell system and oral absorption in vivo. The use and merits of a stirred system for determining the contribution of the unstirred water layer has recently been reported (2). The HT-29 system has also been used to study the uptake of the orally active antibiotic, cefalexin by a dipeptide transport system (26). While it is clear that Caco-2 cells in filter culture display many of the features of normal small intestinal enterocytes it is equally clear that they also show a number of abnormal metabolic and biosynthetic properties (27) probably reflecting their transformed phenotype. Other studies (7) indicate that Caco-2 cell layers consist of morphologically (and probably biochemically) heterogeneous enterocytes. In addition a number of cell types present in the normal intestinal epithelium, e.g. M-cells, goblet cells, have not been detected in these cultures. Thus there are many features of the intact adult intestinal epithelium that are not present in Caco-2 cultures. A challenge for future developments in cell culture methodologies will be to reconstitute an intestinal epithelial barrier, in culture, containing the majority of cell types.

ISOLATED INTESTINAL TISSUES

While much recent attention has focused on the use of cell culture systems, the Ussing chamber technique, which has been widely used to study mechanisms and regulation of ion transport across a variety of isolated epithelia, has found increasing use in studies on the transport and metabolism of a variety of molecules, including peptides, by intestinal preparations (30-34). Viability of intestinal preparations can be routinely assessed by measuring the spontaneous transepithelial potential difference or short circuit current(35,36), and their changes resulting from stimulation of sodium absorption by villus cells (7), or stimulation of sodium secretion by crypt cells (38). These measurements provide a rapid (non-invasive) method for assessing tissue viability during transport experiments. The integrity of the intestinal barrier can be determined by measuring the transepithelial flux of low molecular weight markers such as mannitol. The use of isolated tissues in Ussing chambers has a number of advantages over cell cultures. These include the ease and rapidity by which transport and metabolism by a non-transformed epithelium, with its full cellular repertoire, can be measured, and the ability to examine transport and metabolic processes in tissues from different regions of the intestine. In addition the ability to control the stirring of solutions above and below the tissues results in the presence of a more physiological unstirred water layer than that found in unstirred cell culture systems. Disadvantages include the presence of subepithelial cells which may present an absorption barrier, and the generally short viability times of tissue preparations.

CONCLUSIONS

The exploitation of cell and tissue culture techniques to study the transport of drugs and drug delivery systems across specific cellular barriers is clearly still at an early stage in its development. However, it is already apparent that these techniques offer considerable potential for investigating transport mechanisms to determine their relevance for enhancing the delivery of drugs either through drug modification or through the use of a novel drug delivery system. This brief discussion has focused on in vitro intestinal epithelial systems to illustrate important factors involved in their growth and characterization for transport studies. It is, however, apparent that significant progress is also occurring on the growth and characterization of other cellular barriers, and in particular the epidermis and the endothelium from brain capillaries (see other contributions to this volume).

The use of a variety of novel methodolgies e.g. cellular immortalization, co-culturing, three-dimensional matrices, which are still in their infancy, is likely to yield more sophisticated systems which better resemble their in vivo counterparts. The increasing use of these techniques by scientists in pharmaceutical R&D is being paralleled by their use in basic research in cellular and molecular biology aimed towards elucidating transport pathways and mechanisms. Thus, our use of such systems must currently be viewed against a background of generally poor information on the transport pathways that the in vitro systems are being established to measure and possibly predict. Of crucial importance will be the development of in vitro-in vivo correlations that not only validate the cell culture systems but provide information on the scope of their potential for predicting in vivo absorption.

REFERENCES

1. Von Bonsdorff, C. H., Fuller, S. D., Simons, K. (1985), Apical and basolateral endocytosis in Madin-Darby Canine Kidney (MDCK) cells grown on nitrocelluose filters. EMBO. J., 4, 2781-2792.

2. Hidalgo, I. J., Hillgren, K. M., Grass, G. M., Borchardt, R. T. (1989), Characterization of the aqueous boundary layer in Caco-2 cells using a novel diffusion cell. Pharm. Res. 6:000 (Abstr).

3. Neutra, M., Louvard, D.. (1989), Differentiation of intestinal cells in vitro, Modern Cell Biology: in Functional Epithelial Cells in Culture, (Mattin, K.S., & Valentich, J.D. eds) A. R. Liss, New York, pp. 363-398.

4. Van Meer, G. (1989), Polarity and polarized transport of membrane lipids in a cultured epithelium: in Functional Epithelial Cells in Culture, (Matlin, K.S. and Valentich, J.D. eds) A.R. Liss, new York, pp. 43-70.

5. Caplan, M. and Matlin, K.S. (1989), The sorting of membrane and secretory proteins in polarized epithelial cells: in Functional Epithelial

Cells in Culture, (Matlin, K.S. and Valentich, J.D. eds), A. R. Liss, New York, pp. 71-127.

6. Madara, J.L. and Hecht, G. (1989) Tight (occluding) functions in cultured (and native) epithelial cells: in Functional Epithelial Cells in Culture, (Matlin, K.S. and Valentich, J.D. eds), A.R. Liss, New York, pp 131-164.

7. Wilson, G., Hassan, I.F., Dix, C. J., Williamson, I., Shah, R., Mackay, M., (1990): Transport and permeability properties of human Caco-2 cells: An in vitro model of the intestinal epithelial cell barrier., J. Controlled Release, 11, 25-40.

8. Osiescka, I., Porter, P. A., Borchardt, R. T., Fix, J. A. and Gardner, C. R.(1985), In vitro drug absorption models. Brush border membrane vesicles, isolated mucosal cells and everted intestinal rings: characterization and salicylate accumulation, Pharm. Res., 2, (1983): 284-293.

9. Moyer,M. P. (1983), Culture of human gastrointestinal epithelial cells, Proc. Soc. Exp. Biol. Med., 174, (1986): 11-5.

10. Rousset, M.(1986), The human colon carcinoma lines HT-29 and Caco-2: two in vitro models for the study of intestinal differentiation, Biochimie, 68, 1035-1040.

11. Fogh, J., Fogh, J. M. and Orfeo, T.(1977), One hundred and twenty-seven cultured human tumor cell lines producing tumors in nude mice, J. Natl. Cancer Inst. 59,221-225.

12. Pinto, M., Appay, M-D., Assmann-Simon, P., Chevalier, G., Dracopoli, N., Fogh, J., and Zweibaum, A., (1982), Enterocytic differentiation of cultured human colon cancer cells by replacement of glucose by galactose in the medium, Bio. Cell. 44,193-196.

13. Pinto, M., Leon-Robine, S., Appay, M.-D., Kedinger, M., Triadou, N., Dussaulx, E., Lacroix, B., Assmann-Simon, P., Haffen, K., Fogh, J., and Zweibaum, A. (1983), Enterocyte-like differentiation and polarization of the human colon carcinoma cell line Caco-2 in culture. Bio. Cell 47,323-330.

14. Huet, C., Sahuquillo-Merino, C., Cordier, E., and Louvard, D., (1987), Absorptive and mucus-secreting subclones islolated from a multipotent intestinal cell line (HT29) provide new models for cell polarity and terminal differentiation, J. Cell Biol., 1, 345-358.

15. Hidalgo, I. J., Raub, T. J., and Borchardt, R. T., (1989), Characterization of the human colon carcinoma cell line (Caco-2) as a model system for intestinal epithelial permeability, Gastroenterology 96,736-749.

16. Phillips, T.E., Phillips, T. L., and Neutra, M., (1987), Macromolecules can pass through occluding junctions of rat ileal epithelium during cholinergic stimulation, Cell Tissue Res. 247,547-554.

17. Ellinger, A., and Pavelka, M. (1986), Colchicine-induced tubular, vesicular and cisternal organelle aggregates in absorptive cells of the small intestine of the rat. II.Endocytosis studies, Bio. Cell., 58, 31-42.

18. Grassett, E., Pinto, M., Dussaulx, E., Zweibaum, A., and Desjeux, J. F. (1984), Epithelial properties of a human colonic carcinoma cell line Caco-2: Electrical parameters, Am J Physiol., 247,C260-267.

19. Powell, D. W. Intestinal water and electrolyte transport. (1987), in Physiology of the Gastrointestinal Tract, L. R. Johnson (ed.), Raven Press, New York, pp. 1267-1305.

20. Hidalgo, I. J., and Borchardt, R. T., (1988), Transport of taurocholic acid in an intestinal epithelial model system (Caco-2 cell), Pharm. Res. 5,S110.

21. Hidalgo, I. J., and Borchardt, (1988): Amino acid transport in a novel model system of the intestinal epithelium (Caco-2 cells), Pharm. Res, 5,S110.

22. Dix, C. J., Obray, H. Y., Hassan, I. F., and Wilson, G. (1987), Vitamin B12 transport through polarised monolayers of a colon carcinoma cell line, Biochem. Soc. Trans., 15,439-440.

23. Levine, J. S., Allen, R. H., Alpers, D. H., and Seetharam, B. (1984): Immunocytochemical localisation of intrinsic factor-cobalamin receptors in dog ileum: distribution of intracellular receptors during cell maturation, J. Cell.Biol., 98, 1110-1117.

24. Russell-Jones, G. J., and Aizpurua, H. J. (1988), Vitamin B12:a novel carrier for orally presented antigens, Proc. Int. Symps. Control. Rel.Bioact.Mater., 15, pp. 142-143.

25. Burton, P. S., Hilgers, A. R., and Conradi, R.A. (1988), Human colon intestinal mucosa II:Structure-absorptivity studies with a series of small peptides. Pharm.Res.,5(1988)PD 948.

26. Dantzig, A., and Bergin, L., (1988), Carrier-mediated uptake of cephalexin in human intestinal cells, Biochem. Biophys. Res. Commun. 155:1082-1087 (1988).

27. Hidalgo, I. J., Kato, A., and Borchardt, R. T., 1989, Binding of epidermal growth factor by human colon carcinoma cell (Caco-2) monolayers, Biochem. Biophys. Res. Comm., Vol. 160, 317-324.

28. Hughson, E. J., and Hopkins, C. R., (1990), Endocytic pathways in polarized Caco-2 cells: identification of an endosomal compartment accessible from both apical and basolateral surfaces, J. Cell Biol., pp. 337-348.

29. Neuman, J.T., Kujubu, D., and Fine, L.G. (1989), Regulation of epithelial cell growth in vitro, in Functional Epithelial Cells in Culture (Matlin, K.S. and Valentich, J.D. eds), A.R. Liss, New York, pp.165-191.

30. Smith, P., Mirabelli, C., Fondacaro, J., Ryan, F. and Dent, J. (1988), Intestinal 5-fluorouracil absorption, use of Ussing chambers to assess transport and metabolism. Pharm. Res.5, 598-603.

31. Matuszewska, B., Liversidge, G.G., Ryan, F., Dent, J., and Smith, P.L. (1988), In vitro study of intestinal absorption and metabolism of 8-L-arginine vasopressin and its analogues. Int. J. Pharm. 46, 111-120.

32. Sund, R.B., and Lauterbach, F. (1987), 1-Naphthol and metabolism and metabolite transport in the small and large intestine. II. Effects of sulphate and phosphate ion omission, and of 2,6-dichloro-4-nitrophenol in the isolated guinea pig mucosa, Pharmacol. Toxicol. 60, 262-268.

33. Grass, G.M., and Sweetana, S.A. (1988), In vitro measurement of gastrointestinal tissue permeability using a new diffusion cell. Pharm. Res. 5, 369-376.

34. Ussing, H.H. and Zerahan, K. (1951), Active transport of sodium as the source of electric current in the short-circuited isolated frog skin, Acta Physiol. Scand. 23, 110-127.

35. Schultz, S.G. (1981), Ion transport by mammalian large intestine. In: L.R. Johnson (ed), Physiology of the Gastrointestinal Tract, Raven, New York, pp. 991-1019.

36. Hopfer, U. (1987), Membrane transport mechanisms for hexoses and amino acids in the small intestine. In: L.R. Johnson (Ed), Physiology of the Gastrointestinal Tract, Raven, New York, pp. 1499-1525.

37. Welsh, M.J., Smith, P.L., Fromm, M. and Frizzell, R.A. (1982), Crypts are the site of fluid and electrolyte secretion in mammalian colonic mucosa. Science 218, 1219-1221.

DIFFERENTIATION OF HUMAN COLON CANCER CELLS

Alain Zweibaum

Unité de Recherches sur la Différenciation et la Neuroendocrinologie de Cellules Digestives (INSERM U178), 16 avenue Paul Vaillant Couturier, 94807 Villejuif Cedex (France)

INTRODUCTION

As it is the case for many tissues, there has been an increasing interest over the last decade in the use of cell cultures for studies related to the functions of the intestinal epithelium. These studies have been essentially performed with either organ cultures or primary cultures[1,2]. However there are some limitations in the application of such systems as they are difficult to manipulate, do not allow reproducible dynamic studies, and are not homogenous. An ideal tool would be the use of established differentiated cell lines originating from normal tissues. However, and despite attempts from a number of laboratories[1,2,3], it has not been possible to establish such cell lines so far. In fact, the only cell lines that have been established have regularly failed to express any of the characteristics of terminal differentiation which would make them useful for studies related to the physiological functions of the intestinal epithelium[1,2,3]. This failure has been however circumvented by the finding that cell lines established from human colon carcinomas are able to express in culture most of the differentiation characteristics and functions normally associated with the human intestinal epithelium[1,2,4,5,6]. As appropriate as these cell lines may be for studying a number of functions related to the intestinal epithelium, it must be emphasized that, although these cells are closely similar to, and share a number of physiological properties with intestinal cells, they are not small intestinal, but colonic cells, and are not normal, but malignant cells. In order to further understand why such cell lines would express such differentiation characteristics it appears important to recall some of the cellular differentiation characteristics of colon cancers : by cellular differentiation should be meant the ability of some colon cancer cells to exhibit the ultrastructural morphology of either columnar absorptive cells, or mucus secreting cells, and to express proteins normally associated with the functional differentiation of the corresponding cells. Based on such a definition a variable proportion of differentiated cells are encountered in most colon cancers. Concerning absorptive cell types they may express not only brush border-associated proteins such as those normally associated with the colon, like CEA[7], or villin[8], but also brush border-associated hydrolases[9] (sucrase isomaltase, aminopeptidase N, dipeptidylpeptidase IV, alkaline phosphatase) which in this case appear to be closer to the enzymes present in the fetal colon than to those expressed in the small intestine[10,11,12]. Concerning mucus secreting

Pharmaceutical Applications of Cell and Tissue Culture to Drug Transport
Edited by G. Wilson *et al.*, Plenum Press, New York, 1991

cells they can be of two types depending on the organ specificity of secreted mucins, some being of the colonic type and others being of the gastric type[13,14]. In addition to columnar absorptive and mucus secreting cells, a third type of cells can be found in colon cancers namely cells which are polarized but do not express an apical brush border and do not secrete mucins. With regard to these different cell types it is not surprising that within cell lines established from such cancers some cells would express the differentiation characteristics which preexist in the original tumors. Indeed analysis of established cell lines shows that the same heterogeneity of differentiation which is found in colon cancers also occurs in cultured cell lines. Accordingly, and with the exception of these cell lines in which no differentiated cells can be found (which represent 50% of the cell lines tested so far), colon carcinoma cell lines can be divided into three groups : Group 1 includes cell lines in which 100% cells are differentiated and express a unique differentiation phenotype ; Group 2 includes cell lines which are heterogenous as not all the cells are differentiated and not all the differentiated cells express the same type of differentiation ; Group 3 includes cell lines which express neither a columnar cell absorptive type nor a mucus secreting type but which are able to form a polarized epithelium (references to cell lines cited in this article are reviewed in [6,15]).

GROUP 1 : THE CACO-2 CELL LINE.

The Caco-2 cell line is the only known example of a cell line belonging to this group. It was established in 1974 by Dr. Jorgen Fogh (Sloan Kettering Memorial Cancer Center, Rye, NY)[16,17]. The original tumor had been removed from a 72 year-old male patient of blood group O. It is interesting to note that prior to the removal of the tumor this patient had been treated by 5-fluorouracil and cytoxan (J. Fogh, personal communication). Since the first observation that Caco-2 cells in culture were able to undergo a complete and terminal differentiation of the columnar absorptive cell type[18] (Fig. 1) a considerable number of laboratories have been using this cell line (for review, see [2,4,6]). Because of the amount of literature already available for this cell line, only the main characteristics of differentiation of the cells will be reported here as well as some details which may be of interest for future utilizers of this line.

A first characteristic of the differentiation of the cells is that it is a growth-related process : the cells, although already polarized in the exponential phase[19], are not differentiated ; the process of differentiation starts after the cells have reached confluency and is complete after one week (i.e. on day 16 when cells are seeded at a density of $1x10^4$ cells/cm^2).

A second characteristic which is unexplained so far is that the degree of differentiation of the cells, as appreciated by the level of enzyme activities, increases with the number of passages by 20 : cells obtained from either the Sloan Kettering Memorial Cancer Center or the ATCC are usually at passage 16 ; in our experience the degree of differentiation increases progressively until passage 70 and then remains stable (at least until passage 150, which is the latest passage tested in our laboratory).

The differentiation characteristics of post-confluent cells have been extensively reported[2,4,5,6,18] and will be just summarized : whether cultured on impermeable support (plastic or glass) or on semi-permeable filters post-confluent cultures form a true monolayer of polarized cells which exhibit at their upper surface a typical brush border and at the upper pole of intracellular junctions typical tight junctions (Fig. 1).With the remark already made above that a number of functional or enzymatic characteristics associated with the differentiation of the cells are of a colonic fetal type these cells however share with normal small intestinal cells a number of differentiation features :

(i) <u>proteins associated with the brush border microvilli</u> : they include proteins associated with the core of actin microfilaments such as villin[15,20] and a number of proteins associated with the membrane of the microvilli such as brush border associated hydrolases (sucrase isomaltase, lactase, aminopeptidase N, dipeptidylpeptidase IV, alkaline phosphatase)[18], the sodium dependent glucose transporter[21] or the carcinoembryonic antigen (CEA).

(ii) <u>proteins associated with the basolateral membrane</u> : they include receptors for peptide and hormones such as the vasoactive intestinal peptide (VIP)[19,22], or catecholamines[22].

(iii) <u>other proteins</u> associated with normal intestinal cells such as apolipoprotein B have also been shown to be synthesized by Caco-2 cells[23].

In addition to these structural differentiation characteristics Caco-2 cells exhibit ion transport properties as already suggested by their ability to form domes when grown on impermeable supports[18] and further confirmed by electrophysiological studies[24,25].

The differentiation characteristics of Caco-2 cells and the easiness of their manipulation has allowed a number of studies dealing with biosynthesis and transport to the membrane of brush border enzymes[26,27], the regulation at the mRNA level of the expression of differentiation proteins such as sucrase isomaltase[28,29] or apolipoproteins[23], transcytosis of IgA (see C. Hopkins, this book), bacterial invasiveness[30] etc.

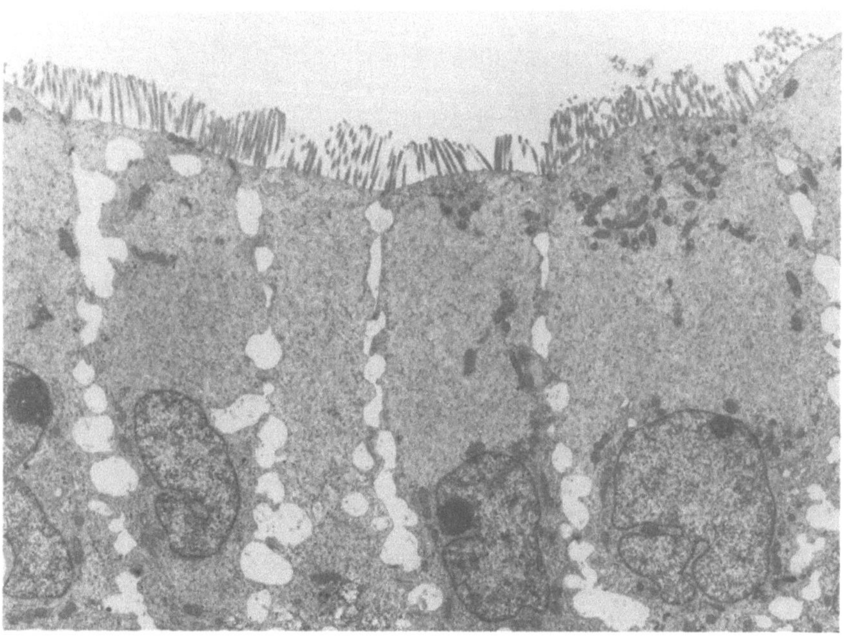

Fig. 1. Transmission electron micrograph of post-confluent Caco-2 cells grown on plastic. Section is perpendicular to the bottom of the flask (bar = 1.1um).

GROUP 2 : HETEROGENEOUSLY DIFFERENTIATED CELL LINES

A few cell lines have been characterized and can be considered as belonging to this group. The best example, and so far the most studied is the HT-29 cell line.

The HT-29 cell line

The HT-29 cell line was established by Jorgen Fogh in 1964 from a colon cancer obtained from a 44 year-old woman of blood group A[17]. Its differentiation characteristics were first reported in 1982 when it was found that replacement of glucose by galactose in the culture medium of HT-29 cells resulted into the emergence of a differentiated subpopulation exhibiting the differentiation characteristics of columnar absorptive cells[31] (Fig. 2). It was further observed that the same type of differentiation could be obtained by culturing the cells either in the absence of glucose (adaptation to a totally hexose-free medium[32] or replacement of glucose by nucleosides like uridine or inosine[32] or by growing the cells in the presence of sodium butyrate[33]. More recently it has been shown that HT-29 cells adapted to growth in the presence of increasing concentrations of 5-fluorouracil or methotrexate also exhibit a total differentiation of the mucus secreting type (A. Zweibaum et al., manuscript in preparation). Using the same nutritional or sodium butyrate treatment conditions, subclones from HT-29 cells have been obtained which express either a columnar absorptive cell type of differentiation[33,34], a mucus-secreting type[34,36] or an ion transporting type[36,37]. In all these works original HT-29 cells (i.e. cells which had been cultured in the presence of glucose since the line was established) have been considered to be undifferentiated and the reason why these cells would differentiate under particular culture conditions had remained unexplained. It was not

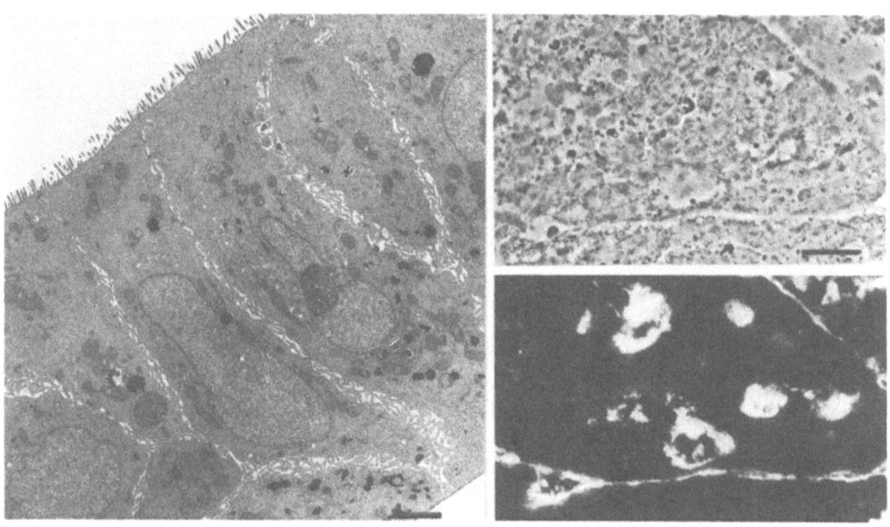

Fig. 2. Columnar absorptive-like HT-29 cells adapted to grow in hexose-free medium. Left : transmission electron micrograph of a section perpendicular to the bottom of the flask (bar = 4 um) ; lower right : cryostat section of the cell layer showing a strong apical immunofluorescent staining of dipeptidylpeptidase IV (compare with Fig. 4) ; upper right: corresponding phase contrast microscopy (bar = 40um).

until very recently that extensive studies of so-called undifferentiated HT-29 cells revealed that post-confluent cultures of HT-29 cells, although constituted of a majority of undifferentiated cells, contained however a small proportion of differentiated cell types (less than 5%) (Fig. 3). These differentiated cells are of three types : columnar absorptive like cells organized either into clusters with an apical brush border directed towards the culture medium, or into intracellular and intercellular lumens ; polarized cells organized around intercellular lumens and devoid of a brush border ; mucus secreting cells. Among the markers associated with the brush border of the very few columnar cells present in these cultures are villin, CEA, and dipeptidylpeptidase IV (Fig. 4) ; however no sucrase isomaltase can be found since these cells are grown in presence of glucose which, as previously reported, is an inhibitor of the expression of this enzyme[38]. As in most colon cancers mucins secreted by mucus secreting cells are of two types, i.e. of the colonic and of the gastric types with different cells secreting each type of mucins (Fig. 4).

That differentiated cell types are present in so-called undifferentiated HT-29 cells sheds some light on the reasons why totally differentiated cell populations can be isolated from these cell lines either by nutritional manipulation or by cloning. Concerning the subpopulation obtained by nutritional manipulation (absence of glucose[32]

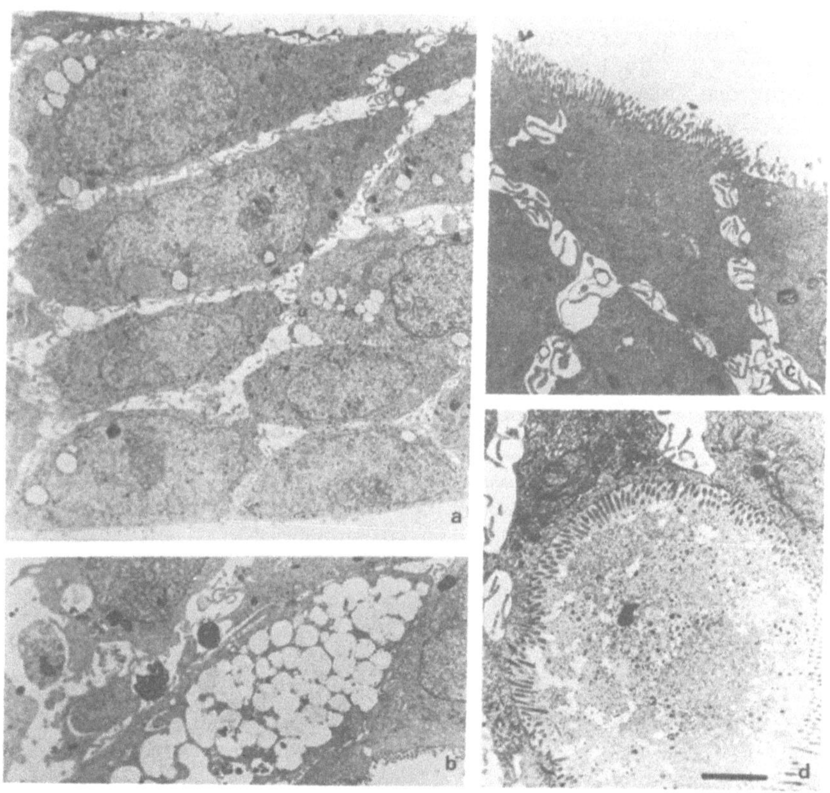

Fig. 3. Transmission electron micrograph of post-confluent so-called undifferentiated HT-29 cells. (a): general appearance of the cell layer ; section is perpendicular to the bottom of the flask ; (b): goblet cell ; (c): detail of a cluster of cells with an apical brush border directed towards the culture medium; (d): intercellular cyst with a brush border (bar = 4 um).

or addition of sodium butyrate[33]) it can be postulated that differentiated, but not undifferentiated cells possess the metabolic pathways associated with their normal differentiated counterparts which include the ability of such cells to develop gluconeogenesis or their ability to use sodium butyrate as a main source of energy[39]. In fact it has been shown that differentiated HT-29 cells grown in the absence of glucose did develop gluconeogenesis as substantiated by an increased activity of phosphoenolpyruvatecarboxykinase[40] a key enzyme of the gluconeogenic pathway. It is therefore most likely that the emergence of differentiated cell subpopulations is the result of a metabolic selection. This may also be the reason for the emergence of differentiated subpopulations from HT-29 cells treated with anticancer drugs. This is further confirmed by the fact that the differentiation of the cells remains stable after they have been switched back to standard culture conditions. It is also most likely that it will be possible, without any metabolic preselection, to isolate by cloning pure populations of each differentiated cell type present in the original cell line i.e. columnar absorptive cells, transporting cells, colonic mucin-secreting cells and gastric mucin-secreting cells.

Whatever the type of differentiation of HT-29 subpopulations they share with Caco-2 cells the same kinetics of differentiation with the set up of differentiation taking place after the cells have reached confluency. As with Caco-2 cells a number of proteins have been characterized in HT-29 cells such as villin[20,41], brush border associated hydrolases[31-34] (Fig. 2), VIP receptors[42-44], HLA antigens[35], and receptors for transferrin[35], neurotensin[45], EGF[46], and alpha-adrenergic receptors[47] ; furthermore absorptive-like differentiated HT-29 subpopulations have been shown to synthesize glycosaminoglycans which are more closely related to normal than undifferentiated HT-29 cells[48].

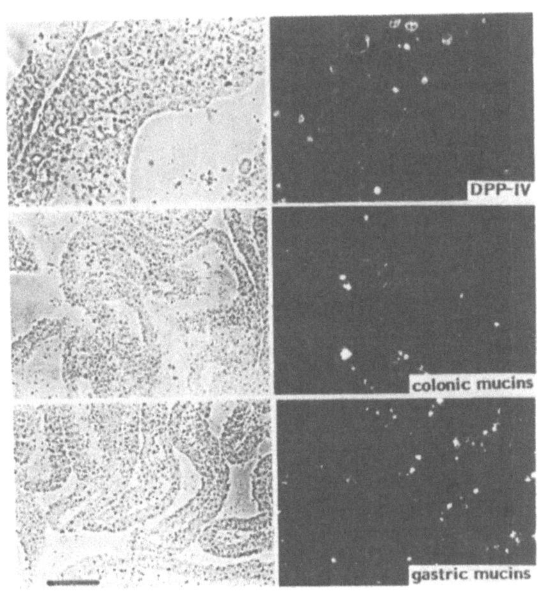

Fig. 4. Cryostat section of "undifferentiated" HT-29 cells showing (right) positive immunoreactivity with antibodies against dipeptidylpeptidase IV, colonic and gastric mucins. Left : phase contrast microscopy of corresponding sections (bar = 200 um).

Other heterogenous cell lines

It is most likely that a number of cell lines which have not been investigated yet as to their differentiation characteristics may behave as HT-29 cells. Because of the very recent availability of antibodies specific for the various differentiation markers and of the necessity of ultrastructural studies very few cell lines have been characterized so far. Preliminary results however already suggest that three cell lines may be good candidates for future differentiation studies : LS174-T[49] which contains a small proportion of mucus-secreting cells ; HCT-EB[15] and HCT-Geo[15] which contain a small proportion of columnar absorptive cells. It can be postulated that subcloning of these cell lines or metabolic manipulations will result, as with HT-29 cells, in the selection of totally differentiated subpopulations.

GROUP 3 : UNDIFFERENTIATED POLARIZED CELLS

This group encompasses a number of cell lines[6,15] which share the same characteristics. After confluency the cells form a true monolayer of polarized cells which show no brush border, no mucus secretion but which exhibit ion transport properties as substantiated by their ability to form domes when grown on an impermeable support. The most studied of these cell lines is T-84[50] which has been extensively studied for its transport properties[51-54]. Other cell lines in this group include SW1116, HRT-18, Co-115, HCA-7, HCT-Gly, ALA, Fri, Fet, CBS (for references see [6,15]). It is most likely that further investigations with this group of cells will provide interesting models for a number of studies dealing with transepithelial transport of molecules.

CONCLUSION

Less than 150 cell lines originating from human colon cancer have been established and published. Of these 150 cell lines very few (less than 25) have been characterized for their differentiation pattern, and except for HT-29, Caco-2, and T84, the cells which have been looked at still need further characterization. It is most likely that, as this field attracts more investigators, other cell lines of interest will emerge. They will help the understanding of both the physiology of intestinal cells and, hopefully, the particular physiology of colon cancer cells. However, real progress will be achieved only when attempts to establish normal intestinal cell lines are finally successful.

ACKNOWLEDGEMENTS

This work was supported by the Association pour la Recherche contre le Cancer (ARC), la Fondation pour la Recherche Médicale Française, le Groupement des Entreprises Françaises de Lutte contre le Cancer (GEFLUC).

REFERENCES

1. M. Kedinger, K. Haffen, and P. Simon-Asmmann, Intestinal tissue and cell cultures. Differentiation, 36:71-85 (1985).
2. M. Neutra, and D. Louvard, Differentiation of intestinal cells in vitro. In : Functional Epithelial Cells in Culture. Alan R. Liss, Inc., p. 363-398 (1989).

3. A. Quaroni, and R.J. May, Establishment and characterization of intestinal epithelial cell cultures. In : Methods in Cell Biology, edited by G. C. Harris, B. F. Trump, and G. D. Stoner, Academic Press, New York, vol. 21B, p. 403-427 (1980).
4. M. Rousset, The human colon carcinoma cell lines HT-29 and Caco-2 : two in vitro models for the study of intestinal differentiation. Biochimie, 68:1035-1040 (1986).
5. C. L. Laboisse, Differentiation of colon cells in culture. In : The Cell and Molecular Biology of Colon Cancer. L.H. Augenlicht ed., CRC Press, pp 28-45 (1989).
6. A. Zweibaum, M. Laburthe, E. Grasset, and D. Louvard, The use of cultured cell lines in studies of intestinal cell differentiation and function. In : Handbook of Physiology, vol. Intestinal Transport of the Gastrointestinal System (M. Field, and R. A. Frizzell, eds), Am. Physiol. Soc. (1989) (in press).
7. F. J. Primus, C. A. Clark, and D. G. Goldenberg, Immunoperoxidase localization of carcinoembryonic antigen in normal human intestinal mucosa. J. Natl. Cancer Inst., 63:1031-1039 (1981).
8. A. Bretscher, and K. Weber, Villin : the major microfilament-associated protein of the intestinal microvillus. Proc. Natl. Acad. Sci. USA, 76:2321-2325 (1979).
9. G. Semenza, Anchoring and biosynthesis of stalked brush border membrane proteins : glycosidases and peptidases of enterocytes and of renal tubuli. Annu. Rev. Cell Biol., 2:255-314 (1986).
10. A. Zweibaum, H. P. Hauri, E. Sterchi, I. Chantret, K. Haffen, J. Bamat, and B. Sordat, Immunohistological evidence obtained with monoclonal antibodies of small intestinal brush border hydrolases in human colon cancers and foetal colons. Int. J. Cancer, 34:591-598 (1984).
11. A. Zweibaum, N. Triadou, M. Kedinger, C. Augeron, S. Robine-Léon, M. Pinto, M. Rousset, and K. Haffen, Sucrase-isomaltase : a marker of foetal and malignant epithelial cells of the human colon. Int. J. Cancer, 32:407-412 (1983).
12. B. Lacroix, M. Kedinger, P. Simon-Assmann, M. Rousset, A. Zweibaum, and K. Haffen. Developmental pattern of brush border enzymes in the human fetal colon. Correlation with some morphogenetic events. Early Hum. Dev., 9:95-103 (1984).
13. M. Rousset, E. Dussaulx, G. Chevalier, and A. Zweibaum, Expression phenotypique des antigènes coliques polymorphes (WZ) dans les adénocarcinomes du côlon humain. C.R. Acad. Sci. (Paris), 286:659-662 (1978).
14. J. Bara, F. Loisillier, and P. Burtin, Antigens of gastric and intestinal mucous cells in human colonic tumours. Br. J. Cancer, 41:209-221 (1980).
15. I. Chantret, A. Barbat, E. Dussaulx, M. G. Brattain, and A. Zweibaum, Epithelial polarity, villin expression and enterocytic differentiation of cultured human colon carcinoma cells : a survey of 20 cell lines. Cancer Res., 48:1936-1942 (1988)
16. J. Fogh, J. M. Fogh, and T. Orfeo, One hundred and twenty-seven cultured human tumor cell lines producing tumors in nude mice. J. Natl. Cancer Inst., 59:221-225 (1977).
17. J. Fogh, and G. Trempe, New human tumor cell lines. In : Human Tumor Cells "In vitro". Edited by J. Fogh, New York, Plenum Press, p. 115-141 (1975).
18. M. Pinto, S. Robine-Léon, M. D. Appay, M. Kedinger, N. Triadou, E. Dussaulx, B. Lacroix, P. Simon-Assmann, K. Haffen, J. Fogh, and A. Zweibaum, Enterocyte-like differentiation and polarization of the

human colon carcinoma cell line Caco-2 in culture. Biol. Cell., 47:323-330 (1983).

19. M. Laburthe, M. Rousset, C. Rouyer-Fessard, A. Couvineau, I. Chantret, G. Chevalier, and A. Zweibaum, A. Development of vasoactive intestinal peptide-responsive adenylate cyclase during enterocytic differentiation of Caco-2 cells in culture : evidence for an increased receptor level. J. Biol. Chem., 262:10180-10184 (1987).

20. S. Robine, C. Huet, R. Moll, C. Sahuquillo-Merino, E. Coudrier, A. Zweibaum, and D. Louvard, Can villin be used to identify malignant and undifferentiated normal digestive epithelial cells ? Proc. Natl. Acad. Sci. USA, 82:8488-8492 (1985).

21. A. Blais, P. Bissonnette, and A. Bertheloot, Common characteristics for Na^+-dependent sugar transport in Caco-2 cells and human fetal colon. J. Membrane Biol., 99:113-125 (1987).

22. M. Laburthe, M. Rousset, G. Chevalier, C. Boissard, C. Dupont, A. Zweibaum, and G. Rosselin, Vasoactive intestinal peptide control of cyclic adenosine 3':5'-monophosphate in seven human colorectal adenocarcinoma cell lines in culture. Cancer Res., 40:2529-2533 (1980).

23. T.E. Hugues, J.M. Ordouas, and E.J. Schaeffer, Regulation of intestinal apolipoprotein B synthesis and secretion by Caco-2 cells. Lack of fatty acid effects and control by intracellular calcium ion. J. Biol. Chem., 263:3425-3431 (1988).

24. E. Grasset, J. Bernabeu, and M. Pinto, Epithelial properties of human colonic carcinoma cell line Caco-2 : effect of secretagogues. Am. J. Physiol. 248 (Cell Physiol. 17):C410-C418 (1985).

25. E. Grasset, M. Pinto, E. Dussaulx, A. Zweibaum, and J. F. Desjeux, Epithelial properties of human colonic carcinoma cell line Caco-2 : electrical parameters. Am. J. Physiol. 247 (Cell Physiol. 16): C260-C267, 1984.

26. H. P. Hauri, E. E. Sterchi, D. Bienz, J. Fransen, and A. Marwer, Expression and intracellular transport of microvillus membrane hydrolases in human intestinal epithelial cells. J. Cell Biol., 101:838-851 (1985).

27. H.P. Hauri, Biogenesis and intracellular transport of intestinal brush border membrane hydrolases. Use of antibody probes and tissue culture. In : Subcellular Biochemistry, G.R. Harris ed., Plenum Press, p. 155-219 (1988).

28. M. Rousset, M. Laburthe, M. Pinto, G. Chevalier, C. Rouyer-Fessard, E. Dussaulx, G. Trugnan, N. Boige, J. L. Brun, and A. Zweibaum, Enterocytic differentiation and glucose utilization in the human colon tumor cell line Caco-2 : modulation by forskolin. J. Cell. Physiol., 123:377-385 (1985).

29. M. Rousset, I. Chantret, D. Darmoul, G. Trugnan, C. Sapin, F. Green, D. Swallow, and A. Zweibaum, Reversible forskolin-induced impairment of sucrase-isomaltase mRNA levels, biosynthesis and transport to the brush border membrane in Caco-2 cells. J. Cell. Physiol. (in press).

30. J.L. Gaillard, P. Berche, J. Mounier, S. Richard, and P. Sansonetti, In vitro model of penetration and intracellular growth of Listeria monocytogenes in the human enterocyte-like cell line Caco-2. Infect. Immun., 55:2822-2829 (1987).

31. M. Pinto, M. D. Appay, P. Simon-Assmann, G. Chevalier, N. Dracopoli, J. Fogh, and A. Zweibaum, Enterocytic differentiation of cultured human colon cancer cells by replacement of glucose by galactose in the medium. Biol. Cell., 44:193-196 (1982).

32. A. Zweibaum, M. Pinto, G. Chevalier, E. Dussaulx, N. Triadou, B. Lacroix, K. Haffen, J. L. Brun, and M. Rousset, Enterocytic

differentiation of a supopulation of the human colon tumor cell line HT-29 selected for growth in sugar-free medium and its inhibition by glucose. J. Cell. Physiol., 122:21-29 (1985).

33. B. M. Wice, G. Trugnan, M. Pinto, M. Rousset, G. Chevalier, E. Dussaulx, B. Lacroix, and A. Zweibaum, The intracellular accumulation of UDP-N-acetylhexosamines is concomitant with the inability of human colon cancer cells to differentiate. J. Biol. Chem., 260:139-146 (1985).
34. C. Huet, C. Sahuquillo-Merino, E. Coudrier, and D. Louvard, Absorption and mucus-secreting subclones isolated from a multipotent intestinal cell line (HT-29) provide new models for cell polarity and terminal differentiation. J. Cell Biol., 105:345-357 (1987).
35. O. Godefroy, C. Huet, L. A. C. Blair, C. Sahuquillo-Merino, and D. Louvard, Differentiation of a clone isolated from the HT-29 cell line: polarized distribution of histocompatibility antigens (HLA) and of transferrin receptors. Biol. Cell., 63:41-55 (1988).
36. Augeron, C., and C. L. Laboisse, Emergence of permanently differentiated cell clones in a human colonic cancer cell line in culture after treatment with sodium butyrate. Cancer Res., 44:3961-3969 (1984).
37. Augeron, C., J. J. Maoret, C. L. Laboisse, and E. Grasset, Permanently differentiated cell clones established from the human colonic adenocarcinoma cell line HT-29 : possible models for the study of ion transport and mucus production. In : Ion-Gradient-Coupled Transport, Edited by F. Alvarado & C. H. Van Os, Amsterdam, Elsevier, (1989) (in press).
38. G. Trugnan, M. Rousset, I. Chantret, A. Barbat, and A. Zweibaum, The post-translational processing of sucrase-isomaltase in HT-29 cells is a function of their state of enterocytic differentiation. J. Cell. Biol., 104:1199-1205 (1987).
39. W. E. W. Roediger, Utilization of nutrients by isolated epithelial cells of the rat colon. Gastroenterology, 83:424-429 (1982).
40. C. Denis, V. Mils, J. C. Murat, M. Rousset, M. Pinto, V. Trocheris, A. Zweibaum, and H. Paris, Evidence for development of gluconeogenesis in the HT-29 human colon adenocarcinoma cell line grown in a glucose-free medium. IRCS Med. Sci., 13:898-899 (1985).
41. E. Pringault, M. Arpin, A. Garcia, J. Finidori, and D. Louvard, A human villin cDNA clone to investigate the differentiation of intestinal and kidney cells in vivo and in culture. EMBO J., 5:3119-3124 (1986).
42. Laburthe, M. Rousset, C. Boissard, G. Chevalier, A. Zweibaum, and G. Rosselin, Vasoactive intestinal peptide : a potent stimulator of adenosine 3':5'-cyclic monophosphate accumulation in gut carcinoma cell lines in culture. Proc. Natl. Acad. Sci. USA, 75:2772-2775 (1978).
43. A. Couvineau, M. Rousset, and M. Laburthe, Molecular identification and structural requirement of vasoactive intestinal peptide (VIP) receptors in the human colon adenocarcinoma cell line, HT-29. Biochem. J., 231:139-143 (1985).
44. M. Laburthe, C. Augeron, C. Rouyer-Fessard, I. Roumagnac, J.J. Maoret, E. Grasset, and C. Laboisse, Functional VIP receptors in the human mucus-secreting colonic cell line Cl.16E. Am. J. Physiol. (Gastrointest. Liver Physiol. 19):G443-G450 (1989).
45. P. Kitabgi, C. Poustis, C. Granier, J. Van Rietschoten, J. Rivier, J. L. Morgat, and P. Freychet, Neurotensin binding to extraneural and neural receptors : comparison with biological activity and structure-activity relationships. Mol. Pharmacol., 18:11-19 (1980).

46. P. Kitabgi, C. Poustis, A. Zweibaum, and P. Freychet, Peptide receptors in colonic tumor cells : specific binding of epidermal growth factor to the HT-29 cell line. In : Hormone Receptors in Digestion and Nutrition. Ed. by G. Rosselin, P. Fromageot & S. Bonfils, Elsevier/North Holland Biomedical Press, p. 255-260 (1979).
47. H. Paris, B. Bouscarel, C. Cortinovis, and J. C. Murat, Growth-related variation of alpha$_2$-adrenergic receptivity in the HT-29 adenocarcinoma cell line from human colon. FEBS Lett., 184:82-86 (1985).
48. P. Simon-Assmann, F. Bouziges, D. Daviaud, K. Haffen, and M. Kedinger, Synthesis of glycosaminoglycans by undifferentiated and differentiated HT-29 human colonic cancer cells. Cancer Res., 47:4478-4484 (1987).
49. B. J. Siddiqui, J. Byrd, F. J. Fearney, and Y. S. Kim, Comparison of metabolically labelled mucins of LS-174T human colon cancer cells in tissue culture and xenograft. Tumor Biology (1989) (in press).
50. H. Murakami, and H. Masui, Hormonal control of human colon carcinoma cell growth in serum-free medium. Proc. Natl. Acad. Sci. USA, 77:3464-3468 (1980).
51. K. Dharmsathaphorn, K. G. Mandel, H. Masui, and J. A. McRoberts, VIP-induced chloride secretion by a colonic epithelial cell line : direct participation of a basolaterally localized Na^+, K^+, Cl^- cotransport system. J. Clin. Invest., 75:462-471 (1985).
52. K. Dharmsathaphorn, K. G. Mandel, J. A. McRoberts, C. A. Cartwright, and H. Masui, Utilization of a human colonic tumor cell line as a model to study electrolyte transport in the intestine. In : Intestinal Absorption and Secretion (Falk Symp. No 36) Ed. by E. Skadhauge & K. Heintze, Lancaster : MTP, p. 325-333 (1984).
53. K. Dharmsathaphorn, K. G. Mandel, J. A. McRoberts, L.D. Tisdale, and H. Masui, A human colonic tumor cell line that maintains vectorial electrolyte transport. Am. J. Physiol. 246 (Gastrointest. Liver Physiol. 9):G204-G208 (1984).
54. A. Weymer, P. Huott, W. Liu, J. A. McRoberts, and K. Dharmsathaphorn, Chloride secretory mechanism induced by prostaglandin E_1 in a colonic epithelial cell line. J. Clin. Invest., 76:1828-1836 (1985).

CRITERIA FOR EVALUATING AND CHOOSING A RELEVANT CELL LINE

Jens Fogh

Novo Nordisk A/S
Niels Steensensvej 1
DK-2820 Gentofte
Denmark

INTRODUCTION

In-vitro test systems, most often based on mammalian cells, are today more and more being used to supplement or even replace in-vivo test systems. There has been an increasing interest in establishing cell lines and the availability of these has stimulated studies which were previously not possible. Investigators are today often seeking cultured lines with predefined characteristics. In most cases though, the presently available lines do not have normal tissue or cell counterparts and this can make correlation to normal controls more difficult. The collection of cell lines is all in all meeting difficulties today, due to the increasing rights of propriety and the fact that cell lines are now being patented.

TISSUE CULTURE SYSTEMS

Tissue culture systems have several advantages over in-vivo testing:

1) Control of the physio-chemical environment (pH, temperature, osmotic pressure, O_2/CO_2 tension).
2) Homogeneity of the substrate after a number of passages and the possibility of characterizing the cell population as well as having an abundance of and being able to quantitate a homogeneous substrate.
3) Being able to screen many variables cheaper and being able to do replicate sampling.

Cells used for in-vitro assays can originate as organ cultures or cell cultures. Organ cultures (Thomas, 1970) retain the differentiated properties of the tissues of origin and the cell proliferation is limited to the periphery of the explant. The disadvantage is that each experiment requires fresh material, resulting in a poorer sample reproducibility.

Cell cultures (Fogh et al., 1975, Jakoby et al. 1979) may be derived from primary explants or dispersed cell suspensions originating from normal tissue, solid tumors or effusions. Many different methods are employed when establishing cell lines from tissue. A simple method is shown in Figure 1. The important thing is that the cultures receive minimal disturbance during early development with only careful pH adjustment and small nutrient supple-

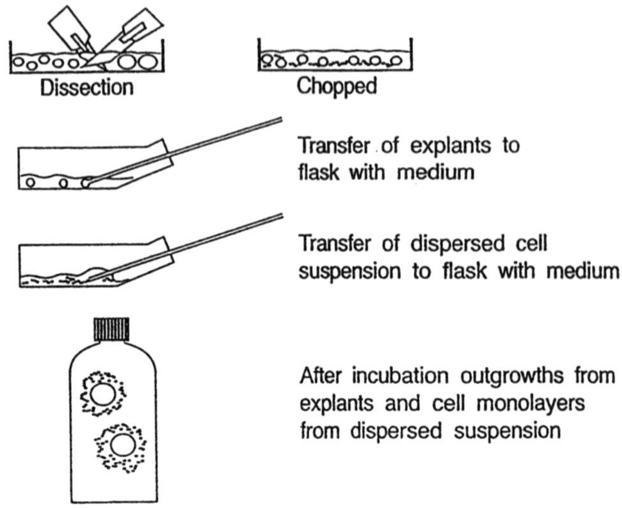

Fig. 1. Culture Establishment Technique

ments. It is also important not to discard cells which do not attach initially, as these may prove to be the important cells.

Tissue used for cell line preparation from oral, gynecological and urinary tract regions as well as skin, will harbor microbial flora. Different combinations of antibiotics (Fogh, 1973) should be used when attempting to culture cells from the different regions. Furthermore, it is important to distinguish between the concentrations of antibiotics for an initial washing of the specimen, the initial culture period and the continuous culture period.

Normal tissue results in mortal cell cultures which have a finite replication possibility. Solid tumors or effusions can on the other hand produce immortal continuous cell lines. The formation of such a cell line requires:

1) An increase in the total number of cells over a period of several generations, although the rate of cell division cannot be considered as an absolute criterion for cell line establishment,
2) that cells or cell subtypes with similar growth capacity predominate, and
3) that this results in a uniform cell population.

Cell lines have normally been designated by the cell line originator. The year of establishment is of interest with regards to the identity of the cell line as well as for historical purposes. The oldest line we have is the HeLa from 1951, but the majority of the cell lines available today have been established in the 1970's and 1980's. The number of cell lines available for each tissue type reflects the particular interest of certain investigators, the availability of tissue and the ease with which certain tissues grow in-vitro.

CHARACTERIZATION AND CONTAMINATION

When establishing cells or cell lines it is important to screen for

Table 1

Presence of contaminants

Mycoplasma
Virus
Fungi
Bacteria

Identification of cultured cells

Cytopathology
Morphology
Growth
Chromosomes
Isoenzymes
HLA-typing
E.M.
Tumor production

the presence of contamination and to fully characterize the cells in order to make certain that the results achieved are correct and only reflect the cell substrate (Fogh et al., 1982) (Table 1).

An important point to note is that mix-ups, mislabellings and contamination are more likely to occur when cells are freely exchanged among investigators who are not primarily concerned with characterization. These mix-ups and mislabellings are clearly demonstrated by the HeLa cell contamination resulting in cultures of unknown identity being used and time-consuming investigations wasted. These interspecies and intraspecies extraneous cell contaminations have been tackled by isozyme analysis (Dracopoli et al., 1983; Fogh et al, 1982; Macy 1978). The HeLa cell was derived from a black patient with G6PD type A. HeLa cell contamination can be excluded when a line is G6PD type B and when the donor is black or has Negro/Caucasian blood. Many cell lines have been screened by isoenzyme analysis and proven to be not HeLa cell contaminated (Fogh, 1974; Povey et al, 1976).

Contamination with bacteria, fungi or mycoplasma must always been checked and controlled. Many of the methods described over the years for mycoplasma testing are time consuming but necessary (Chen, 1977; Fogh, 1973; McGarrity, 1982, Schneider et al., 1975; Stanbridge et al., 1978). The most often used detection methods are cell-culture strains (DNA-specific Orcein stain and Hoechst immunoflourescence) combined with broth and agar culturing. Since the sources of mycoplasma contamination can be nutrients, other cell cultures as well as the technician performing the culturing, it is imperative that good techniques and check systems are implemented rutinely in the laboratory. Unfortunately, there still seems to be some apathy toward mycoplasma contamination and its many detrimental effects on cell cultures and assay results.

A thorough characterization of a cell line (Fogh et al., 1982) should include an in-vitro cytopathology to correlate cultures with donor tissue, isozyme analysis, HLA-typing (Pollack et al., 1981), karyology, tumor producing capacity (Fogh et al., 1978; Tveit et al., 1981), growth pattern, electron-microscopy, virus susceptibility and contamination status. These data should be updated at regular intervals.

Table 2

Finite and Continuous Cell Lines

	Finite	Continuous
Transformation	Normal	Transformed
Growth	Mortal	Immortal
Anchorage dependence	Yes/No	Yes/No
Contact inhibition	Yes	No
Mode of Growth	Monolayer/Suspension/ Semisuspension	Monolayer/Suspension/ Semisuspension
Growth Rate	Slow	Slow - Rapid
Yield	Low	High
Mutation Rate	Low	High, spontaneous
Karyology	Euploid	Aneuploid
Cloning efficiency	Low	High
Serum requirement	High	Low
Growth factor requirement	Yes	No
Special functions	May be retained	Often lost
Tumorgenic	No	Yes

NORMAL AND TRANSFORMED CELLS

Continuous cell lines are distinct from normal, finite cells in that they are transformed (Cameron I.L., 1981) (phenotype modification not necessarily involving uptake of new genetic material) or transfected (change in phenotype dependent on the uptake of new genetic material). Furthermore there are a number of characteristics that distinguish finite cell cultures from continuous cell lines (Table 2), which are essential factors to take into consideration when choosing cells for in-vitro assays. The mode of growth of cell cultures can be as distinct monolayer, suspension or as a semisuspension, where some cells in the culture are monolayer and others in suspension. Monolayer cultures require subculturing with cell dissociation and show a cyclic pattern of growth, while suspension cultures can be maintained at a "steady state". Suspension cultures exhibit a homogeneity and the cell density is affected by nutrient concentration of the medium alone.

Cell cultures in suspension are very suitable for bulk production of cells and easy to harvest. Monolayer cultures on the other hand make serial extractions possible without centrifugation. Most cell types, with the exception of hemopoetic cells, ascites tumors and lymphoblastoid lines (Collins, 1977) will grow as monolayer cultures.

The subculturing of cells gives a number of advantages as well as disadvantages (Table 3). The greatest disadvantage is the selection and overgrowth of specific cells resulting in the loss of differentiated properties, which are of value when trying to interpret an in-vitro result and correlate it with an in-vivo response. All cell lines do not divide at the same rate (transfer factor). If too high a transfer factor is applied, the cells can cease to divide and die.

The mutation rate for normal cells is low compared to continuous cell lines and the karyology reflects this (Biedler 1976; Fogh et al., 1982). Normal fibroblasts will retain their euploid characteristics during subcul-

Table 3

Subculturing

Advantage

Increase the number of cells

Characterize replicate samples

Freeze and store large quantities of cells

Uniform cell populations

Clone specific cell populations

Disadvantage

Disaggregation of cells and possible enzymatic and mechanical damage

Selection and overgrowth of specific cells results in loss of differentiated properties

Genetic instability

ture, while cell cultures from human effusions or solid tumors, which give rise to continuous cell lines as well as many murine cell lines will often exhibit aneuploid karyology after subculturing.

Normal cells are more nutrient demanding than transformed cells and require higher serum concentrations. Furthermore, it has been demonstrated that normal cells will not grow in growth factor deprived serum, while the opposite is true for transformed cells (Gospodarowicz et al, 1978; Jakoby et al., 1979; Klevjer-Anderson et al., 1980).

CELL CHARACTERISTICS

Cells have well defined special functions that are characteristic for a cell. The source of the cells, be it fetal or adult, will reflect the cells possibility for undergoing continuous renewal. The renewal process coupled with the possibilities of transformation can affect the intracellular activity and thereby the production of hormones and metabolites, resulting in changes in cell-cell interactions and cell surface markers (Nicolson, 1976).

Very little is known about the stability of the characteristics of cells or tumors in patients and there is increasing evidence that heterogeneity of cancer cells is expressing a constant shift in tumor population. Certain characteristics may vary among lines derived from histopathologically similar tissue. A conclusion can be that investigations should be based on several lines and not one individual cell line.

Organ cultures are better to retain specific functions and therefore attempts have been made to recreate the organ system. The use of filterwell units, where cells grow on filters allows for the study of transport across an intact cell substrate and attempts to recreate the three-dimensional structure.

Transformed cells will most often be tumorgenic, while normal cells will lack this potential (Fogh et al., 1977; Sharkey et al., 1978). The tumor-producing capacity of the transformed cell lines can be used to propagate in-vivo material for three-dimensional organ cultures exhibiting characteristics of the cell line as well as a means of correlating in-vitro and in-vivo response.

CELL BANKS

An alternative to establishing one's own cell lines is to avail oneself of the different private and commercial cell culture collections that exist. In the US, the American type culture collection has collected and maintains an extensive assortment of well characterized human and animal cells, which are commercially available. In Europe, the European collection of animal cell cultures has a smaller collection that also is commercially available.

An established in-house cell bank is the future for many pharmaceutical companies. The in-house cell bank gives one the advantage of having abundant cells of high quality, well-characterized and documented free from contamination, readily available for research, production or routine assay purposes. An in-house cell bank can also contain newly established relevant cell lines from human and animal tissue, where the ownership and rights to use are indesputably the companies. A criteria for a well-functioning in-house cell bank is a well organized data bank and a continuing follow-up and updating of relevant data pertaining to the cell lines.

FACTORS INFLUENCING CELL ASSAYS

The choice of relevant cells or cell lines and the implementation of these in cell-bioassays has proven to be more complicated than initially entertained. Many different factors have been shown to influence the growth and metabolism of the cells and thereby the results achieved in a bioassay.

Species differences can exist and this should always be taken into consideration. The effect, e.g. binding of a factor to a receptor can be very different in animal verses human cells. It seems though, as if one often sees human products eliciting an effect in animal cells, but not the reverse.

In order to register recombinant derived products produced in mammalian cells today, the products are often correlated with the natural product. Using different cell bioassays to correlate the biological effect, one can observe that the effect of the recombinant product and the natural product can be different in the different cells. This difference in effect can originate in the source of the natural product. A cell supernatant derived natural product can contain factors that can affect a cell population and therefore not elicit the desired, expected effect. The receptors on the cells choosen can also be different but still sensitive to the same factor and thereby elicit a different biological response. Furthermore, the factor tested can have different biological domains and thereby elicit a different biological effect in different cells. Finally one can observe that there can be different cell populations in a cell line used for bioassay and an effect on some cells can cause these to produce a factor that again elicits a response on other cells in the same flask.

In one of our studies we were interested in studying the effects of IL-1 on islet cells in the pancreas (Mandrup-Poulsen et al., 1987). IL-1 inhibits insulin secretion and following prolonged incubation results in cell death. The theory is that macrophages infiltrate the pancreas as the

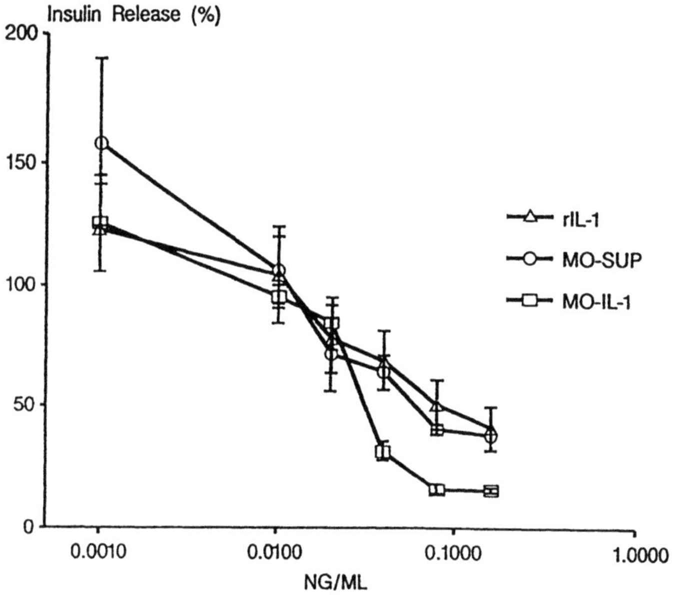

Fig. 2. Comparison of different IL-1 Forms in the Islets Assay

first sign of diabetes and the release of IL-1 from the macrophages is the initial step in insulitis. The desire was to find an antagonist that would inhibit the IL-1 effect on the islet cells.

Three in-vitro assays were set up to study the IL-1 bioactivity. The test material was 1) monocyte supernatant collected from LPS (endotoxin) stimulated monocytes from human patients (MO-SUP); 2) purified monocyte supernatant (same as in 1) resulting in purified monocyte IL-1 (MO-IL-1); 3) E. coli produced, recombinant human IL-1 (rIL-1).

One bioassay was based on rat islets of Langerhans, isolated from newborn rats. In this assay the insulin release from the β-cells was correlated to the IL-1 concentration added to the growth medium. The results showed that IL-1 initially stimulates insulin secretion, but then rapidly inhibits the insulin secretion from the islets and finally results in cell death. Furthermore, all three IL-1 test materials inhibited the insulin secretion.

The inhibition of insulin secretion we see can be due to a difference in receptor type on the islet cells compared to other cells, since there is no cell proliferation which seems to be the normal effect of IL-1, or one could imagine that there are different biological domains on the IL-1 molecule.

A second bioassay was based on a normal human fibroblast cell line (MRC-5) and the use of thymidine incorporation as a means of measuring proliferation. Previous publications had shown that IL-1 exhibited a mitogenic effect on fibroblasts. In our first assays, we noted that monocyte supernatant (MO-SUP) induced a potent mitogenic effect on the fibroblasts, while monocyte IL-1 (MO-IL-1) and recombinant (rIL-1) resulted in a very limited proliferation of the fibroblasts. Using the MRC-5 cell line for our bioassay, we experienced that the proliferative effect was pronounced and reproducible up to passage 45. Thereafter the cells would not respond to different concentrations of IL-1.

The third bioassay was based on mouse thymocytes isolated from 5-6

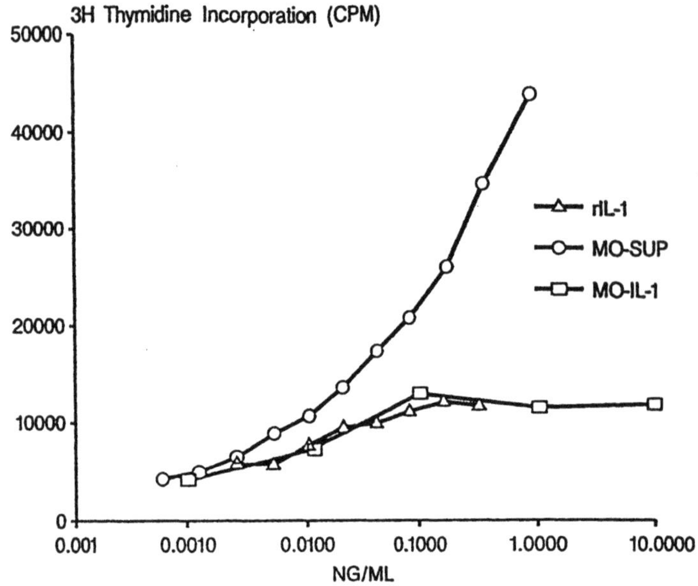

Fig. 3. Comparison of different IL-1 Forms in the Fibroblast Assay

weeks old mice. This assay, the LAF-assay (Leucocyte activating factor) is the most widely used IL-1 assay. The theory is that IL-1 activates the secretion of IL-2 from the thymocytes, which in turn is mitogenic on the thymocytes, resulting in a proliferation measured by thymidine incorporation. Using this assay, we could compare our IL-1's with each other and with the results achieved in other laboratories monitoring the bioactivity of IL-1. In our LAF-assay, the MO-SUP showed the same tendency as in the fibroblast assay, eliciting a potent mitogenic effect. The MO-IL-1 and the rIL-1 elicited only a limited proliferative effect.

We then set out to investigate the difference in the effect of the supernatant MO-SUP and the purified IL-1's MO-IL-1 and rIL-1. From earlier assays, we knew that monocytes produce TNF-α and IL-2.

Initially we then tested MO-SUP, TNF-α and IL-2 in our fibroblast assay. The results showed that the proliferative effect we had noted for the MO-SUP could be due to the TNF-α that was produced by the monocytes and which was removed from the supernatant under purification. The IL-2 had no effect on the fibroblasts.

A colleague of ours performed a study where amounts of rIL-1, rIL-2 and rIL-6 were added to the same culture medium in an amount equivalent to MO-SUP. When thymocytes were incubated with this medium, the proliferative effect was equivalent to the MO-SUP indicating that for this cell type, it was the removal of IL-2 and IL-6 under purification that reduced the mitogenic effect of purified IL-1. When TNF-α was tested in the LAF-assay it gave no response.

The observations on rIL-1 and rIL-6 correspond to the latest publications which indicate that dendritic cells, which are isolated together with the thymocyte population are stimulated by IL-1 to produce IL-6 and that the effect noted on the thymocytes is an indirect effect of IL-6 (Helle et al., 1989; Navarro et al., 1989).

After reviewing the results, our conclusion was that our original choices for assay cultures proved to be less reliable than first thought

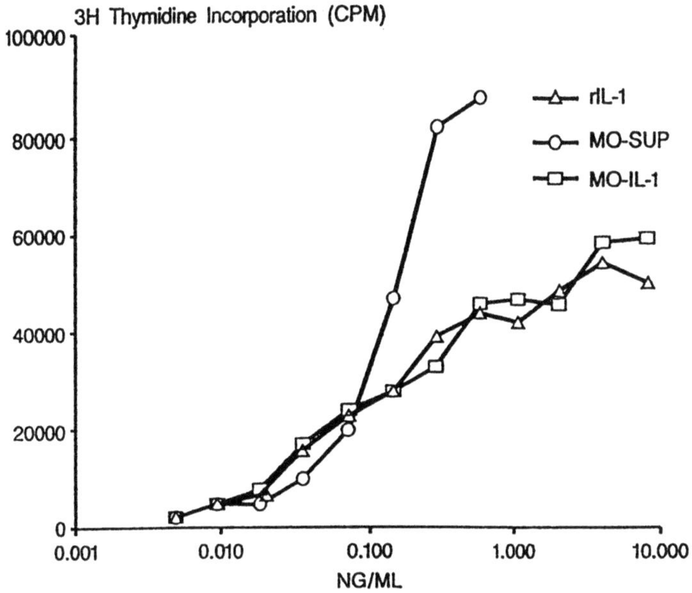

Fig. 4. Comparison of different IL-1 Forms in the LAF Assay

and to be responsive to factors other than the desired test factor. There was only one bioassay out of the three that could give us a reproducible, biologically correct, direct IL-1 effect and that was the islet-cell assay.

CONCLUSION

When evaluating and choosing a cell type or cell line for an assay, one must carefully consider the purpose one hopes to achieve with the assay, how one expects to interpret the response and relate this to the characteristics of the cells one chooses. The existence of many available established cell lines can facilitate the use of more than one cell type for an assay and thereby reduce the possibility of erroneous results caused by unforseeable extraneous factors. One must always keep in mind that cell culture assays can supplement in-vivo assays and can only be used as a substitution if the in-vitro assays are well documented and correlated with an equivalent in-vivo bioassay.

REFERENCES

Biedler, J.L., 1976, Chromosome abnormalities in human tumour cells in culture, in: Human Tumour Cells in Vitro, J. Fogh, ed., Academic Press, N.Y., 359-394.

Cameron, I.L., and Pool, T.B., 1981, The Transformed Cell: Some Introductory Comments, in: The Transformed Cell, I.L. Cameron, and T.B. Pool, eds., Acad. Press., N.Y.

Chen, T.R., 1977, In situ detection of mycoplasma contamination in cell cultures by fluorescent Hoechst 33258 stain, Exp. Cell Res., 104:255.

Collins, S.J., Gallo, R.C., Gallagher, R.E., 1977, Continuous growth and differentiation of human myeloid leukaemic cells in suspension culture, Nature 270:347-349.

Dracopoli, N.C., and Fogh, J., 1983, Polymorphic Enzyme Analysis of Cultured Human Tumor Cell Lines, JNCI, 70:469-476.

Jakoby, W.B., and Pastan, I.H., eds., 1979, Cell Culture, in: Methods in Enzymology, Academic Press, N.Y.

Fogh, J., 1973, Contamination in Tissue Culture, Acad. Press, N.Y.
Fogh, J., Wright, W.C., and Loveless, J.D., 1974, Absence of HeLa Cell Contamination in 169 Cell Lines Derived from Human Tumors, J. Natl. Cancer Inst., 58:209-214.
Fogh, J., and Trempe, G., 1975, New Human Tumor Cell Lines, in: Human Tumor Cells in Vitro, J. Fogh, ed., Plenum Press, N.Y., 115-159.
Fogh, J., Fogh, J.M., and Orfeo, T., 1977, One Hundred and Twenty-seven Cultured Human Tumor Cell Lines Producing Tumors in Nude Mice, J. Natl. Cancer Inst., 59:221-226.
Fogh, J. Bean, M.A., Brüggen, J., Fogh, H., Fogh, J.M., Hammar, S.P., Kodera, Y., Loveless, J.D., Sorg, C., and Wright, W.C., 1978, Comparison of a Human Tumor Cell Line before and after Growth in the Nude Mouse, in: The Nude Mouse in Experimental and Clinical Research, J. Fogh, ed., Academic Press, N.Y., 215-234.
Fogh, J., Dracopoli, N., Loveless, J., and Fogh, H., 1982, Cultured Human Tumor Cells for Cancer Research: Assessment of Variation and Stability of Cultural Characteristics, in: Interaction of Platelets and Tumor Cells, A.R. Liss Inc., N.Y.
Gospodarowicz, D., Greenburg, G, Bialecki, H., Zetter, B.R., 1978, Factors involved in the modulation of cell proliferation in vivo and in vitro: The role of fibroblast and epidermal growth factors in the proliferative response of mammalian cells, In Vitro 14:85-118.
Helle, M., Boeije, L., and Aarden, L.A., 1989, IL-6 is an Intermediate in IL-1 Induced Thymocyte Proliferation, J. Immunol., 142:4335-4338.
Klevjer-Anderson, P., Buehring, G.C., 1980, Effect of hormones on growth rates of malignant and nonmalignant human mammary epithelia in cell culture, In Vitro, 16:491-501.
Macy, M., 1978, Identification of cell line species by isoenzyme analysis, Man. Am. Tissue Cult. Assoc. 4:833-836.
Mandrup-Poulsen, T., Egeberg, J., Nerup, J., Bendtzen, K., Nielsen, J.H., Dinarello, C.A., 1987, Ultrastructural studies of time course and cellular specificity of interleukin-1 mediated islet cytotoxicity, Acta Path. Microbiol. Immunol. Scand. (c), 95:55-63.
McGarrity, G.J., 1982, Detection of mycoplasmic infection of cell cultures, in: Advances in Cell Culture, Vol. 2, K. Maramorosch, ed., Academic Press, N.Y., 99-131.
Navarro, S., Debili, N., Bernaudin, J.-F., Vainchenker, W., and Doly, J., 1989, Regulation of the Expression of IL-6 in Human Monocytes, J. Immunol., 141:4339-4345.
Nicolson, G.L., 1976, Trans-membrane control of the receptors on normal and tumor cells. II. Surface changes associated with transformation and malignancy. Biochim. Biophys. Acta, 458:1-72.
Pollack, M.S., Heagney, S.D., Livingston, P.O., Fogh, J., 1981, HLA-A, B, C & DR alloantigen expression on forty-six cultured human tumor cell lines, J. Natl. Cancer Inst., 66:1003-1012.
Povey, S., Hopkinson, D.A., Harris, H., and Franks, L.M., 1976, Characterization of human cell lines and differentiation from HeLa by enzyme typing, Nature, 264:60-63.
Schneider, E.L., Stanbridge, E.J., 1975, A simple biochemical technique for the detection of mycoplasma contamination of cultured cells, Methods Cell Biol., 10:278-290.
Sharkey, F.E., Fogh, J.M., Hajdu, S.I, Fitzgerald, P.J., and Fogh, J., 1978, Experience in Surgical Pathology with Human Tumor Growth in the Nude Mouse, in: The Nude Mouse in Experimental and Clinical Research, Academic Press, N.Y., 187-214.
Stanbridge, E.J., Doersen, C.-J., 1978, Some effects that mycoplasmas have upon their injected host, in: Mycoplasma Infection of Cell Cultures, G.J. McGarrity, D. Murphy, W.W. Nichols, eds., Plenum Press, N.Y.
Taylor-Robinson, D., 1978, Cultural and serologic procedures for mycoplasmas in tissue culture, in: Mycoplasma Infection of Cell Cultures, G. McGarrity, D.G. Murphy, W.W. Nichols, eds., Plenum Press, N.Y.

Thomas, J.A., 1970, Organ Culture, Academic Press, N.Y.
Tveit, K.M., Pihl, A., 1981, Do cells lines in vitro reflect the properties of the tumours of origin? A study of lines derived from human melanoma xenografts, Br. J. Cancer, 44:775-786.
Van Diggelen, O., Shin, S., Phillips, D., 1977, Reduction in cellular tumorigenicity after mycoplasma infection and elimination of mycoplasma from infected cultures by passage in nude mice, Cancer Res., 37:2680-2687.

USE OF FABRICATED LIVING TISSUE AND ORGAN EQUIVALENTS AS DEFINED HIGHER ORDER SYSTEMS FOR THE STUDY OF PHARMACOLOGIC RESPONSES TO TEST SUBSTANCES

Eugene Bell, Roger Gay, Mark Swiderek, Tom Class, Paul Kemp, Graham Green, Howard Haimes, and Patrick Bilbo

Organogenesis Inc.
83 Rogers Street, Cambridge, Massachusetts 02142

INTRODUCTION

Because animal tissues resemble human tissues, the use of animals as model systems in pharmacologic studies was a natural outcome of laboratory practice. But with the realization that there are significant differences between animal skin and human skin and between responses given by animals as compared with humans,[1,2,3] to substances that come into contact with the skin or tissues of the eye, and because of humane considerations, there has developed a major move away from the use of animals toward alternatives to animal testing.

Although systemic approaches in both animals and humans have been used to study percutaneous absorption and skin irritancy,[4] the criteria used to assess the irritancy potential of substances have been based largely on cutaneous responses in humans and on ocular and cutaneous responses in animals.[5]

Cutaneous responses such as edema, erythema and necrosis are judged subjectively and are, as a consequence, poorly reproducible.[6] Detracting from the value of animal model systems has been their inadequate predictive value, and the difficulty of using them to elucidate mechanisms of injury or favorable changes induced by applied substances because human skin and reactions differ so substantially from animal skin and animal reactions. Recently, many major testing users have been steering away from animal testing and seeking alternatives. A significant alternative is *in vitro* testing. It is hardly a new concept; there have been notable efforts,[7,8,9,10,11] but *in vitro* toxicity, safety and product development testing has yet to have its day.

In searching for new approaches many *in vitro* systems have been explored, ranging from the use of human cell types that are found in skin,[12,13,14,15] to others of non-skin or non-human origin.[16,17,18,19,20] Deciding what changes to assay for in these systems has been a difficult challenge. Beginning with a stimulus *in vivo*, that is, the applied substance(s) or radiation, one can follow the cascade of response signals it provokes as physiologic and immune defenses are activated. There is an advantage in looking near the apex of the defense pyramid for the induction of mediators that represent the earliest signals in the progression that leads to the gross manifestations of damage, edema, erythema and necrosis. Because *in vitro* systems are not equipped with a capillary network and the circulatory system of which it is a part, there is no possibility of simulating erythema and edema *in vitro*. The question is therefore posed, to what extent can the release of a mediator, that can be measured quantitatively, be correlated with the severity of the terminal skin response that is assessed *in vivo* however subjectively? Will it be possible to establish mediator response values that can be used to grade the degree of corrosivity or irritation caused by a substance as a function of applied dose? Are there tests of cytotoxicity, as well, whose results can be correlated with phenomenologic responses *in vivo*? Any new test system should at least be

Pharmaceutical Applications of Cell and Tissue Culture to Drug Transport
Edited by G. Wilson *et al.*, Plenum Press, New York, 1991

able to measure the limit or allowable concentration of a substance alone or in combination with other substances beyond which use would be unsafe. Several approaches are possible. For example, results obtained with a new system can be compared with results obtained with the Draize tests that have been in use for decades and accepted by regulatory agencies. Although the Draize tests, as well as other animal tests, have been used as a basis of comparison for new tests to promote their introduction,[21,22,23] its predictive value for human responses and its reproducibility have been repeatedly questioned.[24,25] Ideally, the ultimate value of a new system's usefulness in predicting human responses should be based on reactions given by people.

In developing systems for testing products or substances that come into contact with human skin, we ask, what are the properties of the ideal system? The ideal system should offer the possibility of discovering reaction mechanisms to help develop a better understanding of how tissues interact with cytotoxic agents or respond to ameliorative ones. To begin to do this effectively, the system should be made to resemble actual skin with maximum faithfulness. The resemblance should be morphological; the model system should possess a degree of organizational complexity and tissue differentiation characteristic of skin. In addition to becoming like skin morphologically in the course of *in vitro* development, and exhibiting a repertoire of biosynthetic properties like those of skin, the expectation is that the model system will provide a broad range of pharmacologic responses including the induced release of cytokines and eicosanoids.

An essential feature of the system of choice is that it provide a structural configuration permitting the application of a test product or substance in a manner similar to its application to actual skin. The susceptibility to damage of the system will be determined in no small measure by the foregoing feature. For example, cells in monolayer culture bathed in a sea of medium are exposed to irritants dissolved or mixed with the medium. The *in vivo* habitat of epidermal cells, keratinocytes in skin for example, is partly arid and partly moist, but not wet; in general, cells in tissues exist in a moist, but not wet environment. As we show in this presentation, monolayered cells on plastic substrates in tissue culture flasks or plates are at least an order of magnitude more susceptible to damage by an irritant of a given dose than the same cells incorporated into a tissue matrix. We think this is due in large part to the unnatural conditions of exposure to applied substances experienced by the cells.

Responses in the ideal system should be quantifiable, and the resolution of dose-response interpolation should be maximized, that is, the slope of the curve should be sufficiently gentle to make interpolation possible. Being able to distinguish among the slopes of dose response curves is a property that could lead to the discovery of qualitative differences among classes of compounds that reflect differences in their mechanism of action. In this paper we present evidence that the Living Skin Equivalent, called TESTSKIN^{tm}, now being manufactured at Organogenesis Inc., has many of the properties desired for an *in vitro* model system.

The interpretation of how a living cellular system responds to an applied substance or mixture of ingredients depends on the character and sensitivity of the assay used to detect and evaluate a response. In looking at a reaction classified as an erythema, only the grossest phenomenologic manifestation of change is estimated. A cascade of chemical and cellular reactions has preceded and led to the manifestation. What hidden damage lies below remains unknown. In estimating what may have happened, there is no possibility of measuring cell or tissue damage in the skin non-intrusively or distinguishing between irreversible and reversible damage. Since the capacity of intact living systems, that is organisms, to repair themselves and to generate adaptive responses that lessen the damage potential of applied or contacted substances is substantial, insufficient attention has been paid to the transient damage inflicted on the system in what have been called tests for irritation. It is highly probable that many irritants at certain concentrations do irreversible damage at the cellular level. We suggest that in given concentration ranges many irritants are cytotoxic. An important property of an *in vitro* system that mimics its actual counterpart in nature, is that it lend itself to the quantitative measurement of cytotoxicity, without giving exaggerated responses. In such a system, it should be possible to measure irritation by measuring cytotoxicity. We report in this paper on early validation studies demonstrating reductions in mitochondrial function as a measure of cytotoxicity and cytokine and eicosanoid induction.

MORPHOLOGICAL AND BIOCHEMICAL SIMILARITIES OF TESTSKIN AND ACTUAL HUMAN SKIN

Methods for preparing living skin equivalents (LSE) have been published elsewhere.[26,27] The model reported in this paper differs from earlier versions in that human neonatal fibroblasts and keratinocytes have been screened for viral and other pathogens and that bovine acid extracted collagen are presently in use. Only strains free of HIV, Herpes, Cytomegalovirus, Hepatitis B, Epstein-Barr virus and human papaloma virus are accepted for manufacturing TESTSKIN. In the course of fabrication, TESTSKIN passes through several stages of development. The precursor of the dermal equivalent (DE), consisting mainly of collagen and fibroblasts, is cast as a fluid mixture into a cylindrical mold in which it gels and undergoes a volume diminution of 20-to 30-fold. The volume decrease is brought about by the collagen condensing activity of the fibroblasts.[26,27] After contraction, the DE is plated with keratinocytes previously cultivated in a proprietary serum free medium. Within a short period, the keratinocytes become confluent, and a stem population gives rise to progeny that undergo differentiation. Complete morphogenesis of the epidermis spans a period of about three weeks from the time of plating. The sequence of changes in a time series of samples spaced four days apart in age is seen in Figs. 1 and 2. From a condition of 2-3 epidermal cell layers (4) days after epidermalization, the epidermis differentiates into a multilayered structure with well defined basal, suprabasal, and granular layers overlaid by a well developed stratum corneum. The morphology of the epidermis is remarkably similar to that of actual skin so it is no surprise that the biochemical markers of epidermal cytodifferentiation are synthesized.[26,28]

Fig. 1. Light micrographs of 2 micron methacrylate embedded Living Skin Equivalents. sections. Samples were fixed in 10% neutral buffered formalin and stained with aniline blue, basic fuchsin and toluidine blue O. Micrographs are 300X: A) 4 days post epidermalization; B) 6 days; and C) 8 days. Note the orderly development of the epidermal layer, that is about 0.1 - 0.15 mm thick, from a simple multi-layered epithelium to a fully matured and cornified epidermis. The stratum corneum is identified by the dark staining at the apical surface of the epidermis. Only a portion of the dermal equivalent, whose full thickness is about 1.0 mm, is shown. Fibroblasts from human neonatal foreskin are surrounded by a well organized collagen containing matrix. (cont.)

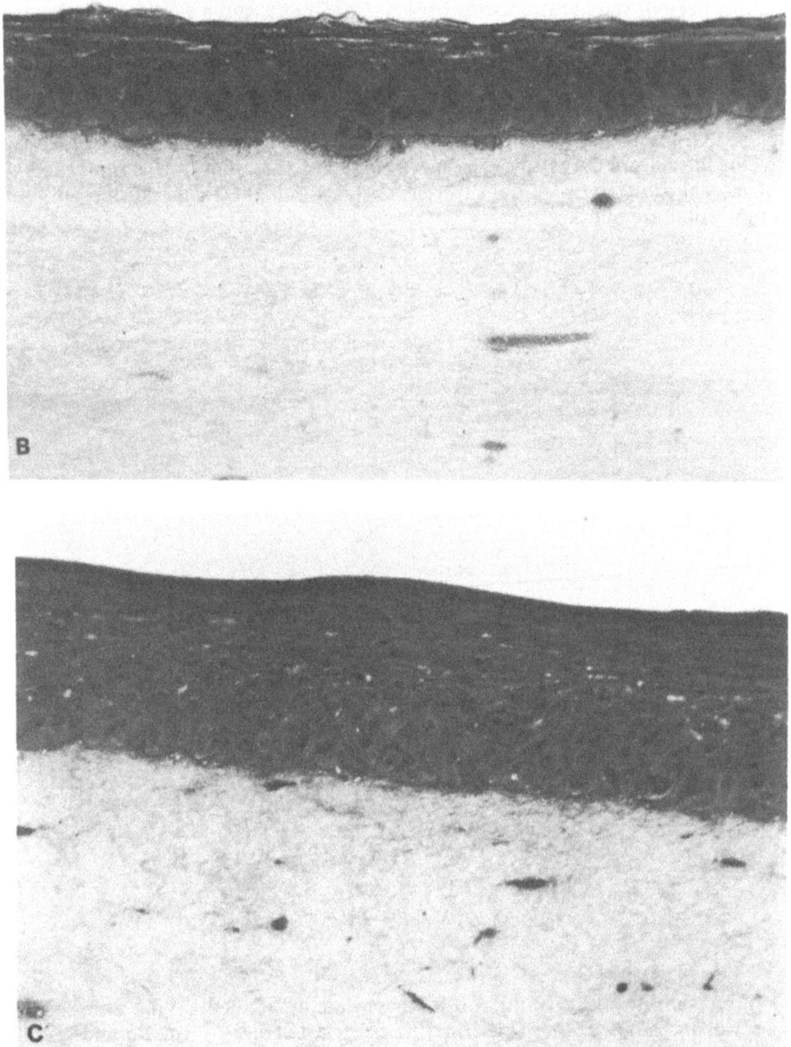

Fig. 1. (cont.)

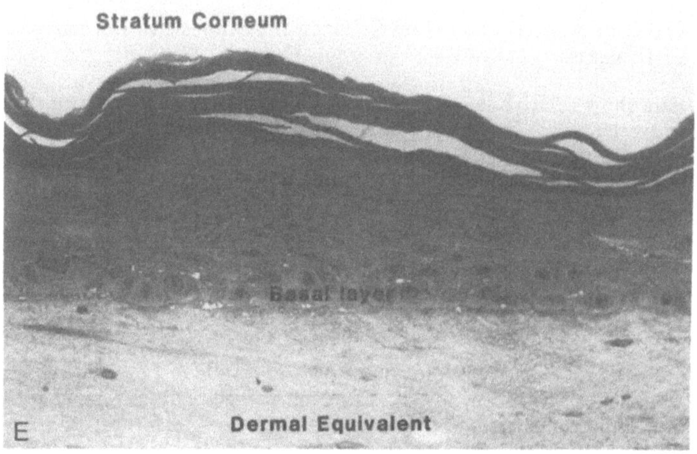

Fig. 2. Light micrographs of Living Skin Equivalents processed as described in Fig. 1: D) 12 days; and E) 25 days.

Both low and high molecular weight keratins are present as we and others have shown; involucrin envelopes are laid down by differentiating keratinocytes as they escalate in the epidermis; keratohyalin granules are found, as is the protein fillagrin with which they come to be filled and which is essential for the organization of the stratum corneum. These features have been documented by immunochemical staining.

Throughout the period of fabrication and maturation, the cells of the DE as well as those of the epidermis are biosynthetically active synthesizing cell type specific proteins. The extent of biosynthesis can be markedly enhanced by the addition to the medium of hormonal or nutritional factors as seen in Fig. 3. We have shown previously that cells in the DE not only synthesize and secrete collagen, but also process it so that the procollagen telopeptides are cleaved enabling the molecules to polymerize.[29] Neither cells grown in monolayers[29,30,31,32] nor cells grown in nylon nets, in pilot experiments, deposit processed collagen; in fact, little or no collagen is found in nylon nets populated with fibroblasts.

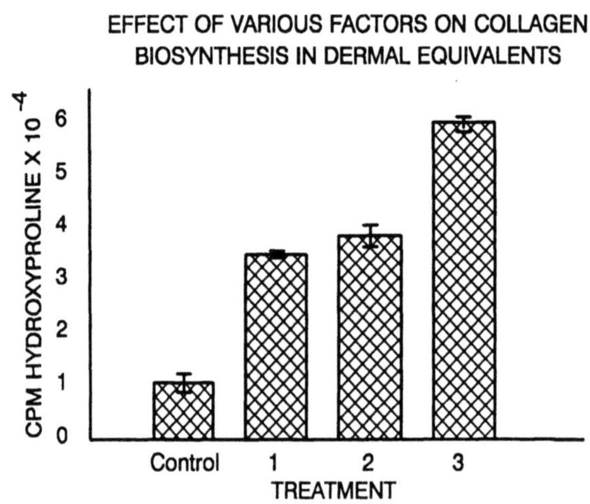

Fig. 3. Effect of Various Factors on Collagen Biosynthesis in Dermal Equivalents.

Bars show relative levels of collagen biosynthesized by dermal equivalents incubated with different factors during pulse labeling with 3H-proline. After incubation, dermal equivalents were extracted and proteins were hydrolyzed to component amino acids. Amino acids were reacted with phenylisothiocyanate for determination of biosynthetically radiolabeled hydroxyproline by reverse-phase HPLC.

While in the foregoing systems, proliferation of cells to a state of confluence is encouraged by the nature of the substrate, biosynthesis is favored and cell bodies remain well separated in the DE, as they do in the dermis *in vivo* even though a two- to three-fold increase in cell number occurs after DE formation. The cells in the DE, however, may become interlinked by cell processes that often extend for tens of micrometers, but the cells nevertheless remain surrounded by large volumes of collagenous intercellular matrix.

THE RESPONSIVENESS OF LIVING TISSUE EQUIVALENTS AS COMPARED WITH MONOLAYERS TO VARIOUS CHEMICALS

Among the criteria listed for a good toxicity assay, the attribute "simulates routes of administration" has been listed as "likely" for an *in vivo* model and "impossible" for an *in vitro* one.[33] Although this paper is only six years old, the author's conclusion at the time it was written was incontestable. With the advent of the LSE and now its standardization in

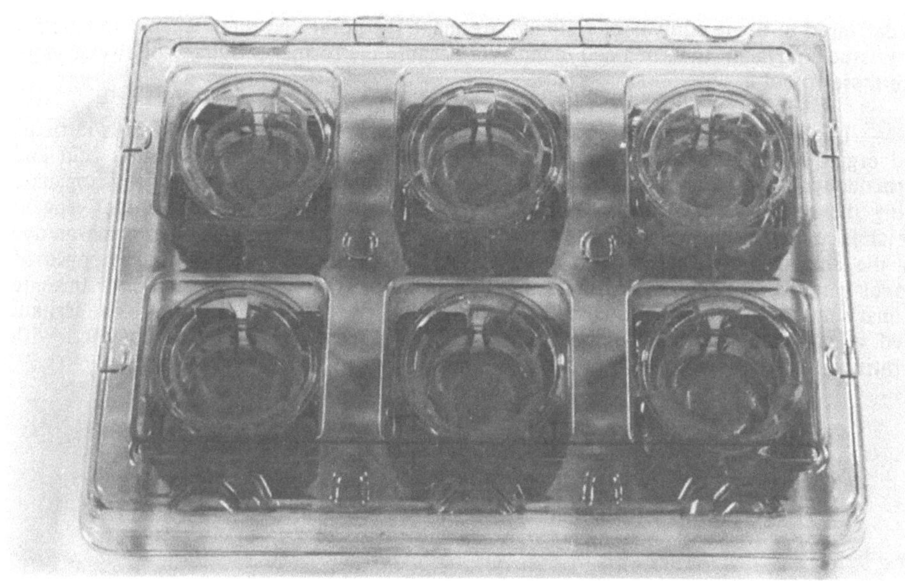

Fig. 4. Six units of TESTSKIN shown in the 6-well shipping tray. Each unit consists of a living skin equivalent composed of a dermal equivalent populated with human dermal fibroblasts and a well-differentiated, stratified epidermis made up of human keratinocytes. The epidermis is exposed to an air-interface while the dermis is bathed in nutrient medium. Each unit of TESTSKIN can be removed from the shipping tray and transferred to a standard 6-well plate for performing assays.

the form of TESTSKIN (Fig. 4), the conclusion needs revision. With an organ equivalent such as TESTSKIN, whose keratinized surface is exposed to the atmosphere and whose dermal underside is in contact with a nutrient pool, but separated from it by a diffusion barrier, there has become available an organotypic *in vitro* system that makes possible a route of administration similar to that provided by the skin of an intact organism. It is important to model the consequences of applying a substance to skin or to an LSE. Consider the results of applying 50-100 microliters of delivery fluid or other vehicle containing a known concentration of a chemical. The applied dose should pass into the hydrated tissue system at some rate determined by the barrier properties of the stratum corneum if one is present. Once past the barrier, taking into account diffusion, active pino or phagocytosis and metabolic conversion, the applied dose should spread in a hemispherical fashion (Fig. 5), creating, because of the resulting dilution of the chemical, a transient dosage gradient that changes with time. The concentration of the applied chemical should vary inversely with the distance from the center of the hemisphere.

Cells in a living tissue will thus see different doses of an applied chemical depending upon where they are in the hemisphere of applied substance distribution, and depending on the time constants of the experiment. In actual skin, clearance and dilution of an applied substance is facilitated by the presence of a capillary network and by lymphatics. In the LSE, which has no circulatory systems to assist in the removal of an applied substance, because of exchange with the nutrient reservoir below the dermis, the concentration of the substance will eventually become uniform throughout the system and will be reduced in concentration in proportion to the total amount of fluid in the reservoir plus that in the hydrated DE. Thus, some cells of the LSE are subjected to a transiently high dose of an applied chemical, while others are subjected to lower doses for some period of time until equilibrium is established, and all cells are subjected to the same dose of reduced intensity until the system is cleared of the substance by a medium change and washing. Because cells in the organ equivalent experience widely different doses, there is an expectation that a response pattern different from that given by cells in a monolayer will result. In a monolayer, all cells are exposed to the same dose of a substance present in the fluid covering the monolayer. Also cells grown in a monolayer differentiate poorly as compared with cells in a tissue equivalent, hence the diversity of cell types is more limited and they exhibit a more limited repertoire of biosynthetic capacity.[34] In addition, primarily aqueous soluble test substances can be applied to monolayers, but salves, emulsions and other vehicles can be delivered to tissue equivalent systems. We make the point that there is more to look for in a tissue equivalent than in a cell monolayer because the former provides conditions more like those found *in vivo*.

In comparing responses of skin cells in a monolayer with responses of cells in tissue and organ equivalents to selected chemicals, namely triethanolamine, benzoic acid and formaldehyde, we have measured changes in the activity of mitochondrial dehydrogenase using the MTT [3-(4,5-dimethylthiazol-2-yl)-2,5-diphenyltetrazolium bromide] reagent which is colored yellow in a neutral aqueous solution and is oxidized to a blue formazan dye by the enzyme. Swisher et al.[14] used the MTT reagent to measure the response of monolayered keratinocytes to irritants and found a "surprisingly good" correlation with acute primary skin irritancy data. In comparing our results (Fig. 6) with theirs for the one irritant used in common, triethanolamine, very similar dose response data was obtained with keratinocytes.

Diffusion of Test Substance from Site of Application through a Tissue

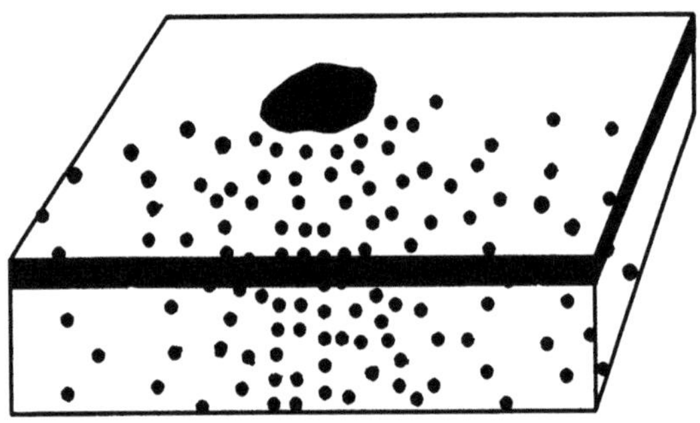

Fig. 5. Theoretical distribution pattern of a substance applied to skin or a tissue equivalent showing local concentration differences that exist transiently until the system reaches equilibrium at some dose much lower than that applied.

However, monolayered keratinocytes and monolayered fibroblasts proved to be hypersensitive in their reactions to triethanolamine as compared with cells in a tissue equivalent. The latter system tolerated doses two orders of magnitude greater than those tolerated by skin cell monolayers being thus a better predictor of both animal and human threshold doses.[1,35] The same magnitude of difference was seen when the applied chemical was formaldehyde (Fig. 7).

Since the spread between monolayer and tissue equivalent responses is different for different substances (Figs. 6, 7 and 8), it is not possible to say that results obtained with tissue equivalents are predictable from those obtained with monolayers. To begin to determine realistic threshold doses for substances that contact the skin it is particularly important to measure susceptibility to damage in a system that neither over or underestimates it. The selection of a test system that can provide realistic limit concentrations, that is, the concentration of a substance in a non-irritant diluent greater than which irritation of the skin may be expected, is described in EEC directives[36] is consonant with current needs. Responses of the DE and LSE to formaldehyde and of the DE to benzoic acid (Fig. 8) all show lower susceptibility to damage than do monolayers of dermal

Cell Systems Exposed to Triethanolamine (24 Hours)

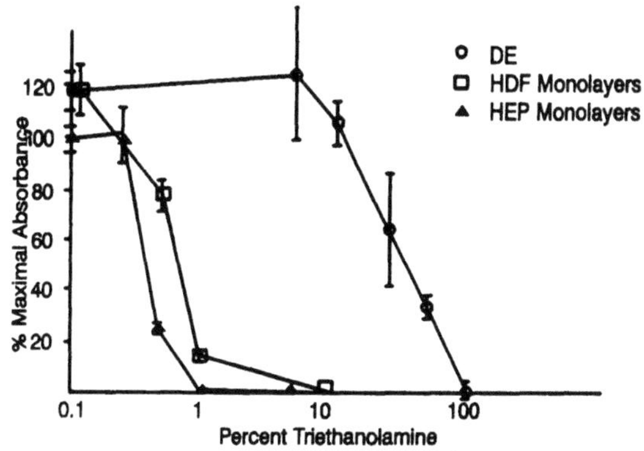

Fig. 6. Dermal equivalents used were 7-10 days old containing 210,000 +/- 20,000 fibroblasts. Monolayers were seeded at 60,000 cells per well in 6-well plates and propagated to confluence. The cell number in the monolayers were 490,000 and 418,000 cells for fibroblasts and keratinocytes, respectively. For sample application to monolayers, the indicated concentrations of triethanolamine (TEA) were solublized in the cell culture medium directly. For DEs, TEA was solublized in water prior to application. DEs had polyethylene washers (circumscribed area of 0.8 cm^2) affixed with silicon grease to the air exposed tissue surface to which 0.08 ml of test or control sample were applied.

All cell systems were exposed to the test substance for 24 hrs. at 37 °C in an incubator gassed with 10% CO_2 and in the presence of 1.5 ml of serum-free cell culture medium (1:1 F-12, DMEM). At the end of the exposure period, medium was aspirated and replaced with medium supplemented with 0.33 mg/ml MTT and incubated for 2 hrs. at 37 °C. Stained monolayers and DEs were extracted in isopropanol, 0.04 N in HCl and absorbances read at 570 nm. Data are expressed as the mean percent +/- the percent standard deviation for four replications. All percentage values are based on 100% MTT conversion determined in controls exposed to the experimental conditions but to which no test substance additions were made.

fibroblasts or keratinocytes. We attribute these differences in susceptibility mainly to the organization of the test tissue or organ system that permits the application of a test substance in a manner that simulates the *in vivo* mode of administration.

EXPOSURE OF TISSUE EQUIVALENTS WITH AND WITHOUT SUNSCREEN PROTECTION TO UVB

Since the surface of the LSE or DE is exposed to air in the test system configuration (Fig. 4), without an overlying layer of fluid, it is possible to study the effects of radiation directly on the tissue. The results of exposing DEs to UVB radiation give a dose response curve with a $Mito_{50}$ of 4.3×10^{-3} Joules/cm^2 which is 1/3 of the Minimal Erythemic Dose(MED) in humans. $Mito_{50}$ is the dose at which mitochondrial function is reduced to 50% of the control value. When DEs were treated with a commercial sunscreen (SPF 36) and then exposed to a dose of 7×10^{-3} Joules/cm^2, which is 1/2 the MED in humans, full protection was afforded to the DE (Fig. 9). It is hoped that the system will be useful for

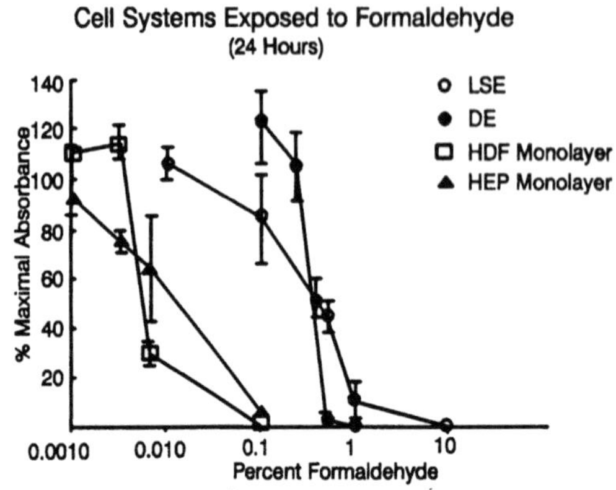

Fig. 7. DEs and LSEs were exposed to 0.08 ml of the indicated concentrations of formaldehyde diluted in water prior to application. For monolayers, formaldehyde was added directly to cell culture medium. All cell systems were exposed to test samples for 24 hrs. at 37 °C prior to assaying for MTT conversion capability. Data are expressed as the mean percent +/- the percent standard deviation for 4 replicate samples. See the legend to Fig. 6 for details.

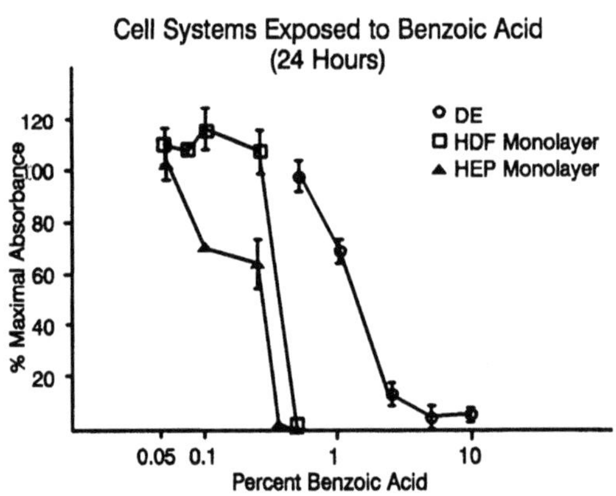

Fig. 8. DEs were exposed to 0.08 ml of the indicated concentrations of benzoic acid solublized in a 50/50 mix of DMSO and water. For monolayers benzoic acid was diluted directly in cell culture medium. All cell systems were exposed to test samples for 24 hrs. at 37 °C prior to assaying for MTT conversion capability. Data are expressed as the mean percent +/- the percent standard deviation for 4 replicate samples. See the legend to Fig. 6 for details.

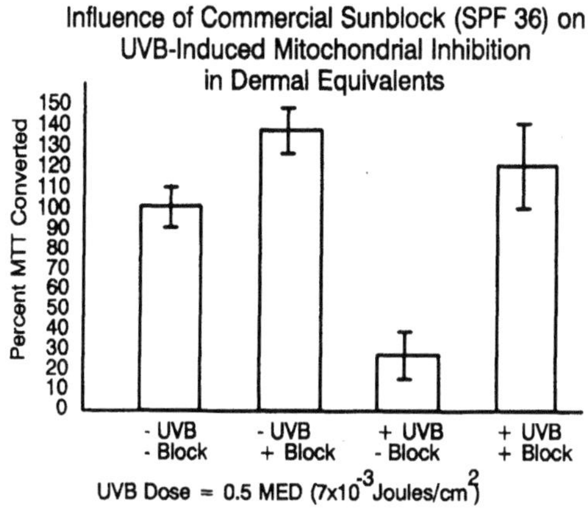

Fig. 9. Dermal equivalents were transferred to 6-well plates containing 1.5 ml of phosphate buffered saline with calcium and magnesium. Polyethylene washers (circumscribed area 0.8 cm^2) were mounted on the air-exposed surface using silicon grease.

DEs were treated with 0.08 ml of sunblock (SPF 36) added to the circumscribed area. Systems were exposed to 7 x 10^{-3} Joules/cm^2 UVB radiation as measured with a research radiometer with a UVB detector. MTT conversion assays were conducted as described in Fig. 6. Data are expressed as the mean percent +/- the percent standard deviation for four determinations normalized to results obtained with DEs which were neither exposed to the sunblock nor to UVB.

developing future sunscreens and for evaluating their efficacy. The LSE on which tests are currently in progress should be even more useful because of the possibility of generating sunburn cells and with added melanocytes, as we have previously reported, stepped up pigment donation to keratinocytes occurs with increasing UVB exposure.

INDUCING IL 1α AND EICOSANOIDS

It is still extremely difficult, if possible, to probe the skin or eye of an intact organism for mediators or other chemical signals induced by an irritant without perturbing the system in the course of making measurements. Thus, the desirability of having an *in vitro* system whose complexity and faithfulness begin to approach nature's counterpart is apparent. In such a system we think it will be possible to accurately measure mediator responses to given concentrations of applied substances known to cause unacceptable damage in humans. While this view is more hypothesis than conclusion, we have conducted experiments with the DE and the LSE showing that cells in the systems are truly responsive to irritants. Interleukin (IL 1α) is induced in LSEs after application of triethanolamine (TEA), benzoic acid orformaldehyde over a range of doses (Fig. 10). As indicated by the response to the lowest dose of TEA, there is no constitutive release of the cytokine. At a high dose of benzoic acid, the response is reduced to zero by the end of 24 hours. To understand the kinetics of decay, a time course study of the first 24 hours is needed. Although not shown, no IL 1α response is given by the DE as expected, since keratinocytes are the source of it. Although a number of eicosanoids are synthesized constitutively by the LSE,[37] including 12-HETE, LTB4, PGE2 and 6-keto-PGF 1α (prostacyclin), thus far with the substances reported above, only the latter two appear to be inducible (Figs. 11 and 12). At the high concentration of formaldehyde the level of prostacyclin is lowered, and PGE2 does not increase relative to the next lower dose. The results correlate moderately well with results given for PGE2 induction and in mitochondrial functions tests, but it is difficult yet to properly interpret the mediator data since it represents the integrated induced responses over a twenty-four hour period, plus the integrated decay of mediator presence as a result

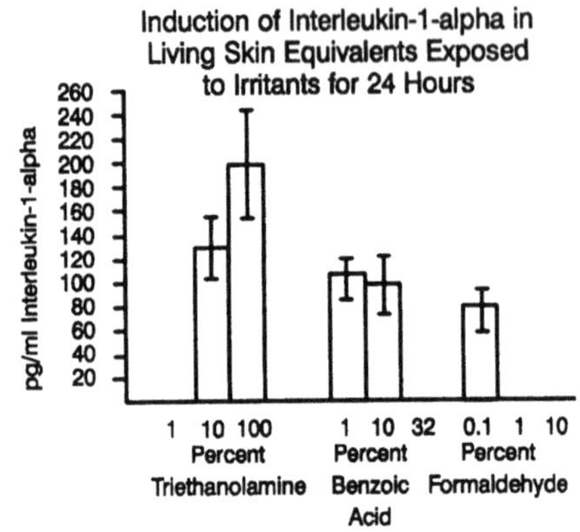

Fig. 10. Dermal equivalents were exposed to the indicated concentrations of triethanolamine, benzoic acid or formaldehyde as described in legends for figures 6, 7 and 8. After the 24 hr. exposure period, medium was collected and stored at -70 °C until thawed for assay. Levels of interleukin-1-alpha were determined using Endogen's ELISA kit for IL-1α. Data are expressed as the mean +/- the standard deviation for 3 determinations.

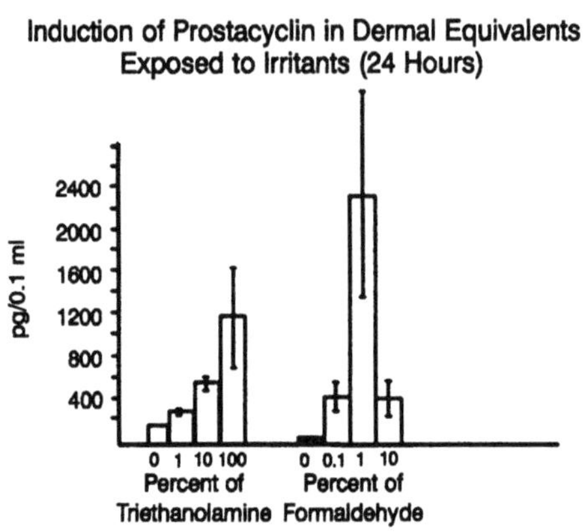

Fig. 11. Dermal equivalents were exposed to the indicated concentrations of triethanolamine or formaldehyde as described in the legends to Figs. 6 and 7. After the 24 hr. exposure period, medium was collected and stored at -70 °C until thawed for assay. Levels of prostacyclin were determined using Advanced Magnetics RIA kit for 6-keto-PGF1-alpha. Data are expressed as the mean +/- standard deviation for 3 determinations.

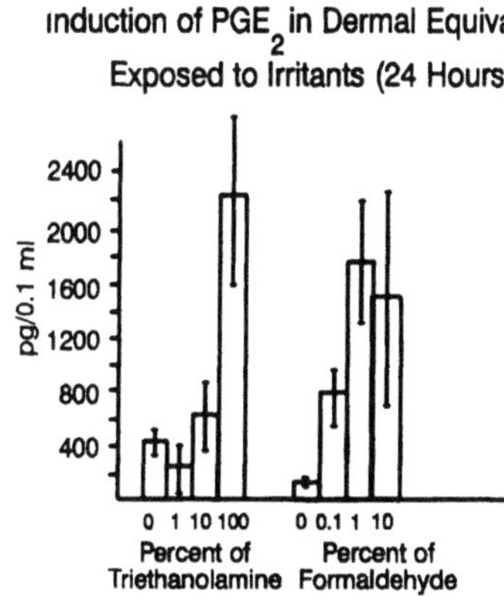

Fig. 12. Dermal equivalents were exposed to the indicated concentration of triethanolamine or formaldehyde as described in the legends to Figs. 6 and 7. After the 24 hr. exposure period, medium was collected and stored at -70 °C until thawed for assay. Levels of PGE2 were determined using Advanced Magnetics RIA kit for PGE2. Data are expressed as the mean +/- the standard deviation for 3 determinations.

of metabolic degradation. Designing experiments that will make fuller interpretation of mediator responses possible is a present goal. There is also the additional issue of understanding how to correlate induction of mediators of response to applied substances with sequelae such as the degree of edema, erythema and necrosis deemed undesirable that occur *in vivo*.

CONCLUSIONS

Substances applied to living systems can damage some components of them reversibly and others irreversibly or not at all; they can also elicit none, few or a broad range of responses of a chemical nature. Seeing what happens depends on how, where and when one looks. A satisfactory *in vitro* test system should lend itself to the measurement of damage and to the measurement of signals with which the system responds to an applied substance or mixture. We have demonstrated that these prerequisites are fulfilled by TESTSKIN. In this paper we have selected reduction of mitochondrial function as a measure of chemical and UV damage and release of the cytokine IL 1α, and the eicosanoids PGE2 and prostacyclin as measures of mediator signaling. We have shown that the TESTSKIN system bears a good resemblance to actual skin morphologically and in its biosynthetic repertoire. It is configured to present *in vivo* conditions for dosing and its resistance to damage by injurious substances is orders of magnitude closer to that given *in vivo* to the same substances.

Although it has not been dwelt upon in this report, the Living Skin Equivalent can be treated as a developing system subject to perturbation by test substances at any point in the course of its maturation. The effect of substances on keratinocyte differentiation which occurs routinely in normal skin and at an accelerated pace during would healing, can be studied in the system. We have previously reported that various steroids affect the capacity of dermal fibroblasts to contract a collagen lattice.[38] The Living Skin Equivalent along with the Living Artery Equivalent, and other defined but complex organotypic *in vitro* systems now being fabricated at Organogenesis Inc. offer new possibilities for testing that begin to resemble their *in vivo* counterparts more closely than do *in vitro* systems available now. The recent report that a living skin equivalent has been successfully grafted in people as a skin replacement[39], along with earlier reports of successful animal trials,[27] attest to the likeness of the LSE to actual skin.

REFERENCES

1. M. Steinberg, W. A. Akers, M. Weeks, A. H. McCreesh and H. I. Maibach, 1. A Comparison of Test Techniques Based on Rabbit and Human Skin Responses to Irritants With Recommendations, Regarding the Evaluation of Mildly or Moderately Irritating Compounds, *in* "Animal Models in Dermatology," H. Maibach, ed., Churchhill, Livingstone (1975).
2. R. Marks and T. Kingston, Acute Skin Toxicity Reactions in Man - Tests and Mechanisms, *Fd. Chem. Toxic.* 23:155-163 (1985).
3. D. W. Swanston, Assessment of the Validity of Animal Techniques in Eye-Irritation Testing, *Fd. Chem. Toxic.* 23:169-173 (1985).
4. D. C. McGill, Cosmetics Companies Quietly Ending Animal Tests, *The New York Times*, August 2:1 (1989).
5. R. J. Feldmann and H. I. Maibach, Percutaneous Penetration *In Vivo* in Man, *in* "Proceedings of International Conference on Cosmetic Sciences," Toilet Goods Assoc., Inc. (1968).
6. J. H. Driaze, G. Woodard and H. O. Calvery, Methods for the Study of Irritation and Toxicity of Substances Applied Topically to the Skin and Mucous Membranes, *J. Pharmacol. Exp. Therap.* 83:377-390 (1944).
7. P. R. Roper and B. Drewinka, Comparison of *In Vitro* Methods to Determine Drug-Induced Cell Lethality, *Cancer Res.* 36:2182-2188 (1976).
8. J. W. Bridges, An Evaluation of the Present Status and Potential Application of *In Vitro* Toxicity Tests, *in* "Towards Better Safety of Drugs and Pharmaceutical Products," D. D. Breimer, ed., Elsevier/North Holland, Amsterdam (1980).
9. A. P. Stammati, V. Silano and F. Zucco, Toxicology Investigations with Cell Culture Systems, *Toxicology* 20:91-153 (1981).
10. W. E. Parish, Relevance of *In Vitro* Tests to *In Vivo* Acute Skin Inflammation: Potential *In Vitro* Applications of Skin Keratome Slices, Neutrophils, Fibroblasts, Mast Cells and Macrophages, *Fd. Chem. Toxic.* 23:275-285 (1985).
11. W. E. Parish, Evaluation of *In Vitro* Predictive Tests for Irritation and Allergic Sensitization, *Fd. Chem. Toxic.* 24:481-494 (1986).
12. M. A. E. Mol, J. Van Genderen and O. L. Wolthuis, Cultured Human Epidermal Cells as a Tool in Skin Toxicology, *Fd. Chem. Toxic.* 24:519-520 (1986).
13. V. DeLeo, S. Scheide, J. Meshulam, D. Hanson and A. Cardullo, Ultraviolet Radiation Alters Choline Phospholipid Metabolism in Human Keratinocytes, *J. Invest. Dermatol.* 91:303-308 (1988).
14. D. A. Swisher, M. E. Prevo and P. W. Ledger, The MTT *In Vitro* Cytotoxicity Test: Correlation with Cutaneous Irritancy in Two Animal Models, *in* "Alternative Methods in Toxicology, Vol. 6, Progress in *In Vitro* Toxicology," A. M. Goldberg, ed., Mary Ann Liebert, Inc., New York (1988).
15. D. A. Swisher, M. Cormier, J. Johnson and P. W. Ledger, A Cytotoxicity Assay Using Normal, Human Keratinocytes: Characterization and Applications, *Models Dermatol.* 4:131-137 (1989).
16. E. Borenfreund and C. Shopsis, Toxicity Monitored with a Correlated Set of Cell-Culture Assays, *Xenobiotica* 15:705-711 (1985).
17. R. B. Kemp, J. Meredith and S. H. Gamble, Toxicity of Commercial Products on Cells in Suspension Culture: A Possible Screen for the Draize Eye Irritation Test, *Fd. Chem. Toxic.* 23:267-270 (1985).
18. M. C. Scaife, An *In Vitro* Cytotoxicity Test to Predict the Ocular Irritation Potential of Detergents and Detergent Products, *Fd. Chem. Toxic.* 23:253-258 (1985).
19. J. Selling and B. Ekwall, Screening for Eye Irritancy Using Cultured HeLa Cells, *Xenobiotica* 15:713-717 (1985).
20. C. Shopsis, E. Borenfreund, J. Walberg and D. M. Stark, A Battery of Potential Alternatives to the Draize Test: Uridine Uptake Inhibition, Morphological Cytotoxicity, Macrophage Chemotaxis and Exfoliative Cytology, *Fd. Chem. Toxic.* 23:259-266 (1985).
21. R. B. Kemp, R. W. J. Meredith, S. Gamble and M. Frost, A Rapid Cell Culture Technique for Assessing the Toxicity of Detergent Based Products *In Vitro* as a Possible Screen for Eye Irritancy *In Vivo*, *Cytobios* 36:153-159 (1983).
22. C. Shopsis and S. Sathe, Uridine Uptake Inhibition as a Cytotoxicity Test: Correlations with the Draize Test, *Toxicology* 29:195-206 (1984).
23. H. E. Kennah II, S. Hignet, P. E. Laux, J. D. Dorko and C. S. Barrow, An Objective Procedure for Quantitating Eye Irritation Based Upon Changes of Corneal Thickness, *Fund. Appl. Toxicol.* 12:258-268 (1989).

24. C. S. Weil and R. A. Scala, Study of Intra- and Interlaboratory Variability in the Results of Rabbit Eye and Skin Irritation Tests, *Toxicol. Appl. Pharmacol.* 19:276-360 (1971).
25. G. A. Nixon, C. A. Tyson and W. C. Wertz, Interspecies Comparisons of Skin Irritancy, *Toxicol. Appl. Pharmacol.* 31:481-490 (1975).
26. E. Bell, B. Ivarsson, C. Merrill, Production of Tissue-Like Structure by Contraction of Collagen Lattices by Human Fibroblasts of Different Proliferative Potential *In Vitro, Proc. Natl. Acad. Sci. U.S.A.* 76:1274-1278 (1979).
27. E. Bell, S. Sher and B. Hull, The Living Skin-Equivalent as a Structural and Immunological Model in Skin Grafting, *Scan. Electr. Micro.* 4:1957-1962 (1984).
28. D. Asselineau, B. A. Bernard, C. Bailly, M. Darmon and M. Pruniéras, Human Epidermis Reconstructed by Culture: Is It "Normal"?, *J. Invest. Dermatol.* 86:181-222 (1986).
29. B. Nusgens, C. Merrill, C. Lapiére and E. Bell, Collagen Biosynthesis by Cells in a Tissue Equivalent Matrix *In Vitro, Collagen Relat. Res.* 4:351 (1984).
30. D. L. Layman, E. B. McGoodwin and G. R. Martin, The Nature of the Collagen Synthesized by Cultured Human Fibroblasts, *Proc. Natl. Acad. Sci. U.S.A.* 68:454-458 (1971).
31. B. Goldberg, E. H. Epstein, Jr. and C. J. Sheer, Precursors of Collagen Secreted by Cultured Human Fibroblasts, *Proc. Natl. Acad. Sci. U.S.A.* 69:3655-3659 (1972).
32. H. Freiberger, D. Grove, A. Sivarajah, S. R. Pinnell, Procollagen I Synthesis in Human Skin Fibroblasts: Effect of Culture Conditions on Biosynthesis, *J. Invest. Dermatol.* 75:425-430 (1980).
33. D. M. Stark and C. Shopsis, Developing Alternative Assay Systems for Toxicity Testing, *Ann. N.Y. Acad. Sci.* 406:92-103 (1983).
34. B. Coulomb, L. Dubertret, E. Bell, C. Merrill, M. Fosse, J. Breton-Gorius, C. Prost and R. Touraine, Endogenous Peroxidases in Normal Human Dermis: A Marker of Fibroblast Function, *J. Invest. Dermatol.* 81:75-78 (1983).
35. K. Y. Chan, Alternative to the Draize Test, *in* "Alternative Methods in Toxicology, Vol. 3, *In Vitro* Toxicology," A. M. Goldberg, ed., Mary Ann Liebert Inc., New York (1985).
36. G. Jacobs, M. Martens and G. Mosselmans, Proposal of Limit Concentrations for Skin Irritation within the Context of a New EEC Directive on the Classification and Labeling of Preparations, *Reg. Toxicol. Pharmacol.* 7:370-378 (1987).
37. E. Bell, M. Rosenberg, P. Kemp, N. Parenteau, H. Haimes, J. Chen, M. Swiderek, F. Kaplan, D. Kagan, V. Mason and L. Boucher, Reconstitution of Living Organ Equivalents From Specialized Cells and Matrix Biomolecules, *in* "Hybrid Artificial Organs", C. Baquey, B. Dupuy, ed., Colloque INSERM, Paris (1989).
38. B. Coulomb, L. Dubertret, E. Bell and R. Touraine, The Contractility of Fibroblasts in a Collagen Lattice is Reduced by Corticosteroids, *J. Invest. Dermatol.* 82:341-344 (1984).
39. J. Nanchahal, W. R. Otto, R. Dover and S. K. Dhital, Cultured Composite Skin Grafts: Biological Skin Equivalents Permitting Massive Expansion, *Lancet* 2:191-193 (1989).

INTRACELLULAR PATHWAYS; EXOCYTIC AND ENDOCYTIC ROUTES RELYING UPON SELECTIVE VESICULAR TRANSPORT

Colin R. Hopkins

Department of Biochemistry
Imperial College
South Kensington
LONDON SW7, 2AZ

INTRODUCTION

For hydrophilic macromolecules to be transported into cells and across tissue barriers the major permeability barriers are the lipid bilayers of the continuous membrane boundaries which surround and compartmentalise all eukaryotic cells. Specialised mechanisms within these boundaries can translocate relatively large molecular species such as polypeptides across the hydrophobic internum of the bilayer but these mechanisms provide major routes of access only into biosynthetic compartments such as the rough endoplasmic reticulum (RER).

Elsewhere within eukaryotic cells the transport of content from one membrane bound compartment to another is accomplished by processes of vesiculation. These processes avoid the need to translocate proteins across the lipid bilayer but involve fusions and fissions within the participating lipid bilayers. They allow one membrane boundary to pinch off free vesicles which can then move to a second boundary and (by what appears to be a reversal of the pinching off process) fuse (Fig.1). These processes of fission and fusion have been described and discussed for more than thirty years but it is only in recent studies using cell free systems that the essential molecular components in the fusion process have begun to be identified (Glick & Rothman, 1987; Shatz, 1989). Despite these recent advances much remains to be understood about the molecular rearrangements which occur within the lipid bilayer during the fission/fusion processes of vesicular transport.

Many integral membrane proteins are mobile within the plane of the lipid bilayers of membrane boundaries and by moving to sites at which vesiculation is taking place they can be transferred from one compartment to another. In this way vesicular transport delivers selected proteins from their site of synthesis in the rough endoplasmic reticulum to destinations throughout the cell. During vesiculation soluble proteins within the compartment lumen can also be transported. Two processes are known to operate; either the protein moves in the fluid phase (a process which is unselective and relatively inefficient) or it is bound to integral membrane proteins trafficking into the vesiculating membrane ('receptor proteins'). The sorting mechanisms which induce trafficking proteins (and if they are receptors, their bound ligand) to concentrate within the vesiculating membrane provide selective, high efficiency transport steps.

Pharmaceutical Applications of Cell and Tissue Culture to Drug Transport
Edited by G. Wilson *et al.*, Plenum Press, New York, 1991

INTRACELLULAR PATHWAYS

Trafficking proteins being carried by vesicular transport through the cell from one compartment to another have closely regulated itineraries, and it is possible therefore to describe several major trafficking routes through the cell. Within these routes regulatory points can be identified at which further progress may be prevented or, alternatively, a variety of destinations selected (Cutler, 1988). Most of the regulatory points which have been defined to date occur at sites of vesiculation.

EXOCYTIC PATHWAYS

In structural and functional terms the best characterised pathways are the exocytic pathways. These include the secretory pathway to the plasma membrane. This route was first identified in the studies of Palade and his co-workers (Palade, 1975) which followed the processing of secretory products (content) through exocrine secretory cells. These proteins were shown to move from the RER through the Golgi to become stored in secretory granules where they accumulated. Movement of the secretory granules to the plasma membrane where they fused and released their content into the extracellular space (exocytosis) only occurred on stimulation. The route through the secretory granule compartment is therefore a 'regulated' pathway. Other routes to the plasma membrane from the Golgi are also known to exist but there is no evidence that there is the equivalent of a long term storage (i.e. hours) compartment within them. In these pathways, as in the earlier stages of the exocytic pathway, transport occurs at a steady rate (i.e. it is "unregulated" or "constitutive")(Kelly, 1985).

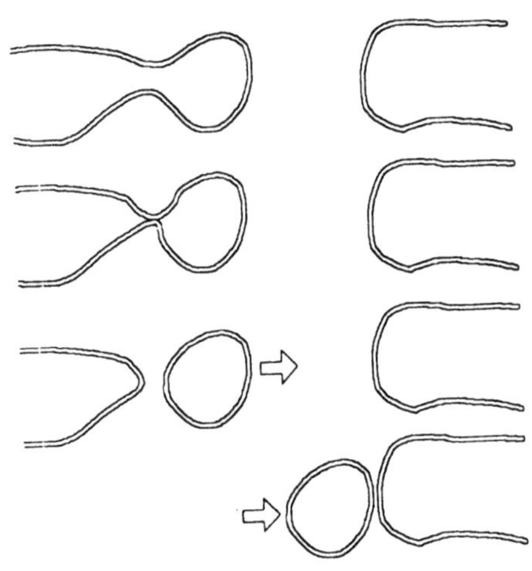

FIG. 1. Vesicular transport from one intracellular compartment to another. By a process of fusion then fission in the lipid bilayer a vesicle is pinched off one cisternum and moves to another.

The RER to Golgi stack is the major biosynthetic pathway of the cell and there are, therefore, routes in addition to those leading to the cell surface which lead to a variety of intracellular destinations. The best characterised of these deliver acid hydrolase enzymes to lysosomal compartments (von Figura and Hasilik, 1986; Griffiths et al., 1988). However, these pathways, like all pathways trafficking through the Golgi cisternae, have yet to be clearly defined and much remains to be elucidated about the routes available to molecules exiting from the RER and Golgi stack (Griffiths and Simons, 1986). One general feature of the pathways leading out of the RER which should be noted is that while these pathways can be considered to be unidirectional for most of the content and many of the membrane proteins they transport, there must also be routes which return membrane components to the RER. These routes have not yet been identified morphologically but must exist if only because the amount of lipid and protein synthesized in the RER is insufficient to provide for the large amounts of membrane which leave it by vesicular transport. The recycling of membrane which can be clearly demonstrated in the endocytic pathway (see below) is therefore likely to be a major component of the vesicular transport which takes place between RER and Golgi and other steps in the exocytic pathway. The very existence of recycling steps implies that there must be molecular mechanisms of recognition controlling the destination of vesicles shuttling between membrane boundaries; otherwise there can be no certainty that an outward bound vesicle will fuse only with the next designated compartment in the pathway.

Some sequential compartments along intracellular pathways (for example; the RER and the "cis" or "receiving" face of the Golgi stack) lie in close juxtaposition to each other. It is reasonable to assume therefore that the vesicles shuttling between them have relatively short half lives and that they fuse within the immediate vicinity of their site of formation. The agency which moves the vesicles through the cytoplasm is unknown but if they have the ability to recognise the membrane with which they are required to fuse, additional vectorial constraints (in the form for example of cytoskeletal elements) may not be necessary. This does not seem to be true of all the steps in exocytic pathways, however, and the movement of vesicular components between the Golgi complex and the plasma membrane (i.e. secretory granules in a regulated pathway) illlustrates such a requirement. Here the vesicular container is required to move from the juxtanuclear area to the cell surface and in the case of the regulated pathway of exocrine cells the secretory granules are required,in addition, to fuse with a specific domain of the cell surface - the apical plasma membrane. Recent reports suggest that an intact microtubular cytoskeleton is required for delivery via the regulated pathway to the apical membrane (or its equivalent)(Eilers et al., 1989; Rivas and Moore, 1989). Where carrier vesicles need to travel over longer distances (axonal transport in neurons would be the most extreme example) there may well be an additional requirement for them to be vectorially constrained so that they can be delivered to the vicinity of the membrane boundary with which they will fuse.

ENDOCYTIC PATHWAYS

Major trafficking routes from the plasma membrane into the cell also exist. The best documented lead from the plasma membrane through a series of endocytic compartments known collectively as the "endosome" to lysosomal elements. Related pathways, which use some of the same endosomal compartments, lead from the plasma membrane through the endosome and back to the plasma membrane either in the same domain as the pathway began (i.e. recycling pathways) or to a different domain of the cell surface (i.e. transcytosis - see below).

The first endosomal compartment reached by vesicles which have pinched off from the plasma membrane (the "early" endosome) lies immediately below the plasma membrane (Ciechanover et al., 1983; Hopkins and Trowbridge, 1983). This compartment is highly pleiomorphic (consisting of tubulovesicles and vacuoles which are often interconnected) and it is clear that there is extensive recycling from this compartment back to the plasma membrane. However, a proportion of the content and membrane components received from the plasma membrane are also transferred deeper into the cell to the juxtanuclear area, adjacent to the Golgi stack and surrounding the centrioles (Hopkins, 1986; Schmid et al., 1988). This "late" endosome compartment is, as yet poorly characterised, but is clearly en route to the lysosomal elements which are also concentrated in this area. It, too, is highly pleiomorphic and consists of vacuolar elements (often containing inner vesicles - multivesicular bodies) and tubulovesicular elements. Trafficking proteins entering this compartment can be transferred to lysosomal elements where they are degraded or they can be returned to the cell surface. The return route, to the plasma membrane is, at the present time, especially obscure. There is some evidence that it may include the biosynthetic compartments of the Golgi stack; both regulated and constitutive exocytic pathways have been implicated (Farquhar, 1978; Orci et al., 1986; Snider and Rogers, 1986). Other, more major routes back to the surface although as yet undefined, cannot, however, be ruled out. The relationships between the biosynthetic and endocytic routes and their relationship to lysosomal compartments are currently areas of intense interst. Temperature manipulation (18.5

vs 24 vs 37°C) and microtubule disrupting agents indicate that several discontinuous steps (presumably shuttling vesicles) and a vectorially constrained delivery system (between the early and late endosomes) exist in the endosome system (Dunn et al., 1980; de Brabander et al., 1988; Matteoni and Kreis, 1988; Hopkins, 1986). There are thus some similarities with the exocytic route. However, because endosomal compartments are more plastic and mobile than the cisternae of the RER and Golgi stack these steps are proving much more difficult to define.

MOLECULAR MECHANISMS INVOLVED IN THE SORTING AND TRAFFICKING OF PROTEINS

The cellular pathways described above consist of series of membrane boundaries connected by shuttling vesicular carriers. Vesicular transport steps imply the transfer of bulk membrane bound from one compartment to the next and while recycling certainly returns some membrane to the donor compartment recent studies (made directly upon living cells using labelled lipid probes) (Koval and Pagano,1989) show that even the constituents of the lipid bilayer can traverse the whole length of the endocytic pathway from plasma membrane to 'late' endosome.

As discussed above the agencies which drive vesiculation and ensure the delivery of a vesicle carrier to its correct destination are not known. However it seems that many of these processes are constitutive. They are independent of the content or integral membrane protein being transferred and provided that the essential requirements of the process (e.g. membrane fluidity, energy) exist their rates of transfer are relatively constant. Only at regulated steps (such as the secretory granule to the plasma membrane step in the exocytic pathway and phagocytic uptake in the endocytic pathway) are the rates of compartment to compartment transfer amenable to experimental manipulation.

From the above, a picture of the compartmented systems of eukaryotic cells begins to emerge in which vesicular transport provides a series of constitutive transfer points along well defined transfer routes. Integral membrane proteins trafficking along these routes need to gain access to the vesiculating membrane if they are to leave a membrane boundary. The molecular recognition systems involved in controlling access to the vesiculating membrane have been studied most intensively in one type of invagination only; the coated pit of the plasma membrane (Pearse, 1975; Pearse and Crowther, 1987). Other selective domains of invaginated membrane certainly exist (and in some membrane boundaries such as in the RER and in the plasma membrane of continuous endothelia,they predominate). However, these other vesiculating domains have not, as yet, been characterised at the molecular level.

Within the clathrin coated pit it is known that a series of accessory proteins (the 100kd adaptins and the clathrin triskelions) are associated with the cytoplasmic surface of the membrane boundary (Robinson, 1989; Kirchausen et al., 1989). There is good evidence to suggest that the cytoplasmic domains of the integral membrane proteins which can sequester within this invaginated membrane directly interact with some of these accessory proteins. It has been shown,for example, that a certain specified sequence on the cytoplasmic tail of the LDL receptor is required for the protein to become concentrated within coated pits and when isolated receptor protein is used to make an affinity column this same sequence is required for the column to bind isolated adaptin proteins selectively (Pearse, 1988). Within the sequence a single residue, a tyrosine, is especially crucial for coated pit localisation since the other flanking residues can be substituted (albeit conservatively) without a very significant loss of function (Davis et al., 1987). Tyrosines have been found to be important in the cytoplasmic tails of other proteins which can concentrate within coated pits (Lazarivits and Roth, 1988) and very recently it has been shown that if the LDL receptor sequence is included in the tail of an integral protein not normally found in pits the protein will behave like LDL-R, concentrating in the pit and internalising.

However other pit sequestering proteins are known in which there is no indication of an LDL-like recognition motif in their primary sequence and in which tyrosine, although required for full function, is not as crucial. In the transferrin receptor for example a ten residue sequence essential for high efficiency clustering in coated pits has been identified (Shuqian Jing et al., submitted for publication). Moving the position of the sequence about within the 62 residue domain of the transferrin receptor's cytoplasmic tail does not significantly alter the ability of the receptor to cluster in pits suggesting that the interactions between receptor tails and pit accessory proteins may be various and of a relatively low order.

In growth factor receptor proteins (such as EGFR, PDGFR and insulin R) the cytoplasmic domains have intrinsic tyrosine kinase activities which are activated by ligand binding (Carpenter, 1987). It has been proposed that this activity may also be involved in the clustering of receptor proteins within coated pits (Glenney et al., 1988). To date, however, most of the evidence suggests that growth factor receptors are able to concentrate within coated pits (and endocytose) without their kinases being activated (Honeggar et al., 1987; Escobedo and Williams, 1988). However, there is also evidence to suggest that the trafficking itinerary of 'kinase silent' receptors is altered at a later stage in their processing - in the late endosome - well beyond the pits of the plasma membrane (Felder et al., submitted).

In addition to signals permitting access to vesicular transport steps signals preventing access also exist. There is, for example, recent evidence from studies of immunoglobulin receptors which suggests that in cells in which the receptor is excluded from coated pits its cytoplasmic domain has a segment of 47 residues not seen in the receptor of cells where clustering in coated pits does occur (Miettinen et al., 1989).

These observations have all been made on coated pits at the cell surface; the most experimentally accessible of the cell's membrane boundaries. However, studies on intracellular pathways on transport out of the RER or into the lysosomal or peroxisome compartments indicate that similar signals must also reside in the cytoplasmic domains of the mobile integral proteins of these membrane boundaries (Nilsson et al., 1989; Lobel et al., 1989; Gould et al.. 1989).

POLARISED CELLS IN LINING EPITHELIA

The cells of the simple lining epithelia which constitute the major tissue barriers of the body are polarised both structurally and fuctionally. The junctional complexes, (and especially the tight junctions) which surround them inhibit transport between neighbouring cells and thus separate the environment to which the cell is exposed on its apical surface from that surrounding its basolateral borders (Diamond, 1977). Such cells have two clearly defined plasma membrane domains, and while similar regionalisations of the cell surface may occur in other cell types (e.g. the neuron or even the sessile macrophage) it is clear that in the lining epithelial cell the spacial organisation of the intracellular compartments is also linearly polarised. Thus for example while most of the RER usually resides in the basal cytoplasm below the nucleus, the Golgi complex and other pericentriolar elements lie in the apical cytoplasm. Additional exocytic pathways serving two plasma membrane domains are thus required and both of these domains can serve as the starting point of endocytic routes into the cell. Studies on the exocytic routes of lining epithelia suggest that there is a common biosynthetic pathway from the RER and through the Golgi stack but that at the "trans" or "releasing face" the Golgi probably gives rise to pathways which serve either the apical or the basolateral domains of the plasma membrane (Simons and Fuller, 1985; Caplan et al., 1986). Where a regulated pathway exists this usually serves only the apical border (Eilers et al., 1989).

ENDOCYTIC PATHWAYS IN LINING EPITHELIA

Studies on endocytic routes in lining epithelia give, as yet, an incomplete picture and to date most of the evidence comes from work on enterocytes. From work described above in cultured monolayers it is to be predicted that if both apical and basolateral plasma membranes are involved in endocytic uptake the vesicles produced will fuse with separate apical and basolateral "early" endosome compartments (Fig. 2). The equivalent of their "late" juxtanuclear endosomes may be expected to reside in the pericentriolar area above the nucleus. Early peripheral and late juxtanuclear endosome compartments are known to recycle components to the plasma membrane. In polarised cells there are more alternatives, proteins trafficking into the early endosome from the apical membrane could recycle to that membrane or alternatively to the basolateral membrane. Similar alternatives apply to recycling routes from the late endosome.

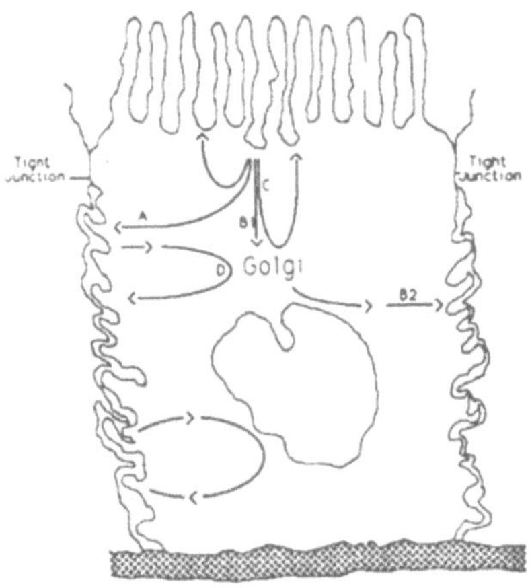

FIG. 2. Endocytic routes which can provide opportunities for transcellular transport. (A) Direct apical to basolateral transcytosis. (B1) Apical border to the lysosome compartment via the late endosome. (B2) Biosynthetic pathway from Golgi to basolateral border. (C) Apical border to late endosome and recycling to the apical border. (D) Basolateral border to late endosome and recycling to basolateral border.If any of these routes (B1/C and B2/D) intersect within the pericentriolar area then indirect transcellular routes may be available. Short circuit recycling pathways to the early endosomes below the apical and basolateral borders are also shown.

In the enterocytes of neonates a direct pathway for trafficking receptors between the apical and basolateral membranes has been shown to exist but there is no evidence that such a pathway exists in adult tissues. Also in the neonate a pathway from the basolateral border to the lysosomes of the apical pericentriolar area has been identified using a fluid phase marker. These same lysosomes can be reached via vesicular routes leading in from the apical border (Fujita et al., 1989).

Studying the trafficking routes of adult enterocytes is difficult in intact tissues and for many recent studies model, in vitro systems have been developed (Cereijido et al., 1978; Huet et al., 1987). The most widely used employ adenocarcinoma cells (HT29 or CaCO-2) growing as tight (200-300 ohms resistance) monolayers on filters. In studies currently in press, Dr. Elaine Hughson working in my laboratory has used the CaCO-2 system to study the uptake of a variety of macromolecular tracers from both basolateral and apical surfaces of these polarised cells. These studies have shown that transferrin receptors which are expressed only on the basolateral border transmit transferrin tracers to a peripheral endosome in the basolateral cytoplasm. From this peripheral endosome most of the transferrin recycles to the basolateral border. After longer incubations some of the internalised transferrin appears in elements in the apical cytoplasm. When chased, all of this transferrin tracer recycles to the basolateral surface. None of the tracers we have applied basolaterally are transferred to the apical surface. Macromolecular tracers introduced across the apical membrane gain access to a peripheral endosomal compartment in the apical cytoplasm. After longer incubations these tracers reach endosomal elements in the pericentriolar area. In double label experiments apically applied tracers are not detectable in the peripheral endosomes (loaded by basolaterally applied tracer) even after prolonged incubations. However, if tracer is applied basolaterally long enough to load endosomal elements in the apical cytoplasm and then a second tracer is applied apically, double labelling is observed within minutes. The late endosomal elements of basolateral routes are thus accessible from the apical border. The failure of basolaterally applied tracer to be transferred to the apical surface even after very prolonged incubations demonstrates the ability of the apical endosomal elements to sort between tracer entering from the separate surfaces with very high efficiency. In the apical cytoplasm connecting pathways thus exist but passive leakage from one to another does not occur.

In a recent study of transcytosis in a different polarised culture preparation a trafficking system with intriguing similarities has been identified (Breitfield et al., 1989). This work followed the uptake of polymeric IgA from the basolateral surface of MDCK cells expressing the human polymeric IgA receptor. It showed that 45% of the IgA at the internalised basolateral border was rapidly recycled to that border but 30% was transferred across the cell to the apical cell surface. Introducing IgA to the apical border resulted in uptake at 70% the level observed at the basolateral border. However all of this tracer was recycled to the apical surface; none was transferred to the basolateral border.

CONCLUSION

Together these in vivo and in vitro studies suggest that the endocytic pathways of polarised epithelial cells offer a wide variety of interconnecting routes reaching to both poles of the cell. Additional exocytic pathways certainly exist and there may well be frequent intersections. However, the preliminary indications are that (except for the neonate) trafficking along these pathways is closely controlled by receptor mediated mechanisms. It seems clear therefore, that while there may well be real opportunities in the enterocyte to exploit existing vesicular pathways for drug delivery more will need to be learned about the behaviour of trafficking membrane proteins before really significant advances can be made.

REFERENCES

Breitfeld, P. P., J. M. Harris, and K. E. Mostov. (1989). Postendocytotic sorting of the ligand for the polymeric immunoglobulin receptor in Madin-Darby Canine Kidney cells. J. Cell Biol. 109: 475-486.

Caplan, M. J., H. C. Anderson, G. E. Palade, and J. D. Jamieson. (1986). Intracellular sorting and polarized cell surface delivery of (Na^+-K^+) - ATPase, an endogenous component of MDCK cell basolateral plasma membranes. Cell. 24: 24-32.

Carpenter, G. (1987). Receptors for epidermal growth factor and other polypeptide mitogens. Ann. Rev. Biochem. 56: 881-914.

Cereijido, M., E. S. Robbins, W. J. Dolan, C. A. Rotuno, and D. D. Sabatini. (1978). Polarized monolayers formed by epithelial cells on a permeable and translucent support. J. Cell Biol. 77: 853-880.

Ciechanover, A., A. L. Schwartz, and H. F. Lodish. (1983). The asialoglycoprotein receptor internalizes and recycles independently of the transferrin and insulin receptors. Cell. 32: 267-275.

Cutler, D. F. (1988). The role of transport signals and retention signals in constitutive export from animal cells. J. Cell Science. 91: 1-4.

Davis, C. G., I. R. van Driel, D. W. Russell, M. S. Brown and J. L. Goldstein. (1987). The low density lipoprotein receptor. Identification of amino acids in cytoplasmic domain required for rapid endocytosis. J. Biol. Chem. 262: 4075-4082.

Diamond, J. M. (1977). The epithelial junction: bridge, gate, and fence. Physiologist. 20: 10-18.

De Brabander, M., R. Nuydens, H. Geerts, and C. R. Hopkins. (1988). Dynamic behaviour of the transferrin receptor followed in living epidermoid carcinoma (A431) cells with nanovid microscopy. Cell Motil. Cytoskeleton. 9: 30-47.

Dudouet, B., S. Robine, C. Huet, C. Sahuquillo-Merino, L. Blair, E. Coudrier, and D. Louvard. (1987). Changes in villin synthesis and subcellular distribution during intestinal differentiation of HT29-18 clones. J. Cell Biol. 105: 359-370.

Dunn, W. A., A. L. Hubbard, and N. V. Aronson. (1980). Low temperature selectively inhibits fusion between pinocytic vesicles and lysosomes during heterophagy of ^{125}I-asialofetuin by the perfused rat liver. J. Biol. Chem. 255: 5971-5978.

Eilers, U., J. Klumperman, and H.-P. Hauri. (1989). Nocodazole, a microtubule-active drug, interferes with apical protein delivery in cultured intestinal epithelial cells (Caco-2). J. Cell Biol. 108: 13-22.

Escobedo, J. A., and L. T. Williams. (1988). A PDGF receptor domain essential for mitogenesis but not for many other responses to PDGF. Nature. 335: 85-87.

Farquhar, M. G. (1978). Recovery of surface membrane in anterior pituitary cell. Variations in traffic detected with anionic and cationic ferritin. J. Cell Biol. 77: R35-R42.

von Figura, K., and A. Hasilik. (1986). Lysosomal enzymes and their receptors. Annu. Rev. Biochem. 55: 167-193.

Fujiki, Y., A. L. Hubbard, S. Fowler, and P. B. Lazarow. (1982). Isolation of intracellular membranes by means of sodium carbonate treatment: application to endoplasmic reticulum. J. Cell Biol. 93: 97-102.

Fujita, M., F. Reinhart, and M. Neutra. (1988). Convergence of apical and basolateral endocytic pathways at the late endosome; double tracer studies in ileal absorptive cells of suckling rats. J. Cell Biol. 107: Abtr. 616.

Glenney, J. R. Jr., W. S. Chen, C. S. Lazar, G. M. Walton, L. M. Zokas, M. G. Rosenfield, and G. N. Gill. (1988). Ligand induced endocytosis of the EGF receptor is blocked by mutational inactivation and by microinjection of anti-phosphoprotein antibodies. Cell. 52: 675-684.

Gould, S. J., G. -A. Keller, N. Hosken, J. Wilkinson, and S. Subramani. (1989). A conserved tripeptide sorts proteins to peroxisomes. J. Cell Biol. 108: 1657-1664.

Griffiths, G., and K. Simons. (1986). The trans Golgi network: sorting at the exit site of the Golgi complex. Science. 234: 438-443.

Griffiths, G., B. Hoflack, K. Simons, and S. Kornfeld. (1988). The mannose 6-phosphate receptor and the biogenesis of lysosomes. Cell. 52: 329-341.

Heuser, J. (1989). Changes in lysosome shape and distribution correlated with changes in cytoplasmic pH. J. Cell Biol. 108: 855-864.

Honeggar, A., M. Dull, T. J. Felder, S. Van Obberghen, E. Bellot, F. Szapary, D. Schmidt, A. Ulrich, and J. Schlessinger. (1987). Point mutation at the ATP binding site of EGF receptor abolishes protein-tyrosine kinase activity and alters cellular routing. Cell. 51: 199-209.

Hopkins, C. R. and I. S. Trowbridge. (1983). Internalization and processing of transferrin and the transferrin receptor in human carcinoma A431 cells. J. Cell Biol. 97: 508-521.

Hopkins, C. R. (1986). Membrane boundaries involved in the uptake and intracellular processing of cell surface receptors. TIBS. 11: 473-477.

Huet, C., C. Sahuquillo-Merino, E. Coudrier, and D. Louvard. (1987). Absorptive and mucus-secreting subclones isolated from a multipotent intestinal cell line (HT-29) provide new models for cell polarity and terminal differentiation. J. Cell Biol. 105: 345-358.

Kelly, R. B. (1985). Pathways of protein secretion in eukaryotes. Science. 230: 25-32.

Kirchausen, T., K. L. Nathanson, W. Matsui, A. Vaisberg, E. P. Chow, C. Burne, J. Keen, J. Hand, and A. E. Davis. (1989). Structural and functional division into two domains of the large (100- to 115-kDa)) chains of the clathrin-associated protein complex AP-2. Proc. Natl. Acad. Sci. USA. 86: 2612-2616.

Koval, M., and R. E. Pagano. (1989). Lipid recycling between the plasma membrane and intracellular compartments: Transport and metabolism of fluorescent sphingomyelin analogues in cultured fibroblasts. J. Cell Biol. 108: 2169-2181.

Lazarovits, J., and M. Roth. (1988). A single amino acid change in the cytoplasmic domain allows the influenza haemagglutinin to be endocytosed through coated pits. Cell 53: 743-752.

Lobel, P., K. Fujimoto, R. Ye, G. Griffiths, and S. Kornfeld. (1989). Mutations in the cytoplasmic domain of the 275kd mannose-6-phosphate receptor differentially alter lysosomal enzyme sorting and endocytosis. Cell. 57: 787-796.

Matteoni, R., and T. E. Kreis. (1988). Translocation and clustering of endosomes and lysosomes depends on microtubules. J. Cell Biol. 105: 1253-1265.

Miettinen,H. M., J. K. Rose, and I. Mellman. (1989). Fc receptor isoforms exhibit distinct abilities for coated pit localization as a result of cytoplasmic domain hetereogeneity. Cell. 58: 3017-3027.

Orci, L., M. Ravazzola, M. Amherdt, D. Brown, and A. Perrelet. (1986). Transport of horseradish peroxidase from the cell surface to the Golgi in insulin-secreting cells: preferential labeling of cisternae located in an intermediate position in the stack. EMBO J. 5: 2097-2101.

Nilsson, T., M. Jackson, and P. A. Peterson. (1989). Short cytoplasmic sequences serve as retention signals for transmembrane proteins in the endoplasmic reticulum. Cell. 58: 707-718.

Pearse, B. M. F. (1975). Coated vesicles from pig brain: purification and biochemical characterization. J. Mol. Biol. 97: 93-98.

Pearse, B. M. F. (1988). Receptors compete for adaptors in plasma membrane coated pits. EMBO J. 7: 3331-3336.

Pearse, B. M. F., and R. A. Crowther. (1987). Structure and assembly of coated vesicles. Ann. Rev. Biophys. Chem. 16: 49-68.

Rivas, R., J. Hsiao-Ping, and H. Moore. (1989). Spatial segregation of the regulated and constitutive secretory pathways. J Cell Biol. 109: 9521-9525.

Robinson, M.S. (1989). Cloning of cDNAs of two related 100kD coated vesicle proteins (alpha adaptins). J. Cell Biol. 108: 833-842.

Schmid, S. L., R. Fuchs, P. Male, and I. Mellman. (1988). Two distinct subpopulations of endosomes involved in membrane recycling and transport to the lysosomes. Cell. 52: 73-83.

Simons K., and S.D. Fuller. (1985). Surface polarity in epithelia. Ann. Rev. Cell Biol. 1: 243-288.

Snider, M., and O. C. Rogers. (1986). Membrane traffic in animal cells: cellular glycoproteins return to the site of Golgi mannosidase 1. J. Cell. Physiol. 115: 199-207.

CARRIER-MEDIATED TRANSPORT OF BILE ACIDS AND AMINO ACIDS IN CACO-2 CELLS

Ismael J. Hidalgo and Ronald T. Borchardt

Department of Pharmaceutical Chemistry
The University of Kansas
Lawrence, Kansas 66045 USA

INTRODUCTION

The small intestinal epithelium comprises a cell population exhibiting continuous growth and differentiation *in vivo*, where the continuous loss of differentiated cells at the villus tip is compensated by cell replication at the villus crypt.[1] The presence of tight junctions, which limits the paracellular diffusion of solutes, together with the numerous enzymes present in the brush border region make the intestinal epithelium the most significant barrier to mucosal absorption of drugs and nutrients.

It is clear that a cell monolayer system consisting of viable and polarized intestinal cells, similar to the epithelial cell layer found in the small intestine, would be a valuable tool in the elucidation of the processes involved in mucosal drug and nutrient transport and metabolism. However, attempts either to culture isolated mucosal cells or to establish cell lines derived from the enterocyte have not been successful.[2,3] For this reason, the use of human colon adenocarcinoma cell lines, some of which have been reported to undergo enterocytic differentiation in culture (i.e., Caco-2, HT-29, SW1116, LS174T),[4-7] appears attractive.

In a recent study, Caco-2 cells were grown on

Pharmaceutical Applications of Cell and Tissue Culture to Drug Transport
Edited by G. Wilson *et al.*, Plenum Press, New York, 1991

collagen-coated microporous polycarbonate membranes and characterized as a model system for intestinal epithelial transport.[8] As indicated by light microscopy and cell growth curves, Caco-2 cells reached monolayer density 6-7 days after seeding. Although occluding junctions were observed as early as day 3, large intercellular spaces were also visible at this time. After day 6, the intercellular spaces disappeared and prominent desmosomes were observed. Concomitant with these observations was the progressive formation of a brush border region of numerous and increasingly organized microvilli.[8] After 16 days in culture, Caco-2 cell monolayers consisted of cells, approximately 30 μm in height, with a morphology similar to that found in the simple, columnar epithelium of the small intestine.[8]

The polarized nature of the cells was evidenced by the asymmetric distribution of the brush border marker enzyme alkaline phosphatase, which after 6 days in culture, was localized only on the apical (AP) surface of Caco-2 cells,[8] and by the predominantly basolateral (BL) expression of human transferrin receptors in confluent monolayers.[9]

A variety of water-soluble, membrane-impermeant fluid-phase markers that ranged in molecular weight from 453 to 70000 daltons were used to evaluate the permeability properties of Caco-2 cell monolayers.[8] Between day 3 and day 17, the amounts of lucifer yellow CH (MW 453) and [^{14}C]inulin (MW 5200) that underwent paracellular leakage decreased with increasing days in culture. In contrast, transepithelial electrical resistance (TEER) values increased from 96.6 $\Omega.cm^2$ at day 3 to 173.5 $\Omega.cm^2$ at day 6 and remained constant through day 17. At day 13, leakage of [^{14}C]polyethyleneglycol (MW 4000) and [^{3}H]dextran (MW 70000) was less than 0.1% of the administered amount per hour. This low leakage was consistent with the presence of prominent junctional complexes (Figure 1). However, it should be noted that although TEER values have been shown to provide a good indication of the resistance of monolayers to ionic flux,[10] the precise meaning of TEER values as it relates to the paracellular diffusion of neutral substances is not clearly established.

The behavior of the monolayer towards the paracellular and

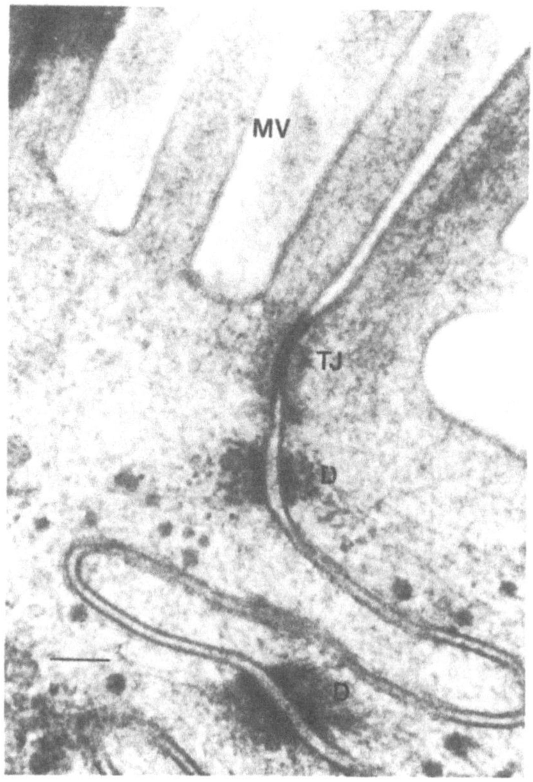

Fig. 1. Transmission electron micrograph showing the different elements of the junctional complexes of Caco-2 cells. Fifteen day old cells were fixed in 2.5% glutaraldehyde in 0.1M cacodylate buffer at room temperature for 1 hr, post-fixed with 1% OsO_4. After dehydration through a graded series of ethanol, the cell-inserts were embedded in a medium-hard recipe of Epon 812. Cross-sections were mounted onto formvar-coated grids, stained with aqueous 7.5% uranyl acetate and lead citrate, and examined with a JEOL 1200EX electron microscope. Key: MV, microvilli; TJ, tight junctions; D, desmosomes. Bar = 100 nm.

transcellular transport of the fluid-phase marker horseradish peroxidase (HRP) provided additional information on the integrity of the monolayers as well as some insight into the pinocytic activity of Caco-2 cells.[8] Presentation of HRP to the AP surface of polarized cell monolayers at 37 °C resulted in fluid-phase pinocytosis of HRP into numerous multivesicular bodies (MVB), 250-300 nm in diameter, that were clustered beneath the terminal web. After exposure to the BL surface, HRP filled the intercellular spaces, stopping at the occluding junctions.[8] These results were in excellent agreement with a previous study in the rat *in vivo*.[11] Following BL administration, HRP was localized within MVB, inside compartments of 500-600 nm diameter, possibly dense bodies or lysosomes, and within smaller diameter vesicles and tubules located near the terminal web and lateral membranes. At 4°C, no endocytosis of HRP was observed by either the AP or BL membrane.

A potential approach to rational drug delivery involves the utilization of the body's own carrier systems to achieve the transport of therapeutic agents across relevant biological membranes. Following the characterization of the Caco-2 cell monolayers as a model transport system of a polarized intestinal epithelium, we have undertaken several studies to determine the suitability of this system for studying specific carriers found in the small intestine. Because bile acids and amino acids are absorbed from the lumenal cavity of the small intestine via active carriers, the transport characteristics of a model bile acid (e.g., taurocholic acid) and a model amino acid (e.g., phenylalanine) were investigated in Caco-2 cell monolayers.

BILE ACID TRANSPORT IN CACO-2 CELL MONOLAYERS

Bile acids (BA's) are produced in the liver and secreted into the small intestine where by virtue of their surfactant activity they play a major role in the intestinal absorption of lipids.[12] After exerting their action, BA's are passively absorbed from the duodenum and jejunum or actively transported across the ileal mucosa and recycled back into the liver. This enterohepatic recirculation ensures minimal loss of BA's into the feces and maximizes their utilization along the entire length of the small intestine. In spite of the significant body of knowledge concerning the enterohepatic recirculation of BA's *in vivo*, little information is available regarding the molecular mechanisms responsible for transepithelial transport of BA's. Therefore, using the Caco-2 cell monolayer system as an example of polarized epithelium, we have conducted the experiments described below that have helped to clarify the mechanisms responsible for transepithelial transport of BA's.

Transcellular Transport of Bile Acids

Since the expression of the marker enzymes sucrase-isomaltase, alkaline phosphatase and aminopeptidases increased with days in culture,[4,8] taurocholic acid (TA) transport studies were performed with Caco-2 cells cultured for different times.[13] Results from these studies indicate that between days 6 and 13, both apical-to-basolateral (AP-to-BL) and basolateral-to-apical (BL-to-AP) transport of TA showed a decrease (Table 1) which most likely was due to a decrease in paracellular leakage across the monolayer. After day 13, however, there was an increase in AP-to-BL transport and a decrease in BL-to-AP transport. At day 28 the rate of transport in the AP-to-BL direction was 14.5

Table 1. Apical-to-basolateral and basolateral-to-apical transport of TA at different times in culture.

TIME IN CULTURE (Days)	TRANSPORT * ($pmol.mg\ protein^{-1}.hr^{-1}$) AP-to-BL	BL-to-AP
6	10.8 ± 0.8	4.9 ± 1.4
9	3.5 ± 1.5	1.5 ± 0.8
13	2.0 ± 0.4	1.0 ± 0.3
24	9.8 ± 0.9	0.8 ± 0.1
28	14.5 ± 1.1	1.1 ± 0.1
32	13.5 ± 0.4	0.9 ± 0.3

*: The monolayers were incubated with 0.1 μM [^{14}C]TA for 60 min at 37°C. Values are means±SD (N=4).

pmol.mg protein^{-1}.hr^{-1}, while that in the opposite direction (BL-to-AP) was only 1.1 pmol.mg protein^{-1}.hr^{-1} (Table 1).

The appearance of TA in the BL side following AP application increased linearly at concentrations up to 10 µM, but saturation of the transport system was observed above this concentration (data not shown). The kinetic parameters, V_{max} and K_m, for transport of TA were calculated to 822.1 pmoles mg protein^{-1}. hr^{-1} and 49.7 µM, respectively, indicating that the carrier system has a high affinity for TA but that its transport capacity is approximately 250-fold lower than that of the ileal BA carrier.[14]

The transport of TA across the monolayer of Caco-2 cells was temperature-dependent. For example, the rates of transport of 0.1 µM TA in the AP-to-BL direction were 0.84, 0.27 and 0 pmoles.mg protein^{-1}.hr^{-1} at 37, 23 and 4°C, respectively. Using the Arrhenius plot, the activation energy (Ea) for transcellular transport of TA was calculated to be equal to 13.2 Kcal.mol^{-1}.[13] This value is consistent with active transport, because Ea's for enzymatic and carrier-mediated processes are commonly between 7 and 25 Kcal.mol^{-1}, while Ea's for single diffusion processes are less than 4.0 Kcal.mol^{-1}.[15]

The transport of TA in the AP-to-BL direction was significantly reduced in the presence of sodium azide plus 2-deoxyglucose (2-DG) or in the absence of glucose, indicating that the transport of BA's in Caco-2 cells is energy-dependent[13]. The Na^+-K^+ ATPase inhibitor ouabain (2.5 mM) reduced the transport of TA by approximately 52%, suggesting the existence of both Na^+-dependent and Na^+-independent transport pathways in Caco-2 cells.[13] TA transport was also significantly reduced by unlabeled TA and by deoxycholic acid (DC).[13]

The evidence described above strongly suggests that Caco-2 cells express a BA carrier, similar in some respects to that found in the enterocyte.[13] Since active transport of BA's has been observed in ileal but not in jejunal cells,[15,16] these results suggest that Caco-2 cells exhibit ileal characteristics with respect to BA transport. The main discrepancies, however,

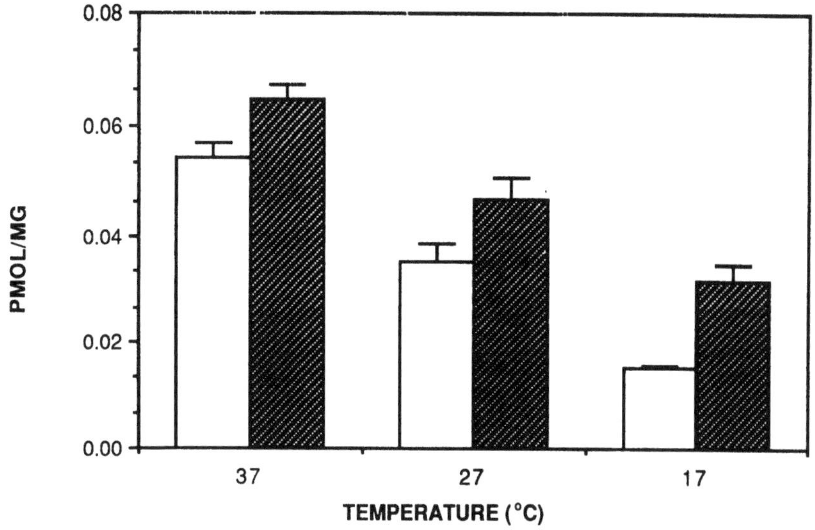

Fig. 2. Temperature-dependency of AP (□) and BL (▨) uptake of TA. Four monolayers were incubated with 0.1 μM [^{14}C]TA either on the AP or BL side at 37, 27 or 17°C for 1 hr. Subsequently, the cells were washed 4 times with HBSS and dissolved with a 1% Triton-X 100 or 1 M NaOH. Mean±SD.

were that BA transport in Caco-2 cells was not as Na^+-dependent as in the enterocyte system and that the transport capacity (i.e., V_{max}) was much lower than in the enterocyte.[14]

<u>Uptake and Efflux of Bile Acids in Caco-2 Monolayers</u>

TA transport was also investigated in Caco-2 cells in terms of uptake and efflux from either the AP or BL membrane of the cell. When administered on the AP or BL side of the monolayers, TA was taken up in a time-dependent manner at approximately equal rates (BL, 27.0 pmol.mg protein^{-1}.hr^{-1}; AP, 17.4 pmol.mg protein^{-1}.hr^{-1}). These results were unexpected, since the transport of TA was polarized almost exclusively in the AP-to-BL direction.[13] Similar to the AP-to-BL transport of TA, the rates of uptake of TA by AP and BL membranes were temperature-dependent (Figure 2). However, the AP uptake exhibited a greater temperature-dependency. Following AP or BL loading of Caco-2 with 10 μM TA for 30 min, the extent of efflux after 30 min was 89% and 95%, respectively (not shown). The fact that the extent of efflux was similar after AP or BL loading and further, that more than 90% of the TA taken up by the AP or BL membrane underwent BL efflux indicates that the BA efflux pathway in Caco-2 cells is independent of the route of entry. The amounts of BA retained in the cells after loading at 37 °C, followed by efflux at 37 or 4°C for 30 min, were 7% and 70%, respectively, indicating that not only the uptake but also the efflux was temperature-dependent.

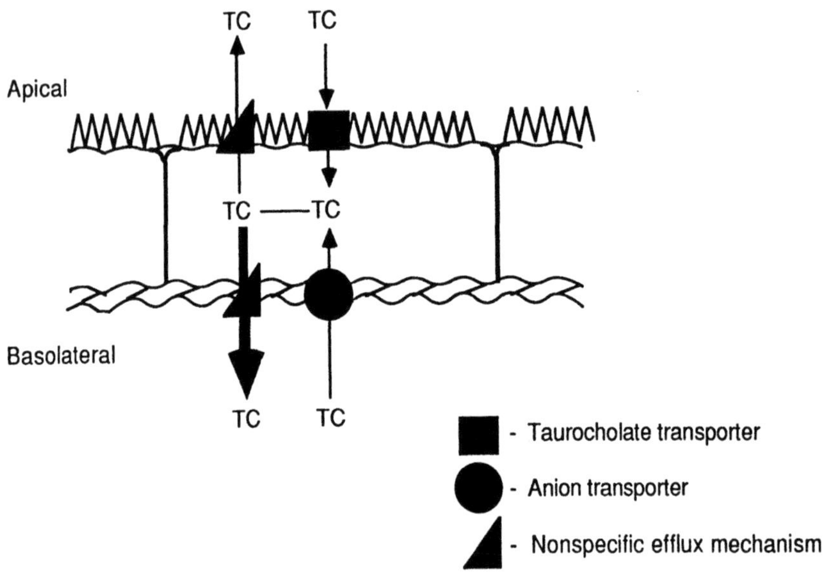

Fig. 3. Uptake and transport of TA in Caco-2 cells. This model indicates that while the AP uptake of TA involves a specific BA carrier, the BL membrane takes up TA via an anion transporter. BA's undergo efflux via a carrier different from that mediating the AP or BL uptake of BA's. Regardless of the route of uptake, TA undergoes efflux through the BL membrane.

Since the AP uptake was inhibited by DC, while the BL uptake was not (not shown), it seems that only the AP uptake is mediated by a BA-specific carrier system. The lack of effect of DC on the BL uptake of TA is consistent with the presence of a different carrier on the BL membrane. The fact that the co-administration of 1 mM p-aminohippuric acid (pAH) inhibited the BL but not the AP uptake of TA suggests the involvement of a relatively nonspecific anion transporter on the BL membrane. Two recent studies have identified an anion transporter in the BL membrane of rat ileal enterocytes.[18,19] A schematic representation of the proposed pathways mediating the uptake and transcellular transport of BA's in Caco-2 cells is shown in Figure 3.

AMINO ACID TRANSPORT IN CACO-2 CELL MONOLAYERS

Caco-2 cell monolayers were also used to investigate the mechanism of amino acid (AA) transport in polarized epithelial cells. Hidalgo and Borchardt[20] have recently shown active transport of the large neutral amino acids (LNAA) (e.g., phenylalanine (Phe)) in Caco-2 cells and the results are summarized below.

To assess the polarity of LNAA transport, Phe was applied to the AP or BL side in separate experiments and the amounts of AA that appeared in the contralateral side (i.e., BL side after AP administration and vice versa) were determined. At day 18, the AP-to-BL transport of Phe was greater than 10-fold faster than that in the BL-to-AP direction (Figure 4). Since the intestinal absorption of amino acids involves their translocation across the epithelial mucosa in the lumenal-to-serosal direction, the polarity of Phe transport exhibited by Caco-2 cells is consistent with the physiologic role of this carrier system in vivo.

The AP-to-BL transport of Phe was time-, concentration- and temperature-dependent (data not shown). The temperature-dependent component of Phe transport was determined by subtracting the transport observed at 4°C, after correcting for the change in diffusion coefficient due to the lower temperature, from the transport observed at 37°C. V_{max} and K_m for transepithelial transport of Phe, calculated by performing linear regression analysis on the double reciprocal plot of the rate of transport vs concentration data, were 3.34 nmol.mg protein^{-1}.hr^{-1} and 0.45 mM, respectively.

The rate of AP-to-BL transport of 10 μM Phe against a steep concentration gradient (100 μM in the BL side) was indistinguishable from that in the absence of such gradient (control). In addition, the presence of an excess amount (10 mM) of cold L-Phe in the AP side reduced the transport of [^{3}H]Phe by approximately 85% (not shown).

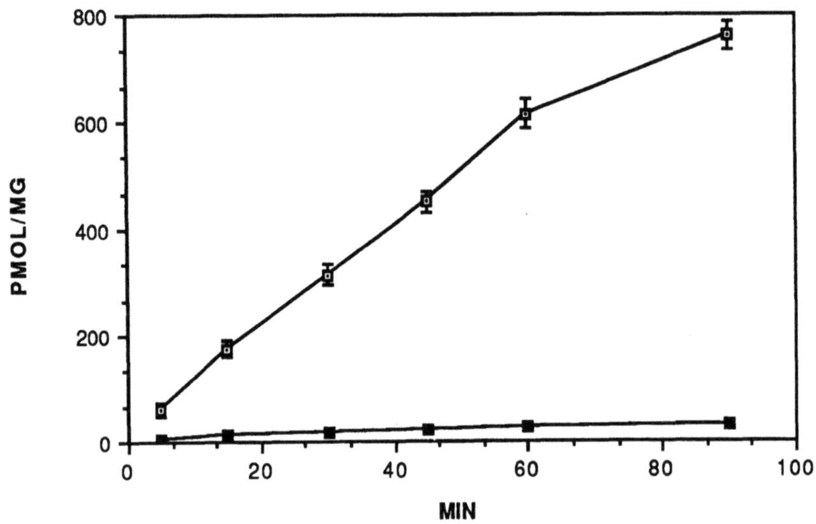

Fig. 4. Polarity of Phe transport in 18 day old Caco-2 cell monolayers. The AP-to-BL (□) and BL-to-AP (■) transport of Phe were determined after applying 0.1 mM Phe to the AP or BL side of 4 monolayers and incubating at 37°C. Values are means±SD.

The LNAA carrier system was further characterized by examining the effect of some experimental conditions on the transport of Phe. The presence of sodium azide plus 2-DG caused a significant reduction in Phe transport, thus confirming the energy requirement of this transport system. Although statistically significant, the effect of azide was not as dramatic as might have been expected. This observation agrees with previous studies showing that the ability of the intestinal epithelium to transport amino acids was not predictable on the basis of the ATP content of the cells[21] and suggest that the energy blockade produced by azide spares small amounts of ATP, which may be sufficient to energize the transport of Phe.[22] This amino acid carrier system showed reduced transport capacity in the presence of 0.1 mM ouabain and in the absence of glucose.[20]

It is known that, with the exception of proline and hydroxyproline, the transporter of LNAA requires a primary amino group in the α-position to the carboxyl group and an uncharged side chain.[22] In order to gain additional insight into the structural requirements of the LNAA transporter in Caco-2 cells, the effects of selected amino acids on Phe transport were investigated. Since in the small intestine Phe is transported by the same LNAA carrier system responsible for the transport of Leu,[23] the inhibitory effect of Leu on Phe transport in the Caco-2 cell monolayer system (Figure 5) was not surprising. In addition, Osiecka et al[24] reported a virtually complete inhibition of Phe uptake by Leu in rat everted intestinal rings.

The lack of effect of Gly on Phe transport was not unexpected since the LNAA transporter requires a bulky substituent in the α-position to the carboxyl group. The pronounced effect of the aromatic amino acids Trp and Tyr further emphasized the importance of side chain bulkiness and agreed with an earlier study showing that Trp and Met inhibited Phe transport by 34% and 55%, respectively[23].

That the cationic amino acids Lys and His inhibited Phe transport by 49.6% and 67.4% most likely indicates that, as in the small intestine, neutral amino acids are also transported by the cationic amino acid carrier system (Figure 5). The inhibition of Phe transport by the L-isomer and the lack of effect of the D-isomer attest to the stereospecificity of the carrier (Figure 5).

The amino acid-type drug L-Dopa was the most potent inhibitor of Phe transport (Figure 5). Indeed, the effect of L-Dopa was similar to that of unlabeled L-Phe. The antihypertensive drug α-methyldopa caused a significant

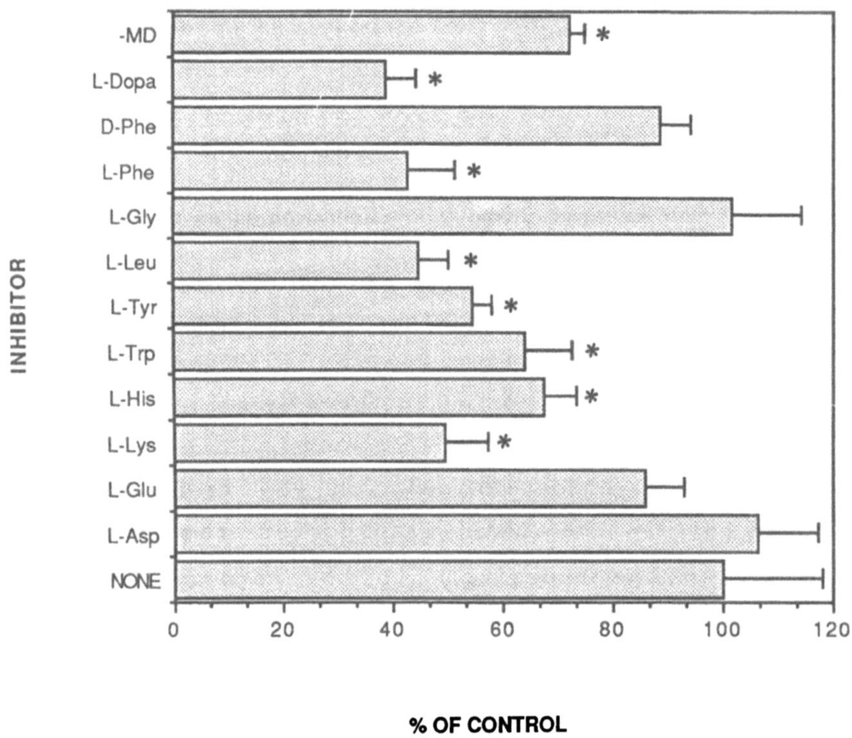

Fig. 5. Differing effects of various amino acids on the AP-to-BL transport of Phe across 18 day old Caco-2 monolayers. The transport of 10 µM [^{3}H]Phe was determined in the absence (control) and presence of 1 mM concentration of selected amino acids, and incubated for 30 min at 37°C. The values represent the Means±SD of 4 monolayers. *: $P<0.05$, compared to the control group.

inhibition of Phe transport. These results are consistent with data reported by Osiecka et al[24] in rat everted intestinal rings.

CONCLUSIONS

Caco-2 cells possess carrier systems for BA's and LNAA's. The transport of both BA's and LNAA's exhibit full polarity in the AP-to-BL direction. Although the cells take up BA's from both the AP and BL sides, virtually all of the BA that is internalized undergoes BL efflux. While on the AP membrane the uptake of BA appears to be mediated by a BA carrier, the BL uptake most likely involves an anion transport system. The evidence suggests that the BL efflux of BA's involves a carrier system which is neither the BA carrier that mediate the AP uptake of BA's nor the anion transporter which mediate the BL uptake of BA's. The carrier system responsible for the transcellular transport of LNAA's is shared by neutral and cationic amino acids. Besides showing preference for branched and aromatic amino acids, this transport system appears to be stereospecific towards L-isomers. The morphologic properties of Caco-2 cell monolayers and their ability to undertake the transcellular transport of BA's and AA's in a pattern reminiscent of that found *in vivo* may make this cell culture system very helpful in improving our understanding of intestinal mucosal transport at the cellular level.

ACKNOWLEDGEMENT

This work was supported by a grant from The Upjohn Company, Kalamazoo, Mich.

REFERENCES

1. A. B. Cairnie, L. F. Lamerton and G. G. Steel, Cell proliferation studies in the intestinal epithelium of the rat. I. Determination of the kinetic parameters, Exp. Cell Res. 39:528-538 (1965).
2. A. Quaroni, J. Wands, R. L. Trelstad and K. J. Isselbacher, Epithelioid cell cultures from rat small intestine, J. Cell Biol. 80:248-265 (1979).

3. F. Raul, M. Kedinger, P. Simon, J. Grenier and K. Haffen, Behaviour of isolated rat intestinal cells maintained in suspension or monolayer cultures, Biol. Cell 33:163-168 (1978).
4. M. Pinto, M.-D. Appay, P. Simon-Assmann, G. Chevalier, N. Dracopoli, J. Fogh and A. Zweibaum, Enterocytic differentiation of cultured human colon cancer cells by replacement of glucose by galactose in the medium, Biol. Cell 44:193-196 (1982).
5. M. Pinto, S. Robine-Leon, M.-D. Appay, M. Kedinger, N. Triadou, E. Dussaulx, B. Lacroix, P. Simon-Assmann, K. Haffen, J. Fogh and A. Zweibaum, Enterocytic-like differentiation and polarization of the human colon carcinoma cell line Caco-2 in culture, Biol. Cell 47:323-330 (1983).
6. A. Leibovitz, J. C. Stinson, W. B. McCombs III, C. E. McCoy, K. C. Masur and N. D. Mabry, Classification of human colorectal adenocarcinoma cells lines. Cancer Res. 36:4562-4569 (1976).
7. B. H. Tom, L. P. Rutzky, M. M. Jakstys, R. Oyasu, C. I. Kaye and B. D. Kahan, Human colon adenocarcinomas I. Establishment and description of a new line, In Vitro 12:180-191 (1976).
8. I. J. Hidalgo, T. J. Raub and R. T. Borchardt, Characterization of the human colon carcinoma cell line (Caco-2) as a model system for intestinal epithelial permeability, Gastroenterology 96:736-749 (1989).
9. T. J. Raub, I. J. Hidalgo, S. L. Kuentzel and R. T. Borchardt, Characterization of the human colon carcinoma cell line (Caco-2) as a model transport system of the intestinal epithelium, FASEB J. 2:2538 (1988) (Abstr.).
10. M. Cereijido, E. S. Robbins, W. J. Dolan, C. A. Rotunno and D. D. Sabatini, Polarized monolayers formed by epithelial cells on a permeable and translucent support, J. Cell Biol. 77:853-880 (1978).
11. T. E. Phillips, T. L. Phillips and M. Neutra, Macromolecules can pass through occluding junctions of rat ileal epithelium during cholinergic stimulation, Cell Tissue Res. 247:547-554 (1987).
12. S. Erlinger, in: Physiology of the Gastrointestinal Tract", L. R. Johnson, ed., Raven, New York, 1987, p. 1557-1580.

13. I. J. Hidalgo and R. T. Borchardt, Transport of taurocholic acid in an intestinal epithelial model system (Caco-2 cell), Pharm. Res. 5(Suppl):949 (1988) (Abstr.).
14. F. A. Wilson and J. M. Dietschy, Characterization of bile acid absorption across the unstirred water layer and brush border of the rat jejunum, J. Clin. Invest. 51:3015-3025 (1972).
15. M. Hofer and J. G. Hoggett, "Transport across biological membranes," Pitman, Boston, 1981, p. 96-102.
16. E. Krag and S. F. Phillips, Active and passive bile acid absorption in man: perfusion studies of the ileum and jejunum, J. Clin. Invest. 53:1686-1694 (1974).
17. L. Lack and I. M. Weiner, Intestinal absorption of bile acids and some biological implications, Gastroenterology 22:1334-1338 (1963).
18. S. L. Weinberg, G. Burckhardt and F. A. Wilson, Taurocholate transport by rat intestinal basolateral membrane vesicles. Evidence for the presence of an anion exchange transport system, J. Clin. Invest. 78:44-50 (1986).
19. M. C. Lin, S. L. Weinberg, W. Kramer, G. Burckhardt and F. A. Wilson, Identification and comparison of bile acid-binding polypeptides in ileal basolateral membrane, J. Membr. Biol. 106:1-11 (1988).
20. I. J. Hidalgo and R. T. Borchardt, Amino acid transport in a novel model system of the intestinal epithelium (Caco-2 cell), Pharm. Res. 5(Suppl):949 (1988) (Abstr.).
21. J. R. Bronk and H. J. Leese, Changes in the adenine nucleotide content of preparations of the rat small intestine in vitro, J. Physiol. (Lond.) 235:183-196 (1973).
22. B. G. Munck, Intestinal absorption of amino acids, in: "Physiology of the Gastrointestinal Tract," L. R. Johnson, ed., Raven, New York, 1981, p. 1097-1122.
23. R. P. Sencer and A. H. Samiy, Intestinal absorption of L-phenylalanine in vitro, Am. J. Physiol. 200:501-504 (1961).
24. I. Osiecka, M. Cortese, P. A. Porter, R. T. Borchardt, J. A. Fix and C. R. Gardner, Intestinal absorption of α-methyldopa: In vitro mechanistic studies in rat small intestinal segments, J. Pharmacol. Exp. Ther. 242:443-448 (1987).

PASSIVE ABSORPTION OF DRUGS IN CACO-2 CELLS

Per Artursson and Johan Karlsson

Department of Pharmaceutics
Uppsala University
Uppsala, Sweden

Introduction

In vitro studies on the intestinal absorption of drugs are usually performed on isolated intestinal segments from various experimental animals (1). The intestinal segments are mounted in Ussing chambers or are alternatively used as everted intestinal sacs or rings (2, 3). The available *in vitro* models have made it possible to perform relatively detailed studies on intestinal drug transport. However, these models have a number of limitations. Firstly, they are not of human origin. In addition, they are often difficult to perform and only a limited number of experiments can be performed on each occasion. The viability of isolated intestinal segments is also limited which at some occasions reduces the duration of the experiments to only a few minutes (3). However, experiments that extend over a few hours can usually be performed in Ussing chambers.

The recent characterization of a number of differentiated human adenocarcinoma cell lines now allows the possibility of investigation of the use of human intestinal cells to study drug absorption (4-10). In these models a large number of experiments can be performed under long time periods. One of these cell lines, Caco-2 (11), has been shown to form well differentiated monolayers that appear to be similar to small intestinal enterocytes (reviewed in ref. 7). Thus, the monolayers are joined by tight junctions and have a polarized ultrastructural appearance. Several microvillar enzymes have been identified in quantities similar to those of small intestinal enterocytes, and the cells transport ions, water and different endogenous products (7, 9, 12-18, see other chapters in this volume). However, it should be noted that a closer examination of the properties of the Caco-2 cells indicates that they are more similar to normal 15 week human fetal colon than to small intestinal enterocytes (5, 7, 19). For instance, the electrical parameters of Caco-2 cells are more similar to colonic than to small intestinal epithelium, see below.

Pharmaceutical Applications of Cell and Tissue Culture to Drug Transport
Edited by G. Wilson *et al.*, Plenum Press, New York, 1991

Caco-2 cells can be cultivated in permeable filter chambers, fig. 1. Different compounds can be added to the apical or the basolateral chamber and the transport can be followed with time. This methodology has been used to study the active transport of a variety of endogenous compounds such as amino acids, bile acids and vitamins and should be useful for studies on actively transported drugs as well (9, 12-15). However, almost all drugs are absorbed through the intestinal wall by passive diffusion. The present study reports therefore the investigation of whether Caco-2 monolayers can be used as a model for passive diffusion of drugs over the intestinal epithelium (10, 20-21).

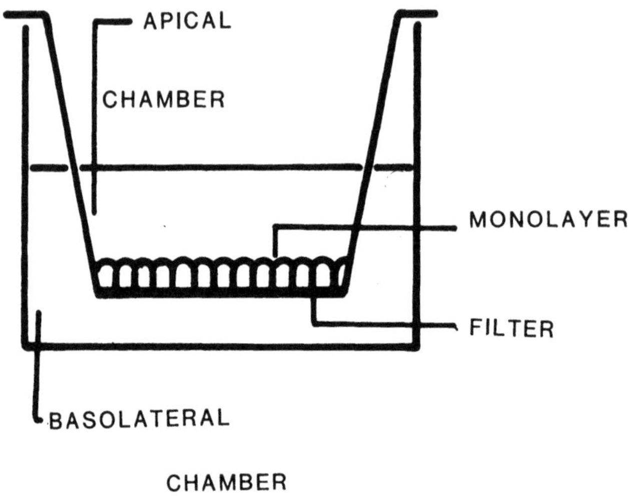

Fig. 1. Caco-2 cells are cultivated directly on permeable polycarbonate filters (Nucell; Nuclepore or Transwell; Costar culture plate inserts).Reproduced from ref. 10 with permission of the copyright owner, The American Pharmaceutical Association.

Methods

Cells

Caco-2 cells (11) were obtained from American Cell Culture Collection, Rockville, Maryland. The cells were cultivated on polycarbonate filters (Nuclepore or Costar cell culture inserts; mean pore diameter 0.45 μm) as described elsewhere (10). Our cell culture procedure is comparable to that recently reported for cultivation on cellulose filters (9) but differs from that recently described for polycarbonate filters (8) since the addition of extracellular matrix components, such as collagen, has been excluded. The integrity of the monolayers was routinely checked by measurements of transmembrane resistance and by determination of the permeability to the two hydrophilic markers ^{3}H-mannitol and ^{14}C-polyethyleneglycol (m.w. 4,000) (10).

Drug transport studies

A suitable concentration (1×10^{-8} - 1×10^{-4} M) of the marker drug was added to the apical or basolateral chamber and samples were withdrawn at predetermined intervalls (10). The resistance of the monolayers was checked at the end of each experiment. A relatively high drug concentration (up to 1×10^{-4} M) had to be used in order to avoid non-specific adsorption to the polycarbonate filters when lipophilic drugs were studied. The drug adsorption to these filters is many times lower than to cellulose based filter chambers such as Millicell (Millipore). However, the potential toxicity of the drug must always be investigated separately when high drug concentrations are used. The use of too high drug concentrations may give erroneous results if saturable processes such as intracellular accumulation, active transport or drug metabolism are to be studied.

Results and discussion

Epithelial permeability

A drug that is absorbed by passive diffusion can pass the epithelial monolayer by two parallel pathways; the transcellular pathway (across the cell membrane) and/or the paracellular pathway (through the intercellular spaces). The transcellular pathway is believed to be the dominating transport route for most drugs since the membranes occupy a much larger surface area than do the intercellular spaces (1). In addition the intercellular spaces are sealed by the components of the junctional complex. (The junctional complex is comprised of tight junctions, intermediate junctions and desmosomes (22)). The mean pore diameter of the intercellular aqueous pores is a few Angstrom units only and consequently much smaller than the hydrodynamic diameter of most drugs (22, 23). However, recent findings indicate that the permeability of the tight junctions is regulated by various endogenous compounds such as nutrients (24, 25), hormones (26) and lymphokines (27). The contribution of the paracellular pathway to the overall absorption of drugs may therefore vary in different tissues and under different physiological conditions.

In order to obtain reproducible determinations of drug absorption, it is important to routinely check the integrity of the monolayers. Two different methods were used for this purpose. The transmembrane resistance is generally used as an approximate measure of the passive ion permeation through the paracellular spaces. This is based on the assumption that the transcellular ion conductivity is smaller than the paracellular, in low or intermediate resistance epithelia. However, it has been shown that the transcellular ion-permeability may account for up to 50% of the total epithelial conductance in the colon while the corresponding value for more leaky epithelia such as the duodenum is only 5% (28). The contribution of the transcellular ionic conductance to the total ion conductivity is therefore significant in colonic epithelium.

The transmembrane resistance across Caco-2 monolayers is higher than that of small intestinal epithelium, fig. 2. The resistance values are similar to those of colonic crypt cells (5). Thus, the passive paracellular ion permeability is lower in

Caco-2 monolayers than in small intestine. If this observation can be extended to drugs it follows that drugs that are passively transported through the paracellular pathway may be absorbed at a slower rate in the Caco-2 model than in the small intestine.

Different values of transmembrane resistance have been reported for Caco-2 monolayers. Thus, values ranging from 154 ±6 ohm x cm^2 (5) to 300-400 ohm x cm^2 (10) have been observed. The transmembrane resistance shows a tendency to vary with passage number. A possible explanation to the differences in transmembrane resistance is that different passages of the cells may have been used in the different studies.

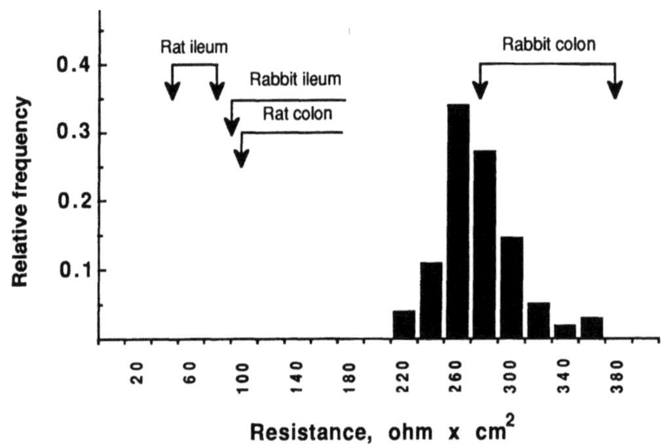

Fig. 2. Distribution of transepithelial resistance for 103 monolayers of Caco-2 cells cultivated on permeable polycarbonate filters. The resistance values of rat and rabbit intestines were taken from reference (28) and (29). Reproduced from ref. 10 with permission of the copyright owner, The American Pharmaceutical Association.

Another test of the integrity of the monolayers is to study the transport of a hydrophilic marker molecule that is transported by passive diffusion and has a very low distribution to the cell membranes. The absorption of such a marker will be equal to its absorption through the paracellular pathway. We routinely use ^{3}H-mannitol (m.w. 180) and ^{14}C-polyethyleneglycol (PEG; m.w. 4,000) to check the permeability of Caco-2 monolayers. Intact Caco-2 monolayers are almost impermeable to these compounds. Thus, the permeability to ^{3}H-mannitol and ^{14}C-PEG has been determined to be approximately 0,05%/h and 0,02%/h of the dose respectively (10), tab.2. These markers have the additional advantage of being metabolically inert. It has recently been shown that another common marker, inulin, is partly degraded to low m.w. sugars by Caco-2 monolayers (9).

Drug absorption studies

Initial studies of passive drug absorption were performed with a series of ß-blocking agents (10). These drugs were chosen since they represent a homologous series of clinically relevant drugs with different lipophilicities, tab. 1. In addition, absorption data from animal studies were available (30). Thus, the influence of lipophilicity on absorption could be studied and the results could be compared with *in situ* data.

Table 1 Comparation of $t_{0,1}$ values from the Caco-2 model and rat ileum[1]

Drug	log D[2]	$t_{0,1}$, h		Ratio[3]
		Caco-2	rat ileum	
Propranolol	1,19	0,29	0,046	0,16
Alprenolol	1,00	0,29	0,045	0,16
Metoprolol	-0,28	0,44	0,066	0,15
Practolol	-1,40	13	0,73	0,056
Atenolol	-2,14	58	0,86	0,015

1 Data compiled from reference 10
2 D = Octanol/water distribution coefficients
3 Ratio = $t_{0,1}$ Caco-2/$t_{0,1}$ rat ileum

The transport of metoprolol from the apical to the basolateral side of Caco-2 monolayers is shown in fig. 3. Accurate rate constants could not be obtained from these curves since the contribution of drug diffusing back from the recieving chamber was significant after some time. Instead rate constants were obtained from the obtained linear graph under 'sink' conditions, insert in fig. 3. Apparent permeability coefficients (Papp) were calculated according to (31):

$$P_{app} = dQ/dt \;\; 1/AC_0 \qquad /1/$$

where dQ/dt is the permeability rate, Co the initial concentration in the donor chamber and A the surface area of the monolayer. When Papp is plotted as a function of lipophilicity (expressed as octanol/water distribution coefficients; D) on a log/log scale a curve that appears to have a sigmoidal shape is obtained, fig. 4. A similar curve has been previously presented from *in situ* rat absorption data (30).

The absorption of the ß-blockers in rat ileum and in Caco-2 cells is compared in tab. 1 (no data was available for rat colon). The rate constants of the two models were not directly comparable since the absorption in the rat intestine followed first order kinetics while the absorption in the Caco-2 model was apparent zero order. The time it took for 10% of the drugs to be absorbed were compared instead. The absorption in the Caco-2 model was several times slower than in rat ileum. A possible explanation is that the surface area of the rat intestine was assumed to be similar to that of a smooth tube (32). Recent studies indicate that the mixing in rat intestine is better than that reported previously (33) and it is therefore possible

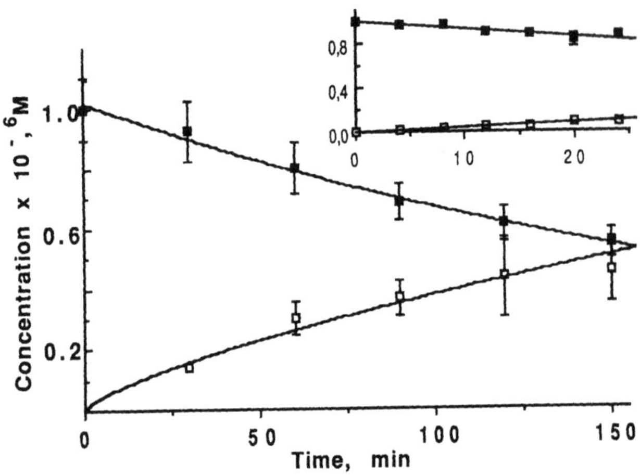

Fig. 3. Transport of metoprolol from the apical (luminal) to the basolateral side of the monolayers. The initial concentration was 1 x 10^{-6} M. Reproduced from ref. 10 with permission from the copyright owner, The American Pharmaceutical Association.

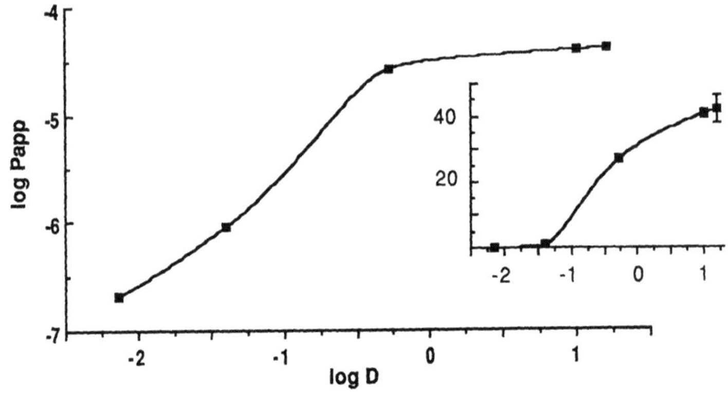

Fig. 4. Apparent permeability coefficients for five beta-blocking agents as a function of their distribution coefficients. Redrawn from ref. 10 with permission from the copyright owner, The American Pharmaceutical Association.

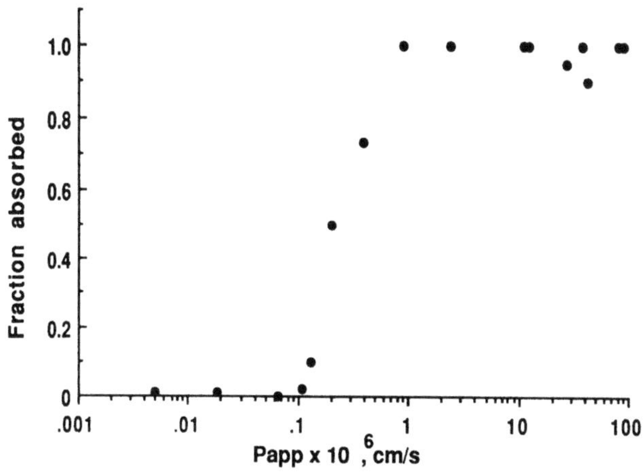

Fig. 5. Fraction of drugs absorbed in humans as a function of their apparent permeablity coefficients in the Caco-2 model.

that the available surface area is larger than that of a smooth tube. In addition, a better mixing will decrease the thickness of the aqueous boundary layer which will result in an increased permeability for more lipophilic drugs (34). If the total surface area of rat ileum is used for comparision, the surface compensated absorption will be slightly faster in the Caco-2 model than that in rat ileum. However, the ratios for the permeabilities of the drugs are comparable even if the correct absorptive surface area of the rat intestine is unknown. Thus, the three lipophilic drugs propanolol, alprenolol and metoprolol were absorbed at comparable rates in the two models while the more hydrophilic drug practolol was absorbed at a slightly slower rate in the Caco-2 model. Atenolol (the most hydrophilic drug) was absorbed at a much slower rate in the Caco-2 model.

We have recently investigated the absorption of a larger number of drugs that have different physical/chemical properties and molecular weights (Artursson et al., manuscript in preparation). Some of the results are summarized in fig. 5 where literature values of the absorbed fraction after a peroral dose *in vivo* is shown as a function of the apparant permeability coefficient in the Caco-2 model. Drugs that are completely absorbed in humans have apparent permeability coefficients that are in the region of 1 x 10^{-6} cm/s or higher. Drugs that are incompletely absorbed or not absorbed at all have apparent permeability coefficients that are lower. The results obtained so far indicate that the Caco-2 model can be used to differentiate drugs that are incompletely absorbed *in vivo*.

Paracellular and Transcellular permeability

It is possible to compare the differences in absorption of the hydrophilic ß-blocking agents in rat intestine and in Caco-2 monolayers, with the differences in electrical parameters, tab. 1, fig. 2. The hydrophilic drugs have a low lipid solubility and will consequently be transported at a slow rate across the cell membranes. When the transcellular transport rate is reduced, the transport through the alternative paracellular route becomes more important. If the data on passive ion permeability can be extrapolated to the larger drug molecules, it follows that the paracellular route may be more accessible in rat ileum than in Caco-2 monolayers.

Table 2 Effect of EGTA treatment on Caco-2 monolayers[1]

Treatment	Resistance ohm x cm^2		Mannitol transport, %/h	
	directly after treatment	2h after treatment	directly after treatment	2h after treatment
None	279±7	287±11	0,05±0,01	0,06±0,01
EGTA, 45 min	63±7	270±11	0,6±0,2	0,08±0,02
EGTA, 90 min	<30	<30	n.d.	n.d.

[1] The cells were exposed to cell culture medium containing 2,5 mM EGTA n.d. = not determined. Data are partly compiled from reference 20

This question was further addressed in a study where the intercellular spaces were selectively opened and closed (20). The integrity of tight junctions is dependent on the extracellular concentration of calcium ions (Ca^{2+}) (35-38). The extracellular calcium concentration can be reduced to a varying degree by the use of specific calcium chelators such as EGTA. When the extracellular calcium concentration was reduced to the micromolar range in the Caco-2 model, the intercellular spaces were widened and the tight junctions were resolved at the ultrastructural level (20). A corresponding decrease in transmembrane resistance and increase in the transport of hydrophilic marker molecules was observed, tab. 2. The process was reversible only if the calcium levels were normalized within a certain time. Prolonged exposure to low calcium concentrations resulted in irreversible changes including cell detachment, tab. 2. It was therefore important to assure that the EGTA treatment was reversible and that no cell losses can be detected in the electron microscope. Similar observations have been reported in other epithelial cell culture models (e.g. 35, 38).

In different experiments, the transport of atenolol was increased 4-9 fold after pretreatment with EGTA, fig. 6. Similarly, the transport of the other hydrophilic ß-blocker, practolol, was increased 2-6 times while the treatment had a limited effect on the more lipophilic drugs, fig. 7. Thus, when the transmembrane resistance was decreased to a level similar to that of rat ileum, the transport rates of all five drugs, ranging 2000-fold in lipophilicity, was comparable in the two models.

We have also investigated the absorption of an experimental ß-blocking agent (H216/44) that is very lipophilic but only slowly absorbed in rats. This drug is also slowly absorbed in the Caco-2 model and its absorption is enhanced by EGTA treatment (10, 21). These observations increase the predictive value of the Caco-2 model since they indicate that more complex structure-absorption relationships than lipophilicity can be detected. Further, a selective widening of tight junctions

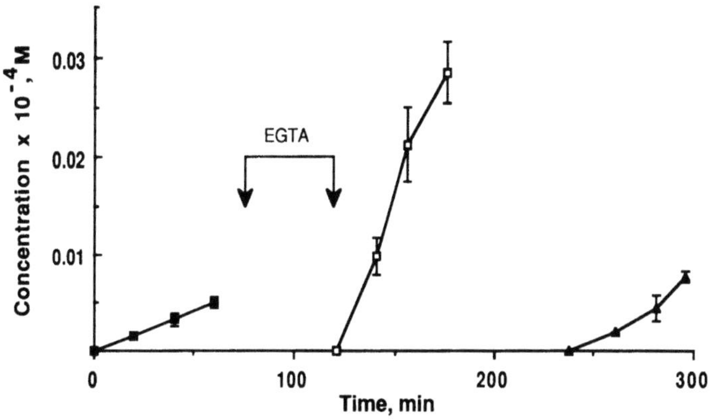

Fig. 6. Transport of atenolol before, immediately after and 2 h after exposure to the selective calcium ion chelator, EGTA. The same monolayer was used in all three experiments. Reproduced from ref. 20 with permission from the copyright owner, The American Pharmaceutical Association.

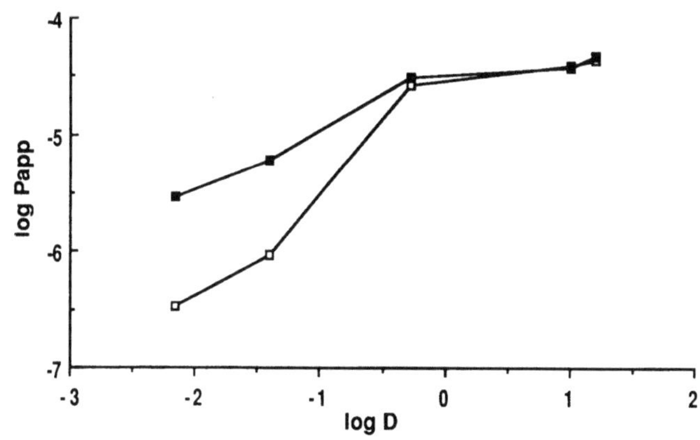

Fig. 7. Apparent permeability coefficients for five beta-blocking agents as a function of their distribution coefficents. The permeability coefficients were determined before (open symbols) and after (closed symbols) exposure to EGTA. Redrawn from ref. 20 with permission from the copyright owner, The American Pharmaceutical Association.

will not only enhance the transport of hydrophilic but also that of lipophilic drugs that are slowly absorbed. The EGTA experiments also indicate that it is possible to study the relative contribution of the transcellular and paracellular pathways to the overall transport of a certain drug in the Caco-2 model. The effects of other compounds such as various pharmaceutical additives and absorption enhancers on passive drug absorption has also been studied in the Caco-2 model and will be reported at a later date. (E.K. Anderberg and P. Artursson, manuscript in preparation).

The unstirred water layer and the epithelium

Adjacent to the apical cell surface is an aqueous boundary layer (unstirred water layer). The unstirred water layer has been shown to be rate limiting for the absorption of (lipophilic) drugs that are transported rapidly across the cell membrane (34, 39). In different studies, the thickness of the unstirred water layer has been determined to be 200-800 μm and it has been shown to be dependent on the luminal stirring (41-50). However, these studies have usually been performed in anesthetized experimental animals. Recent experiments on conscious rats have indicated that the unstirred water layer is approximately 100 μm and that the luminal stirring is more efficient than reported previously (33). It was therefore important to study to what extent the unstirred water layer influenced the apparent permeability coefficients in the Caco-2 model (J. Karlsson and P. Artursson, manuscript in preparation).

The total permeability for a given drug in the Caco-2 model can be considered as a series of permeabilities. Thus, the apparent permeabililty coefficient (Papp) can be related to the permeabilities of the aqueous boundary layers (unstirred water layers) adjacent to the apical surface of the Caco-2 cells and to the basolateral surface of the polycarbonate filters, (Paq), the polycarbonate filters (Pf) and the epithelial cells (Pc) so that from equation /2/:

$$1/Papp = 1/Pc + 1/Paq + 1/Pf \qquad /2/$$

it follows that the resistance (1/P) over polycarbonate filter without Caco-2 cells (1/Pw) is:

$$1/Pw = 1/Pf + 1/Paq \qquad /3/$$

Thus the permeability of the epithelial cells (Pc) can be calculated if the permeability of the empty filters (Pw) is determined and it is assumed that the unstirred water layer has the same thickness at the cell and filter surface. In addition, the thickness of the unstirred water layer can be calculated if the thickness and the porosity of the polycarbonate filter are known (51, 52). We have determined the unstirred water layer in the Caco-2 model to approximately 475 μm under our standard culture conditions. This value can be reduced if the stirring rate is increased. The relative contribution of the non-cellular (1/Pw) and the cellular resistance (1/Pc) to the overall resistance to the ß-blocking agents (1/Papp) is shown in fig. 8. The unstirred water layer has a significant influence on the permeability of the more lipophilic ß-blocking agents while it is of limited importance for the membrane restricted hydrophilic drugs. Thus, the unstirred water layer in the Caco-2 model has a thickness that is similar to that observed in most *in vitro* and *in vivo* models but is thicker than that recently obtained in conscious rats (33).

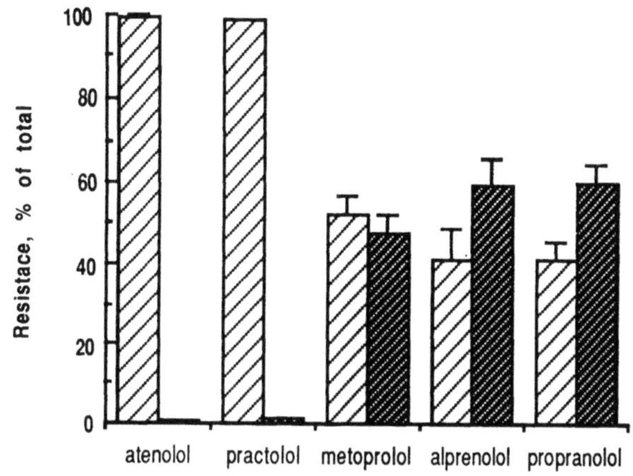

Fig. 8. Relative contribution of 1/Pc (light bars)and 1/Pw (dark bars) to the overall monolayer resistance to the passive diffusion of the ß-blocking agents (1/Papp).

Conclusions

In summary, the results indicate that the Caco-2 model is a useful alternative to other models for intestinal absorption. The model has been used in various mechanistic studies such as structure-absorption studies, studies on the transport across the paracellular pathway and studies on the relative importance of various permeability barriers. It should also be applicable in studies of other factors of importance for drug absorption such as water transport or 'solvent drag' (24, 53, 54), hydrostatic and osmotic pressure (55, 56), pH (57, 58) and regulation of tight junctions (24-27). Accumulating findings suggest that the model can be used to predict the absorbed fraction of an orally administered drug in humans. In addition, the procedure has improved reproducability, is less time consuming and avoids the unnecessary use of experimental animals.

Acknowledgements

This work was supported by grants from the IF's Foundation for Pharmaceutical Research, the Swedish Foundation for Scientific Research without Animal Experiments and O.E. and Edla Johanssons Scientific Foundation.

References

1. Jackson, M.J. in Physiology of the Gastrointestinal Tract, Second Edition (Johnson, L.R. Ed.), Raven Press, New York 1987, chapter 59, p. 1597-1621.
2. Ussing, H.H. and Zehran, K., Acta Physiol. Scand. 23, 110-127 (1951).
3. Gardner, C.R. in Directed Drug Delivery (Borchardt, R.T., Tepta, A.J. and Stella, V.J. Eds.), Humana Press Inc., Clifton, NJ, 1985, p. 61-82.
4. Pinto, M. et al., Biol. Cell 47, 323-330 (1983).

5. Grasset, E. et al., Am. J. Physiol. 247, C260-C267 (1984).
6. Chantret, I.et al., Cancer Res. 48, 1936-1942 (1988).
7. Neutra, M. and Louvard, D. in Modern Cell Biology, vol. 8: Functional Epithelial Cells in Culture (Matlin, K.S. and Valentich, J.D. Eds.), Alan R. Liss, New York, 1989, p. 363-398.
8. Hidalgo, I.J., Raub, T.J. and Borchardt, R.T., Gastroenterology 96, 736-749 (1989).
9. Wilson, G.et al., J. Controlled Release 11, 25-40 (1990).
10. Artursson, P., J. Pharm. Sci. 79(5) 1990, in press.
11. Fogh, J., Fogh, J.M. and Orfeo, T., J. Natl. Cancer Inst. 59, 221-226 (1977).
12. Hidalgo, I.J. and Borchardt. R.T., Pharm. Res. 5 (suppl.), S-110 (PD949) (1988).
13. Hidalgo, I.J. and Borchardt. R.T., Pharm. Res. 5 (suppl.), S-110 (PD950) (1988).
14. Muthia, R. and Seetharam, B., J. Cell Biol. 105, 235a (1987).
15. Dix, C.J.et al., Biochem. Soc. Trans. 15, 439-440 (1987).
16. Grasset, E., Bernabeu, J. and Pinto, M., Am. J. Physiol. 248, C410-C418 (1985).
17. Mohrmann, I.et al., Am. J. Physiol. 250, G323-G330 (1986).
18. Burnham, D.B. and Fondacaro, J.D., Am. J. Physiol. 256, G808-G816 (1989).
19. Hauri, H.P. et al., J. Cell Biol. 101, 838-851 (1985).
20. Artursson, P. and Magnusson, C., J. Pharm. Sci. 79(6) 1990, in press.
21. Karlsson, J. and Artursson, P., Proc. Int. Union Physiol. Sci. XVII, P1306 (1989).
22. Madara, J.L. and Dharmsathaporn, K., J. Cell Biol. 101, 2124-2133 (1985).
23. Madara, J.L. and Trier, J.S. in Physiology of the Gastrointestinal Tract, Second Edition (Johnson, L.R. Ed.), Raven Press, New York 1987, chapter 44, p. 1209-1249.
24. Pappenheimer, J.R. and Reiss, K.Z., J. Membr. Biol. 100, 123-136 (1987).
25. Pappenheimer, J.R., J. Membr. Biol. 100, 137-148 (1987).
26. Lowe, P.J.et al., Am. J. Physiol. 255, G454-G461 (1988).
27. Madara, J.L. and Stafford, J., J. Clin. Invest. 83, 724-727 (1989).
28. Powell, D.W. in Physiology of the Gastrointestinal Tract, Second Edition (Johnson, L.R. Ed.), Raven Press, New York 1987, chapter 46, p. 1267-1305.
29. Ungell, A.-L.et al., Proc. 3rd Int. Conf. Drug Absorption, Edinburgh, 27-30 September 1988, p. 34.
30. Taylor, D.C., Pownall, R. and Burke, W., J. Pharm. Pharmacol. 37, 280-283 (1985).
31. Schoenwald, R.D. and Huang, H.-S., J. Pharm. Sci. 72, 1266-1271 (1983).
32. Wilson, F.A. and Dietschy, J.M., Biochem. Biophys. Acta 363, 112-126 (1974).
33. Anderson, B.W. et al., Am. J. Physiol. 254, G843-G848 (1988).
34. Komiya, I. et al., Int. J. Pharm. 4, 249-262 (1980).
35. Gumbier, B., Stvenson, B. and Grimaldi, A., J. Cell Biol. 107, 1575-1587 (1988).
36. Martinez-Palomo, A. et al., J. Cell Biol. 87, 736-745 (1980).

37. Sedar, A.W. and Forte, J.G., J. Cell Biol. 22, 173-188 (1964).
38. Cereijido, M. et al., 77, 853-880 (1978).
39. Burnette, R.R. in Controlled Drug Delivery, Fundamentals and Applications 2nd Ed. (Robinson, J.R. and Lee, V.H. Eds.) Marcel Dekker, Inc., New York, 1987, p. 95-138.
40. Ho, N.F.H. et al. in Animal Models for Oral Drug Delivery in Man: In Situ and In Vivo Approaches (Crouthamel, W. and Sarapu, A.C. Eds.) American Pharmaceutical Association, Washington DC, USA, 1983, p. 27-107.
41. Cerda, J.J. et al., J. Parenter. Enteral Nutr. 11, 63-66 (1987).
42. Gerencser, G.A. et al., Proc. Soc. Exp. Biol. Med. 176, 183-186 (1984).
43. Westergaard, H. and K.H. and Dietschy, J.M., J. Clin. Invest. 54, 718-732 (1974).
44. Westergaard, H., Holtermuller, K.H. and Dietschy, J.M., Am. J. Physiol. 250, G727-G735 (1986).
45. Flourie, B. et al., Gut. 25, 936-941 (1984).
46. Frase. L.L. et al., Gastroenterology 88, 478-484 (1985).
47. Levitt, M.D. et al., J. Clin. Invest. 74, 2056-2064 (1984).
48. Levitt, M.D. et al., Am. J. Physiol. 252, G325-G332 (1987).
49. Read, N.W. et al., Gut. 18, 865-876 (1977).
50. Sparso, B.H., Luke, M. and Wium, E., Scand. J. Gastroenterol. 19, 568-574 (1984).
51. Barry, P.H. and Diamond, J.M., Physiol. Rev. 64, 763-872 (1984).
52. Pedley, T.J., Q. Rev. Biophys. 16, 115-150 (1983).
53. Lynch, J. et al., Proc. 3rd Int. Conf. Drug Absorption, Edinburgh, 27-30 September, 1988, p. 34.
54. Johno, I. et al., Int. J. Pharm. 25, 255-263 (1985).
55. Madara, J.L., J. Cell Biol. 97, 125-136 (1983).
56. Paulmichl, M. et al., Pflügers Arch. 413, 456-462 (1989).
57. Griepp, E.B. et al., J. Cell Biol. 96, 693-702 (1983).
58. Kiviluoto, T., Mustonen, H. and Kivilaakso, E., Gastroenterology 96, 1410-1418 (1989).

RECEPTOR-MEDIATED TRANSPORT OF COBALAMIN IN CACO-2 CELLS: INTRACELLULAR LOCALISATION OF TRANSCOBALAMIN II

IAN F. HASSAN, IAN J. PARSONS and MARTIN W. MACKAY

Advanced Drug Delivery Research
Ciba-Geigy Pharmaceuticals
Horsham, West Sussex, RH12 4AB (UK)

INTRODUCTION

A number of in vitro models have been developed to study the absorption of drugs across the intestinal epithelium (Osiescka et al., 1985). The majority of these models do not have the morphological and functional properties of normal adult human epithelial cells. Recently, a human adenocarcinoma cell line (Caco-2) that reproducibly displays properties of differentiated distal ileal epithelial cells has been characterised (Wilson et al., 1989; Hidalgo et al., 1989). When cultured on permeable supports (nitrocellulose filters), in chambers, Caco-2 cells form a confluent monolayer which is morphologically polar with a well developed brush border at their apical surface. The monolayers are

---

Abbreviations used in this paper

ATP, adenosine 5′-triphosphate; BSA, bovine serum albumin; Cbl, cobalamin; DPM, disintegrations per minute; HEPES, N-2-hydroxyethyl piperazine-N-2-ethane sulphonic acid; IF, intrinsic factor; MES, 2-[N-morpholino]ethanesulfonic acid sodium salt; NADH, β-nicotinamide adenine dinucleotide; PBS, phosphate buffered saline; PTA, phosphotungstic acid; RCF, relative centrifugal force; TCA, trichloroacetic acid; TCII, transcobalamin II; TEM, transmission electron micrograph; Tris, Tris(hydroxymethyl)aminomethane-HCl; UDP-galactose, uridine diphospho-d-galactose; VSV, vesicular stomatitis virus.

Pharmaceutical Applications of Cell and Tissue Culture to Drug Transport
Edited by G. Wilson *et al.*, Plenum Press, New York, 1991

functional in the specific and unidirectional transport of taurocholic acid, (Wilson et al., 1989), and the active transport of phenylalanine (Hidalgo and Borchardt, 1988).

An essential nutritional requirement of all mammals is cobalamin (Cbl, vitamin B_{12}). Cbl is only synthesised by microorganisms. Consequently, man and other animals have developed a complex transport mechanism for intestinal absorption (Ellenbogen, 1975). We have demonstrated that Caco-2 cells also represent an excellent model of the receptor-mediated transport of Cbl in the intestine. Caco-2 cells will bind and internalise Cbl carrier complexes specifically on the apical membrane and after 28 days in culture, transport Cbl from the apical to basolateral side (Dix et al., 1989).

The absorption of Cbl through the intestine is an intricate process. Cbl is released in the stomach from its dietary sources and binds to intrinsic factor (IF), a glycoprotein secreted by gastric parietal cells (Allen et al., 1978). When the IF-Cbl complex reaches the distal ileum, it binds to specific receptors that are located in the intermicrovillous pits of the epithelium (Levine et al., 1984). After a delay of several hours, Cbl appears in the portal circulation (Doscherholmen and Hogen, 1959) bound to transcobalamin II (TCII) (Chanarin et al., 1970). TCII then mediates the transport and cellular uptake of Cbl throughout the body. TCII is synthesised by various tissues and organs, including the liver, kidney, spleen, heart, bone marrow, gastric mucosa and ileal mucosa (Frater-Schroder et al., 1985). Several cell lines have been shown to secrete TCII. These include hepatocytes (Savage and Green, 1976), fibroblasts (Green et al., 1976), mononuclear phagocytes (Frater-Schroder et al., 1983), endothelial cells (Hall, 1979) and Caco-2 cells (Dix et al., 1987). We have also demonstrated that Cbl transported to the basolateral side of Caco-2 monolayers is bound to TCII (Dix et al., 1989).

Information is available concerning the formation of IF-Cbl complexes (Allen et al., 1978) and the binding of IF-Cbl to its receptor (Levine et al., 1984). However, subsequent events such as those responsible for the transfer of Cbl from the apical membrane into and through epithelial cells are not well understood. For instance, it is not known whether TCII binds Cbl in the intracellular compartment or post-secretion.

We have isolated specific secretory vesicles containing TCII as a means of elucidating in which compartment binding occurs.

SUBCELLULAR FRACTIONATION OF CACO-2 CELLS

Fractionation studies were used to determine if the TCII secreted with polarity from Caco-2 cells is either stored or located in intracellular compartments. The fractionation was achieved with a ficoll gradient (40 ml, 1-22%) over a nycodenz cushion (45%) using a vertical rotor (MSE, Europa 65M, rotor MVT 50.35) centrifuged at 200,000 RCF for 1

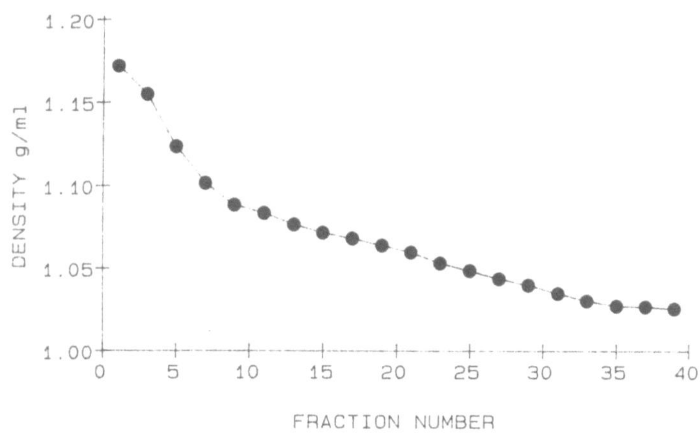

Fig. 1 Density of fractions on a 1-22% ficoll gradient, determined by refractive index (RI) and % sucrose using: density = (RI x 556.5 - 739.9) + 221.9 / 224.6

h (Branch et al., 1987). The gradients were pumped from the base of the tube (1 ml min^{-1}) and collected in 1 ml fractions. Prior to fractionation, single cells were isolated and disrupted using sheer force. Nuclear material was removed by centrifuging the homogenate (1500 RCF, 10 min).

Buoyant density was derived from refractive index measurements. Figure 1 shows that the gradient is linear and will fractionate material with a buoyant density in the range of 1.15474 to 1.02683 g ml^{-1}.

Constitutively secreted proteins are not thought to be stored intracellularly, unlike proteins of the regulated secretory pathway (Chung et al., 1989). Little is known about the mechanism which ensures that newly synthesised proteins, specific for either the apical or basolateral domain, achieve their distribution. However, it is widely understood that sorting to each domain does not occur until the protein has budded into a vesicular compartment at the trans-golgi membrane (Rindler et al., 1985).

In order to indicate the position of compartments prior to trans-golgi involved in de novo synthesis of TCII in Caco-2 cells, the location of subcellular organelles in the fractionating gradient was determined. Marker enzymes of the biosynthetic compartments have been assayed across the gradient. The golgi has been located using sialyltransferase and galactosyltransferase and the endoplasmic reticulum (ER) identified using NADH-cytochrome c oxidoreductase and viral infection studies.

Sialyltransferase (EC 2.4.99.1)

The assay was essentially that of Minnifield et al. (1986) with asialofetuin as acceptor. The reaction mixture contained 50 μl of fraction with 50 μl of reaction medium (3.75 mM MES, pH 6.8, 0.01% Triton X-100, 0.2 mg asialofetuin, cytidine 5′-monophospho-[^{14}C]-sialic acid, 1.1×10^5 DPM, 0.4 mg BSA). The mixture was incubated for 3 h at 37°C and terminated by precipitating the protein with 300 μl of ice cold 5% TCA, 1% PTA. The reaction mixtures were individually pipetted into a microtitre filter plate, 0.22 μm filters (Millipore), and were filtered under vacuum. The wells were washed six times with 250 μl of ice cold 5% TCA, 1% PTA before extracting filters into scintillation vials. The levels of bound [^{14}C] in each fraction were determined. Results indicate a peak of enzymatic activity between fractions 16 to 23 (Fig. 2).

Galactosyltransferase (EC 2.4.1.13)

The assay was a modification of the Bretz and Staubli (1977) method with ovomucoid as acceptor. The reaction mixture contained 50 μl of

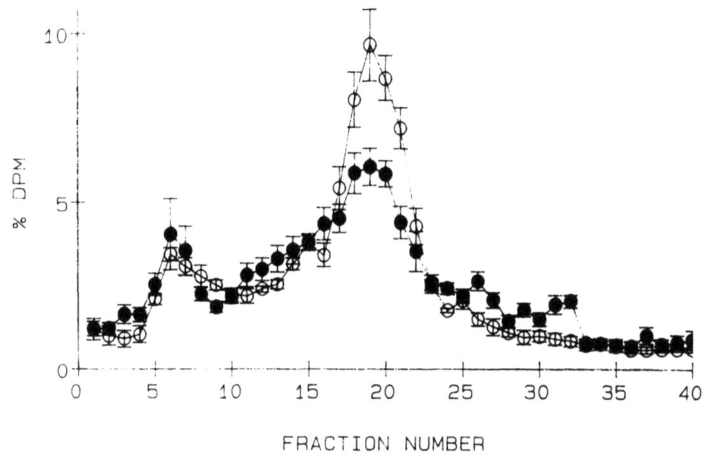

Fig. 2 The mean sialyltransferase , n=3 (-O-), and galactosyltransferase , n=3 (-●-), activities in subcellular fractions of Caco-2 cells on a 1-22% ficoll gradient.

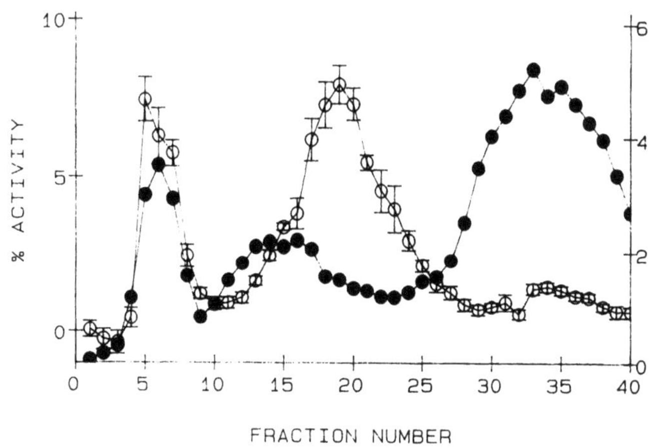

Fig. 3 The level of incorporated [^{35}S]-methionine (-●-) after viral infection with TS045, and the mean NADH-cytochrome c oxidoreductase activity, n=4 (-O-), in subcellular fractions of Caco-2 cells on a 1-22% ficoll gradient.

fraction with 50 μl of reaction medium (50 mM HEPES, pH 7.0, 40 mM β-mercaptoethanol, 0.4% Triton X-100, 40 mM $MnCl_2$, 17 mg ml^{-1} ovomucoid, [^3H]-UDP-galactose, 2.2×10^6 DPM, 1 mM UDP-galactose, 2 mM ATP, 5 mg ml^{-1} BSA). The mixture was incubated for 1 h at 37°C and the reaction was stopped by precipitating the protein with 300 μl of ice cold 10% TCA. The protein was separated from the reaction mixture using a microtitre filter plate (Millipore) and washed six times with 250 μl of ice-cold 10% TCA, before extracting the filters into a scintillation vial. The levels of bound [^3H] in each fraction were determined. Figure 2 indicates that the peak of activity for galactosyltransferase (fractions 10 to 23) is broader than that for sialyltransferase (fractions 16 to 23) and may represent a wider distribution of this enzyme in the golgi stack.

NADH-cytochrome c oxidoreductase (EC 1.6.99.3)

The assay was essentially that of Sottocasa et al. (1967) incorporating the metabolic inhibitors cyanide and rotenone to inhibit the mitochondrial source of the enzyme. The reaction mixture contained 50 μl of fraction with 50 μl of reaction medium (0.1 nM KCN, 0.3 mg NADH, 1.8 mg cytochrome c, 43.8 mM KH_2PO_4 pH 7.6, 4 ng ml^{-1} rotenone, 0.004% Triton X-100). The substrate was added and the oxidation of NADH-cytochrome c was measured within 5 min of the start of the reaction on a Dynatech MR600 Microtitre plate reader at 570 nm.

Figure 3 shows a broad peak of activity between fractions 12 and 26 with a second peak of activity between fractions 5 to 9. This denser peak of activity may be due to the rotenone-insensitive NADH-cytochrome c reductase system present in mitochondria (Sottocasa et al., 1967). This is corroborated using electron microscopy which demonstrates the presence of mitochondrial bodies in these denser fractions (Fig. 4).

Vesicular Stomatitis Virus (VSV) Infection

Viruses have been used as models to investigate the biosynthetic and intracellular pathways of epithelial cells. VSV has been shown to infect Caco-2 cells and bud with polarity at the basolateral membrane (Rindler and Traber, 1988). Several temperature sensitive mutants of VSV have been isolated e.g. TS045. When Caco-2 cells are infected by TS045 and incubated at the non-permissive temperature, the spike glycoprotein, G-protein, locates exclusively to the rough endoplasmic reticulum (RER).

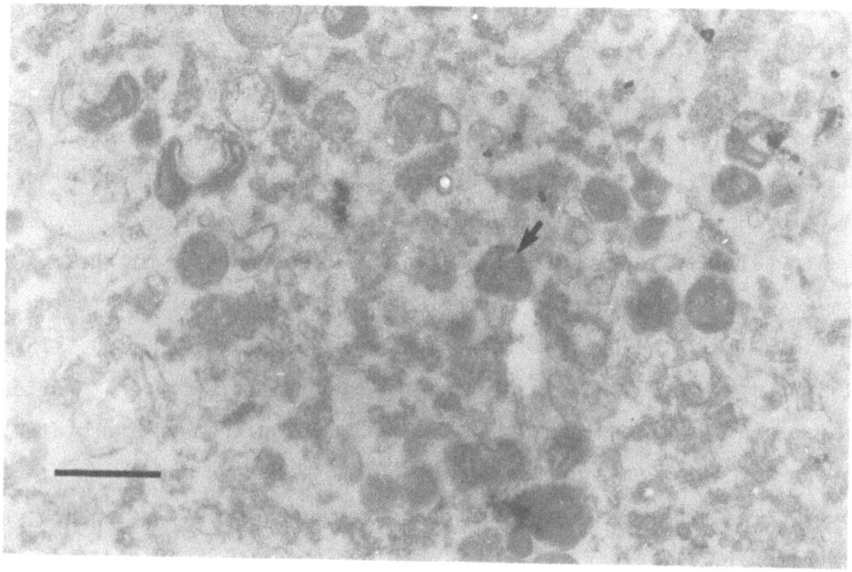

Fig. 4 Transmission electron micrograph, TEM, of fractions 5 to 6 from 1-22% ficoll gradient. Mitochondria are indicated (→). Bar = 1 μm

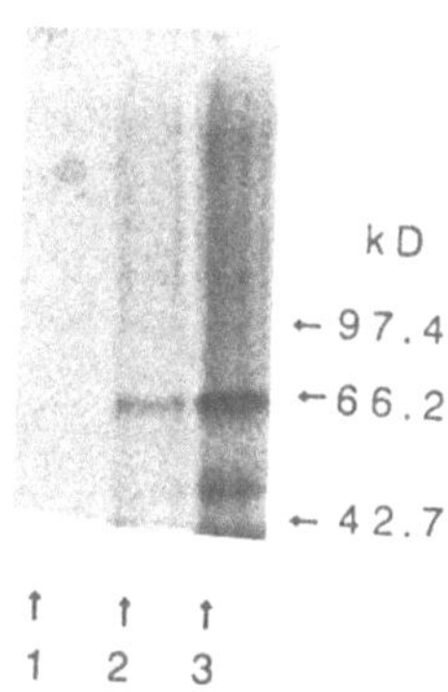

Fig. 5 Autoradiograph of SDS-PAGE of immunoprecipitated G-protein from fractions of a 1-22% ficoll gradient. Lane 1, fractions 14-16; Lane 2, fractions 22-24; Lane 3, fractions 32-34.

Figure 3 shows the amount of incorporated [35S]-methionine (after pulse labelling for 30 min) in Caco-2 cells infected with TS045 (1.5 h at 32°C and then incubated at the non-permissive temperature of 39°C for 4 h). At this stage of infection the majority of synthesised proteins are viral. The five proteins encoded by VSV are L, N, NS, G and M with G-protein being translocation competent. The results indicate three peaks of activity (Fig. 3). To determine which peak is G-protein, immunoprecipitation using a polyclonal antiserum to G-protein (kindly donated by Dr M. Marsh, Chester Beatty, London) was performed. Briefly, aliquots (0.5 ml) were pooled from two fractions on each of the peaks. These were subjected to preclearing using non-specific rabbit serum (4 µl, 30 min) followed by protein-A sepharose 4B (50 µl, 30 min). This was centrifuged at 15,000 RCF, 5 min and the supernatant removed. Antiserum to G-protein was added (4 µl, 2 h). The bound antigen was separated by reaction with protein-A sepharose 4B (50 µl, 2 h) and the supernatant removed. The protein-A sepharose beads were washed (x2) using a high salt buffer (0.0125 M KH_2PO_4, pH 7.4, 0.3 M NaCl, 0.2% azide) and washed (x2) using a mixed micelle buffer (10 mM Tris, pH 8.6, 0.3 M NaCl, 0.1% SDS, 0.5% Triton X-100, 0.2% azide). The beads were washed finally in dH_2O before being added to the sample buffer prior to being separated on SDS-PAGE. G-protein could only be visualised as a prominent single band, 66kD, in fractions 14 and 16, a faint band in fractions 22-24 and not seen in fractions 32-34 (Fig. 5). There is good correlation between the NADH-cytochrome c oxidoreductase activity, the G-protein peak (Fig. 3) and the dominant G-protein immunoprecipitation result (Fig. 5). The broader peak of reductase activity may correspond to the amount of smooth ER isolated in the cells, whereas G-protein is located exclusively in the RER.

TCII LOCALISATION

Positive identification of TCII was achieved using two antibodies; one raised to purified TCII (kindly provided by Dr E. Quadros, Veterans Administration, New York) and one raised to a peptide, prepared using the 19 amino acid sequence of the N-terminus of TCII determined by Quadros and Rothenberg (1986). Both antibodies have been used to immunoprecipitate TCII from either culture medium, which has been exposed to Caco-2 cells, or a cell lysate prepared from Caco-2 cells using 0.1% Triton X-100. Single bands were obtained with both antibodies on SDS-PAGE (Fig. 6) corresponding to 43 kD, the reported molecular weight (Quadros and Rothenberg, 1986).

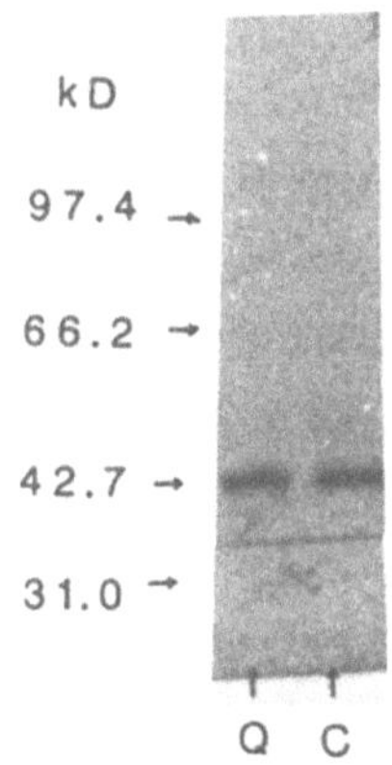

Fig. 6 Autoradiograph of SDS-PAGE of immunoprecipitated TCII from culture medium exposed to Caco-2 cells for 48 h. Bands were precipitated using antibodies to native TCII (Q), and N-terminal peptide (C). Single bands are present at 43kD in both lanes.

TCII was detected on the gradient using immunolocalisation. Briefly, a 50 µl aliquot of each fraction was added to a well of a 96 well plate (Immuno-maxisorp, NUNC) to which 50 µl of 0.2 mM $Na_2CO_3/NaHCO_3$, pH 9.5 was added and incubated for 16 h at 25°C. The wells were drained and blocked using 200 µl of 0.5% gelatine in phosphate buffered saline (PBS), for 1 h at 37°C. The plate was washed (x3) with buffer (PBS, 0.05% Tween 20, 1.0% butan-1-ol). All subsequent washings were carried out using the same buffer. The antibody raised to whole TCII was added to the wells (50 µl, 1:500 dilution) and incubated at 25°C for 3 h. The plate was then washed and the developing antibody added (50 µl, 1:500 of goat anti-rabbit peroxidase conjugated, Sigma). The plate was incubated for a further 2 h at 25°C. The plate was washed and the peroxidase substrate added (75 µl of 0.1 M Na_2HPO_4, 0.05 M citric acid, pH 5.5, 0.04% o-phenylenediamine, 0.04% H_2O_2) and the colour was allowed to develop. The reaction was terminated using 1 M HCl (75 µl) and absorbance determined at 490 nm. Results in Figure 7 indicate a broad distribution of TCII on the gradient. Fractions 15 to 20 correspond to the presence of TCII in both the ER and golgi. The peak of activity between fractions 21 to 24 cannot be accounted for by these two organelles of the biosynthetic pathway. Electron microscopy indicate the presence of numerous vesicular bodies in fractions 22 and 23 (Fig. 8). The buoyant density of these fractions (1.059 to 1.047 g ml^{-1}), is consistent with the light endosomal fraction isolated from liver on this gradient system (Branch et al., 1987).

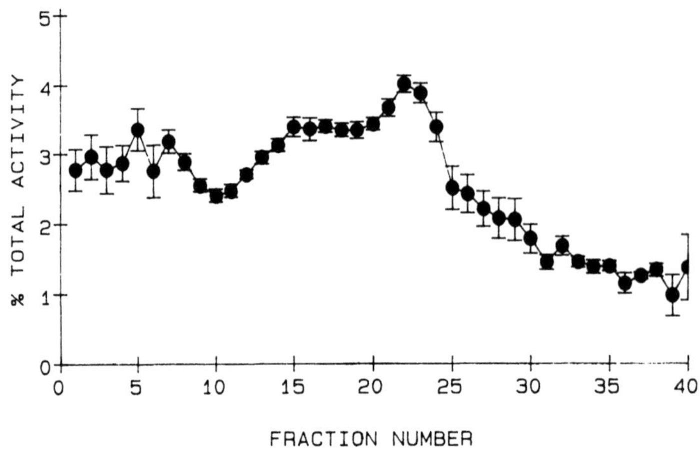

Fig. 7 Detection of TCII in subcellular fractions of Caco-2 cells on a 1-22% ficoll gradient.

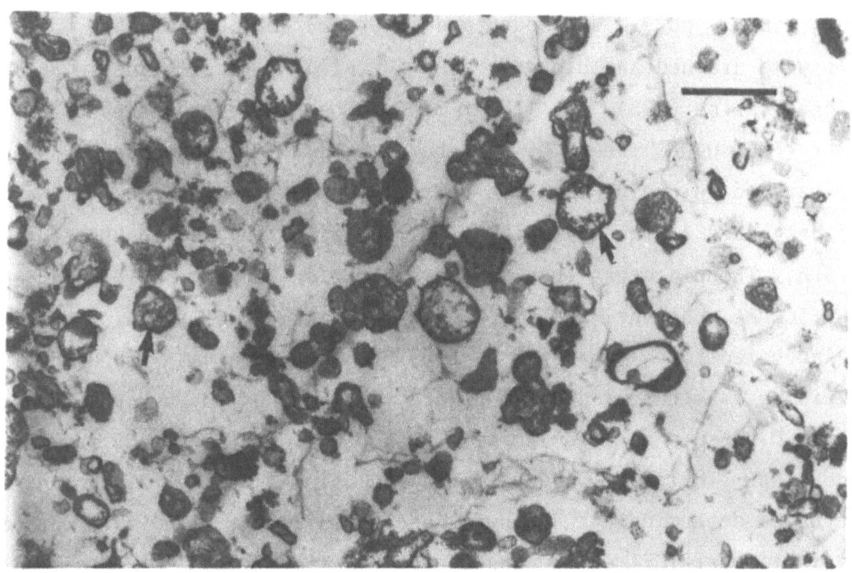

Fig. 8 TEM of fractions 22 to 23 from a 1-22% ficoll gradient. Numerous vesicular structures (→) are present. Bar = 1 μm

CONCLUSIONS

Little is understood about the route which Cbl is transported across the epithelial cell. We believe that TCII has an important role in the polar secretion of Cbl from epithelial cells. In order to understand the role of TCII in the transport of Cbl, fractionation studies were undertaken to discover in which intracellular compartment TCII is located. The study has clearly identified the major biosynthetic compartments, ER and golgi, but has also located a lighter vesicular compartment containing TCII. Recent evidence concerning the transport and sorting of membrane and secreted proteins, indicates that the golgi, and in particular the trans-golgi network, play a vital role in the sorting and processing of proteins into different vesicular populations which are destined for either the basolateral or apical membrane (Achler et al., 1989). Therefore, as TCII is secreted with polarity in Caco-2 cells (Dix et al., 1987), we believe that this lighter vesicular compartment may represent secretory vesicles transporting TCII from the golgi to the basolateral domain. Further work is in progress to investigate whether Cbl is bound to TCII within these vesicles and identify the putative signals directing them to the basolateral membrane.

ACKNOWLEDGEMENTS

The authors warmly acknowledge Siân Woodcock for the culture of Caco-2 cells, Ian Williamson for electron microscopy, Kate Xavier for preparing the manuscript and Eric Tomlinson for his support throughout this study. A special thanks to Dr M Marsh a the Chester Beatty Laboratories, London for providing the facilities for performing the VSV experiments. This contribution is publication number 98 from the Advanced Drug Delivery Unit of Ciba-Geigy Pharmaceuticals.

REFERENCES

Achler, C., Filmer, D., Merte, C., and Drenckhahn, D., 1989, Role of microtubules in polarised delivery of apical membrane proteins to the brush border of the intestinal epithelium, J. Cell Biol., **109**:179-189.

Allen, R.H., Seetharam, B., Allen, N.C., Podell, E., and Alpers, D.H., 1978, Correction of cobalamin malabsorption in pancreatic insufficiency with a cobalamin analogue that binds with high affinity to R protein but not to intrinsic factor. In vivo evidence that a failure to partially degrade R protein is responsible for cobalamin malabsorption in pancreatic insufficiency, J.Clin.Invest., 61:1628-1634.

Branch, W.J., Mullock, B.M., and Luzio, J.P., 1987, Rapid subcellular fractionation of the rat liver endocytic compartments involved in transcytosis of polymeric immunoglobulin A and endocytosis of asialofetuin, Biochem.J., 244:311-315.

Brezt, R., and Staubli, W., 1977, Detergent influence on rat liver galactosyltransferase activities toward different acceptors, Euro.J.Biochem., 77:181-192.

Chanarin, I., Muir, M., Hughes, A., and Hoffbrand, A.V., 1970, Evidence of intestinal origin of transcobalamin II during vitamin B_{12} absorption, Brit. Med. J., 1:1454-1455.

Chung, K-N., Walter, P., Aponte, G.W., and Moore, H-P.H., 1989, Molecular sorting in the secretory pathway, Science, 243:192-197.

Dix, C.J., Hassan, I.F., Obray, H.Y., Shah, R., and Wilson, G., 1989, The transport of Vitamin B_{12} through polarised monolayers of Caco-2 cells, Gastroenterology, (accepted for publication).

Dix, C.J., Obray, H.Y., Hassan, I.F., and Wilson, G., 1987, Vitamin B_{12} transport through polarised monolayers of a colon carcinoma cell line, Biochem. Soc. Trans., 15:439.

Doscherholmen, A., and Hogen, P.S., 1959, Delay in absorption of radiolabelled cyanocobalamin in the intestinal wall in the presence of intrinsic factor, J. Lab. Clin. Med., 54:434-439.

Ellenbogen, L., Absorption and transport of cobalamin: intirinsic factor and the transcobalamin in: "Cobalamin: biochemistry and patho physiology," Babior, B.M., ed., John Wiley & Sons, Inc., New York, 1st Edn., 215-286 (1975).

Frater-Schroder, M., Haller, O., Gmur, R., Kierat, L., and Anastasi, S., 1983, Allelic forms of mouse transcobalamin II expressed in serum and by peritoneal macrophages, Hereditas, **98**:151-152.

Frater-Schroder, M., Porck, H.J., Erten, J., Muller, M.R., Steinmann, B., Kierat, L., and Artwert, F., 1985, Synthesis and secretion of human B_{12}-binding protein, transcobalamin II, by cultured skin fibroblasts and by bone marrow cells, Biochim. Biophys. Acta., **845**:421-427.

Green, P.D., Savage Jr., C.R., and Hall, C.A., 1976, Mouse transcobalamin II: Biosynthesis and uptake by L-929 cells, Arch. Biochem. Biophys., **176**:683-689.

Hall, C.A. The plasma transport of cobalamin (cbl), in: "Proceedings of the Third European Symposium on Vitamin B_{12} and Intrinsic facrtor," Zagalak, B., and Friedrich, W., eds., 725-742, Walter de Gruyter, Berlin, 725-742 (1979).

Hidalgo, I.J., and Borchardt R.T., 1988, Amino acid transport in a novel model system of the intestinal epithelium (Caco-2 cell), Pharm. Res., PD 950.

Hidalgo, I.J., Raub, T.J., and Borchardt, R.T., 1989, Characterisation of the human carcinoma cell line (Caco-2) as a model system for intestinal epithelial permeability, Gastroenterology, **96**:736-749.

Levine, J.S., Allen, R.H., Alpers, D.H., and Seetharam, B., 1984, Immunocytochemical localisation of intrinsic factor-cobalamin receptors in dog ileum: distribution of intracellular receptors during cell migration, J. Cell Biol., **98**:1110-1117.

Minnifield, N., Creek, K.E., Navas, P., and Morre, D.J., 1986, Involvement of cis and trans golgi apparatus elements in the intracellular sorting and targeting of acid hydrolases to lysosomes, Euro. J. Cell Biol., **42**:92-100.

Osiescka, I., Porter, P.A., Borchardt, R.T., Fix, J.A., and Gardner, C.R., 1985, In vitro drug absorption models. Brush border membrane vesicles, isolated mucosal cells and everted intestinal rings:

characterisation and salicylate accumulation, Pharm. Res., 2:284-293.

Quadros, E.V., and Rothenberg, S.P., 1986, Purification and molecular characterisation of human transcobalamin II, J. Biol. Chem., 261:15455-15460.

Rindler, M.J., Ivanov, I.E., Plesken, H., and Sabatini, D.D., 1985, Polarised delivery of viral glycoproteins to the apical and basolateral plasma membranes of MDCK cells infected with temperature-sensitive viruses, J. Cell Biol., 100:136-151.

Rindler, M.J., and Traber, M.G., 1988, A specific sorting signal is not required for the polarised secretion of newly synthesised proteins from cultured intestinal epithelial cells, J. Cell Biol., 107:471-479.

Savage Jr., C.R., and Green, P.D., 1976, Biosynthesis of transcobalamin II by adult rat liver parenchymal cells in culture, Arch. Biochem. Biophys., 173:691-702.

Sottocasa, G. L., Kuylenstierna, B., Ernster, L., and Bergstrand, A., 1967, An electron-transport system associated with the outer membrane of liver mitochondria, J. Cell Biol., 32:415-438.

Wilson, G., Hassan, I.F., Dix, C.J., Williamson, I., Shah, R., Mackay, M., and Artursson, P., 1989, Transport and permeability properties of human Caco-2 cells: an in vitro model of the intestinal epithelial cell barrier, J. Cont. Rel., (in press).

AN IN VITRO ABSORPTION MODEL SYSTEM BASED ON CELL MONOLAYERS (INTESTINE 407, MDBK)

H. Leuenberger, S. Buchmann*, C. Reinke and B. Schmid

School of Pharmacy
University of Basel
Totengaesslein 3
CH-4051 Basel
Switzerland

*Present address: Hoffmann-La Roche
Grenzacherstrasse 4002 Basel, Switzerland

INTRODUCTION

Human in vivo studies on the physicochemical properties of drugs are difficult to perform, first of all because of ethic reasons. For similar reasons the number of animal tests should be reduced if possible. Therefore alternative methods are desired such as biological in vitro experiments based on cell cultures. Thus, the development of a suitable in vitro absorption model system based on defined cell monolayers may offer some advantages if the experiments are performed in very controlled conditions. This is an important precondition, because in vivo as well as in vitro biological experiments usually exhibit important variances in their results, i.e. relative standard errors around 30 - 40% are rather common. Many papers even present results without an estimate of the experimental error involved. With the in vitro absorption model system presented here, two goals can be pursued simultaneously: a possible reduction in the number of animal tests, such as e.g. in situ rat intestine perfusion experiments and hopefully a possible reduction of experimental error.

Pharmaceutical Applications of Cell and Tissue Culture to Drug Transport
Edited by G. Wilson *et al.*, Plenum Press, New York, 1991

DEVELOPMENT OF THE IN VITRO ABSORPTION MODEL SYSTEM BASED ON CELL MONOLAYERS

Concept

An efficient control of the experimental conditions during drug transport through a cell monolayer, which separates the donator and acceptor compartment, needs to take into account the following considerations:

(a) For an easy and reproducible handling the cell monolayer needs to be cultivated in situ on a suitable supporting membrane. This supporting membrane is preferentially transparent, i.e. there is a possibility to check the confluency of the cell monolayer by inverse microscopy. The membrane has to be suitable for a sterile treatment, biocompatible and should exhibit a high permeability coefficient for the drug substances used for the permeation experiments, i.e. the quality of the supporting membrane is extremely important. Additional tests should be possible to control the quality of the cell monolayer, e.g. amount of biomass present, viability, electrical resistivity, etc.

(b) The in vitro absorption model system should permit the performance of relative experiments, i.e. experiments with the cell monolayer on the supporting membrane and with the supporting membrane alone. In addition it is necessary to use low drug concentrations in the donator compartment because of possible toxic effects on the cell monolayer and, as a consequence, a need to detect low drug concentrations in the acceptor compartment. The drug concentration in the acceptor compartment is preferentially measured in a continuous way.

(c) It is an important precondition to be able to measure the influence of unstirred aqueous boundary layers associated with biological or artificial membranes. Thus, it is necessary to control hydrodynamic conditions, but also the effects of hydrostatic and osmotic pressure between the donator and acceptor compartment.

(d) Finally the same apparatus should allow the testing of other types of cell monolayers, as well as tissues such as patches of freshly excised animal or human intestine or other tissues. The donator compartment should be suitable as recipient for an already solubilized drug or a dosage form of interest such as micellar solutions, mixed micellar solutions, nanoparticles, liposomal solutions or other drug carriers. Thus, it should be possible to test e.g. absorption enhancers or new dosage forms in relation to the pure substance solubilized.

Apparatus

The measuring system (see fig. 1) consists of an acceptor compartment A (340 ml) and a cylindrical donator compartment D (10 ml) which can be rotated at defined speeds in order to establish controlled hydrodynamic conditions [1]. The donator compartment can be immersed to a definite depth into the acceptor compartment. The bottom of the donator compartment consists of an artificial membrane, i.e. the supporting membrane for the cell monolayer. This membrane M (surface = 16.6 cm^2) separates both compartments. The solution in the acceptor compartment is mixed by a magnetic stirrer. Above this stirrer a steel plate is fixed to prevent any turbulence of the liquid close to the separating membrane. Thus, a better control of the hydrodynamic conditions can be achieved and a build-up of a hydrostatic pressure difference across the separating membrane can be reduced if not avoided. The temperature was kept at 37°C by a water bath. The liquid of the donator and acceptor compartment was kept isoosmotic and isohydric.

The liquid of the acceptor phase is pumped (5.5 ml/min) through a solid state scintillator (Ramona D, Raytest Isotopenmessgeraete, D-7541 Straubenhardt, Federal Republic of Germany) in order to detect low concentrations of carbon-14 or tritium-labelled drug substances. Thus, the concentration $C_A(t)$ in the acceptor compartment is continuously measured and registered as a function of time.

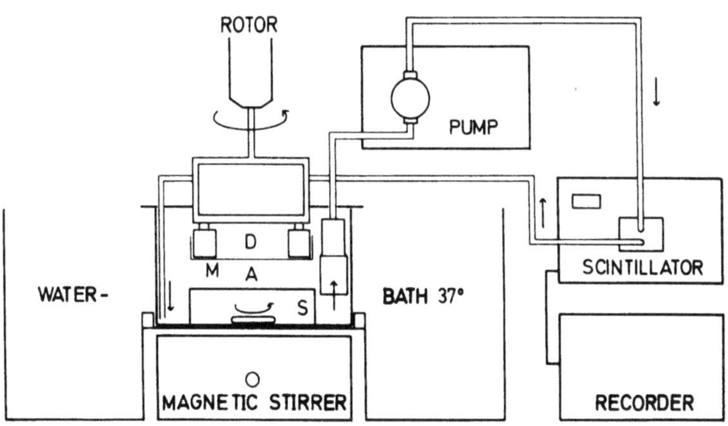

Fig. 1. Schematic representation of the in vitro absorption model system. D = donator compartment, A = Acceptor compartment, M = Supporting membrane with cell monolayer, S = Shield for magnetic stirrer.

The rate constant k, the concentration $C_A(t)$ in the acceptor compartment and the time lag t_{LAG} of drug transport can be calculated on the basis of equation (1) using a non-linear regression analysis (program No. 09835-15044, Hewlett-Packard Co., Fort Collins, Colorado, U.S.A.).

$$C_A(t) = C_A(t{=}\infty)\{1 - \exp(-k\,[t-t_{LAG}])\} \qquad (1)$$

The apparent permeability coefficient P_s is related to the volumes V_A, V_D of the acceptor respective donator compartment, and the surface A of the membrane as follows [2]:

$$P_s = \frac{k}{(1/V_A + 1/V_D)A} \qquad (2)$$

The apparent permeability coefficient P_s depends on both the properties of the membrane and the adjacent unstirred boundary layers, which may be the rate determining barriers for the more lipophilic drug. The unstirred boundary layers depend on the hydrodynamic conditions and are controlled by the rotating speed of the cylindrical donator compartment. An estimate of the thickness h_w of the aqueous boundary layer in dependence of the rotating speed f is given according to Levich [3] by equation (3)

$$h_w = 0.643\, D_w^{1/3}\, \nu^{1/6} f^{-\frac{1}{2}} \qquad (3)$$

The permeability coefficient P_s is equal to the reciprocal of the total transfer resistance R_T of the different resistances of the barriers in series such as the aqueous boundary layer at the donator (D) side $R_w(D)$, the monolayer R_Z, the supporting membrane R_M and the aqueous boundary layer at the acceptor (A) side $R_w(A)$:

$$\frac{1}{P_s} = R_T = R_w(D) + R_Z + R_M + R_w(A) \qquad (4)$$

An estimate of R_Z can be calculated by the determination of the total transfer resistance R_T for the supporting membrane alone, i.e. R_{TM}, and for the supporting membrane with the cell monolayer, i.e. R_{TZ}:

$$R_Z \approx R_{TZ} - R_{TM} \qquad (5)$$

It is evident from equation (5) that the results of R_Z are more reliable if R_{TZ} is $>>$ R_{TM}. The effect of the aqueous boundary layers are cancelled out in a first approximation. However, a higher variance has to be expected for the more lipophilic drugs where the aqueous boundary layers represent the rate-determining transfer resistance. The diffusion coefficient D_M of the drug molecule in the membrane M can be obtained if the permeability coefficient P_M or the resistance R_M, the membrane thickness h_M and the partitioning coefficient K_M (membrane/ solution) are known:

$$D_M = \frac{P_M h_M}{K_M} = \frac{1}{R_M} \frac{h_M}{K_M} \tag{6}$$

In an analogous way diffusion coefficients can be calculated for the cell monolayer with the thickness h_Z and the aqueous boundary layers with the thickness h_W

$$D_Z = \frac{P_Z h_Z}{K_Z} = \frac{1}{R_Z} \frac{h_Z}{K_Z} \tag{7}$$

and $$D_W = P_W h_W = \frac{h_W}{R_W} \tag{8}$$

The total transfer resistance R_T is for $R_w = R_w(A) = R_w(D)$ combined with equation (3) equal to

$$R_T = 1.286\ \nu^{1/6}\ f^{-\frac{1}{2}}\ D_W^{-2/3} + \frac{h_M}{K_M D_M} + \frac{h_Z}{K_Z D_Z} \tag{9}$$

If the diffusion process is mainly controlled by $R_T = 2R_w$ the latter two terms in expression (9) can be neglected.

In case of an important thickness, e.g. membrane thickness h_M, permeation experiments permit the measuring of a lag time t_{LAG} due to the passage of the drug through the barrier (e.g. unstirred aqueous boundary layers, cell monolayer, supporting membrane). If the concentration C_D of the drug in the donator compartment is kept constant and there are sink conditions in the acceptor compartment, i.e. there is a constant gradient $dc/dx = C_D/h_M$, the lag time t_{LAG} is equal to

$$t_{LAG} = \frac{h_M^2}{6D_M} \tag{10}$$

In our experimental situation the gradient dc/dx is only approximately constant for the time period $t << 1/k$ where the exponential function $(1-e^{-kt})$ shows a linear increase of the drug concentration in the acceptor compartment:

$$C_A(t) = C_A(t=\infty)\ kt \quad \text{for} \quad t << 1/k \tag{11}$$

Equation (11) represents the "zero order" flux situation where a constant amount of drug is released per unit time. Thus it has to be kept in mind that equation (10) is an estimation for the lag time t_{LAG}.

If the membrane exhibits a laminated structure and/or has pores the calculation of lag times becomes more tedious [2,4]. In case of an idealized trilaminate system, which takes into account the thickness h_w of the unstirred aqueous boundary layer and the thickness h_M of the membrane, the lag time t_{LAG} is equal to:

$$t_{LAG} = \left[\frac{4h_W^3}{3D_W^2} + h_W^2 \left\{\frac{h_M}{D_W D_M K_M} + \frac{h_M K_M}{D_W^2}\right\} + \frac{h_W h_M^2}{D_W D_M} + \frac{h_M^3}{6 D_W D_M}\right] / RT \tag{12}$$

$$\text{with } R_T = \frac{2h_W}{D_W} + \frac{h_M}{D_M K_M}$$

For membrane controlled diffusion equation (6) and (10) can be used:

$$t_{LAG} = \frac{h^2_M}{6D_M} \qquad \text{with } R_T = \frac{h_M}{D_M K_M}$$

Thus

$$t_{LAG} = \frac{h_M\ K_M}{6} R_T \tag{13}$$

Because of the complex structure of the biological membrane it is often only possible to analyse whether the diffusion process is controlled by the usually lipophilic membrane or by the unstirred aqueous boundary layers adjacent to the membrane. In the first case t_{LAG} can be estimated by equation (10) and in the latter case of an idealized trilaminated structure (unstirred boundary layer thickness h_w, membrane thickness h_M, unstirred boundary layer thickness h_w) the following equation holds if $K_M >> 1$ [4]:

$$t_{LAG} = \frac{K_M\ h_M\ h_W}{2\ D_W} \tag{14}$$

The combination of equation (8) and (14) with $R_T = 2\ R_w$ yields

$$t_{LAG} = \frac{K_M h_M}{4} R_T \tag{15}$$

Equation (15) can be used to determine experimentally the apparent partitioning coefficient $K_M = K_{app}$ for the more lipophilic drugs.

Madin-Darby Bovine Kidney (MDBK) cells: This cell line (ATCC CCL 22) is described in detail by [6]. Medium for cultivation: Dulbecco's modified Eagle's MEM with 10% fetal calf serum, 100 I.U./ml penicilline and 100 μg/ml streptomycine.

Incubation: The cells were cultivated directly on the supporting membrane which was fixed in advance to the transportable donator compartment. Without additional growth promotors, confluent monolayers were achieved after 48 hours for MDBK-and after 78 hours for Intestine 407-cells.

Supporting membrane

PAN membrane: PAN is a sterile polyacrylonitrile membrane used in hemodialysers (RP 607 HOSPAL, F-69330 Meyzien). Thickness of the moist and hydrophilic membrane: (31 $\pm$ 1) μm. The membrane is translucent. Thus, the growth of the confluent monolayer can be controlled in situ by the inverse light microscope technique.

Polyamide 66: This membrane polyamide 66 type Scrynel NY 5 HC (Zurich Bolting Cloth Mfg. Co. Ltd. CH-8803 Rüschlikon) is not translucent enough for in situ growth inspection of the cell monolayer. The thickness of this nylon tissue is equal to 100 μm; the width of the mesh i.e. "pore size" is equal to 5 μm. It is assumed that the confluent cell monolayers cover the pores completely.

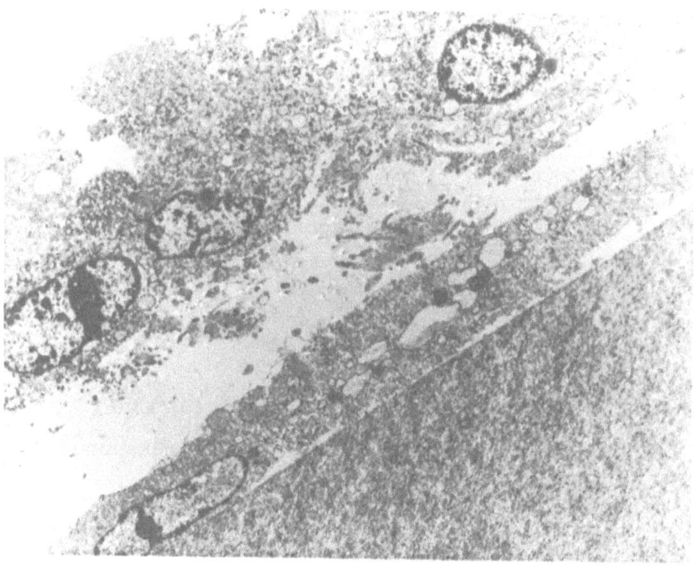

Fig. 2. Transmission electron microscopic view of the interface between supporting membrane and confluent cell monolayer (PAN; Intestine 407, Courtesy of Y. Litzistorf, Pathological Institute, Department of Medicine, University of Basel).

Because of the assumption $K_M >> 1$ the meaning of $K_M = K_{app}$ should not be stressed. However, K_M may be taken as a possible fitting parameter and be compared with the result of K_M in equation (12 and 13). The important point is that t_{LAG} is proportional to the resistance R_T for hydrophilic and lipophilic drugs.

The aqueous diffusion coefficient D_W can be calculated according to equation (8) if the diffusion process is rate-limited by the aqueous boundary layer as it is the case with lipophilic drugs.

Error Analysis

Due to the important variances found in biological experiments, a subtile analysis of experimental errors is essential. Unfortunately, in many cases it is difficult or impossible to detect the presence of systematic errors. However, the analysis of the statistical error can yield useful information. Thus, it may be important to reduce the consumer's risk (second type of error), i.e. to facilitate the detection of existing real differences in two results. If the experiments are suitably standardized, mean values and the standard error of the mean may be applied. In an experimental design which provides redundant information i.e. results, existing correlations can be used to reduce the statistical error. It is, however, important to critically analyse the physical or mathematical model which is the basis of these correlations. The design of suitable relative measurements can improve substantially the reliability of results.

Concerning the determination of permeability values P_Z for the cell monolayer, using different supporting membranes, an error analysis according to Gauss was performed. The use of standard errors of the single value (S_X) or the mean value ($S_{\bar{X}}$) are indicated in the tables of experimental results (equations 16, 17)

$$S_{\ddot{X}} = \left[\frac{\Sigma(x-\bar{x})^2}{n-1} \right]^{\frac{1}{2}} \qquad (16)$$

$$S_{\bar{X}} = \left[\frac{\Sigma(x-\bar{x})^2}{n(n-1)} \right]^{\frac{1}{2}} \qquad (17)$$

The standard error of the mean of the permeability coefficient of the supporting membrane is equal to $S_{\bar{X}}(P_{TM})$. Thus, the corresponding standard error $S_{\bar{X}}(R_{TM})$ of the resistance R_{TM} is equal to

$$S_{\bar{X}}(R_{TM}) = \frac{S_{\ddot{X}}(P_{TM})}{P_{TM}^2} \qquad (18)$$

In an analogous way the following standard errors can be calculated

$$S_{\bar{x}}(R_{TZ}) = \frac{S_{\bar{x}}(P_{TZ})}{P_{TZ}^2} \qquad (19)$$

with R_{TZ} = resistance of the supporting membrane with the cell monolayer

P_{TZ} = permeability coefficient of the supporting membrane with the cell monolayer

$$S_{\bar{x}}(R_Z) = \{S_{\bar{x}}(R_{TM})^2 + S_{\bar{x}}(R_{TZ})^2\}^{\frac{1}{2}} \qquad (20)$$

and ultimately

$$S_{\bar{x}}(P_Z) = \frac{S_{\bar{x}}(R_Z)}{R_Z^2} \qquad (21)$$

Important variances in R_Z and P_Z may result if R_{TZ} and R_{TM} have the same order of magnitude [compare equation (5)]. This problem should be kept in mind for the more lipophilic drugs where the unstirred aqueous boundary layers represent the major transfer resistance. Thus, the determination of P_Z may become unrealistic.

EVALUATION AND VALIDATION OF THE IN VITRO ABSORPTION MODEL SYSTEM

Materials and Methods

Apparatus and Methods

The apparatus developed, the theoretical background and the methods applied are described in detail in section: *In vitro model developed for permeation experiments*.

Cell cultures and characterization

Intestine 407: This is an established cell line (ATCC, human origin) described in detail by Henle and Deinhardt [5]. Medium for cultivation: Eagle's Minimum Essential Medium (MEM) with 10% fetal calf serum, 100 I.U./ml penicilline and 100 μg/ml streptomycine.

Solutions for donator and acceptor compartment during permeation experiment

Isotonic phosphate buffered saline (PBS) (Dulbecco) with pH 7.4, density (20°C) = 1.005 g/cm^3 and viscosity (20°C) = 1.03 cP.

Isotonic buffer TRISC with 2% PVP with pH = 7.4, density (20°C) = 1.011 g/cm^3 and viscosity (20°C) = 1.15 cP. The TRISC buffer solution with 2% PVP is used in case of the more lipophilic drugs to reduce a possible memory effect in the solid state scintillation counter.

Patches of intestine

For each experimental run an excised patch of intestine (jejunum) of freshly slaughtered Swiss noble pig was used for comparison with the cell monolayer experiments. No supporting membrane was necessary to fix the patch. Swelling experiments showed, that the thickness of the patch in increased linearly with time [7].

Control of the growth viability and morphology of the cell monolayer

Concerning the translucent membrane the confluency of the cell monolayer was inspected in situ by inverse light microscopy. The viability test was performed with the Trypan blue method [8]. A viability of 85-90% of the cells was found after a 4 hours' exposition to PBS and TRISC buffer solution with 2% PVP.

Other controls included the counting of the number of cells present per unit area with Videoplan 2, image processing system, Kontron AG, CH-8048 Zurich and the determination of the biomass, i.e. mass of protein per unit area [9] after the permeation experiment. The interface between the supporting membrane and the confluent cell monolayer was inspected by transmission electron microscopy (see fig.2). The mean thickness of the confluent cell monolayer is equal to 28 μm.

Radiolabelled substances for the permeation experiments

As a hydrophilic model drug (7-14 C) salicylic acid (New England Nuclear, 0.25 mCi in 2.5 ml ethanol water mixture), and as a lipophilic model drug (1,2(N)-3H)-progesterone (Amersham, 1 mCi in 1 ml toluol) were used. Other test substances included C-14 labelled ascorbic acid, butyric acid, (sodium salt), octanoic acid (sodium salt), testosterone and biotin. Concerning the model drugs, 40-80 μl of the radioactive solution was sufficient to assure a convenient detection of the drug concentration in the acceptor compartment by the continuous flow measurement of the solid state scintillation counter (Ramona D, Raytest Isotopen-messgeraete, D-7541 Straubenhardt 1, Federal Republic of Germany). Because of the low drug concentration no toxic effects on the cell culture had to be expected. To avoid any adsorption effects only stainless steel tubes and glass were used in the measuring equipment.

Results of the validation experiments

Effect of the hydrodynamic conditions

Position of the donator compartment: In order to measure precisely the concentration of the labelled drug in the acceptor compartment it is an absolute prerequisite to stir the buffer solution of the acceptor phase sufficiently. Thus, a build-up of pressure gradients between donator and acceptor compartment can be expected. A special stainless steel plate was mounted above the magnetic stirrer to protect the thin membrane separating the donator compartment from the acceptor compartment, from the effect of the whirl. To validate the system permeation experiments included different positions of the donator compartment relative to the acceptor compartment. For this purpose a high precision positioning of the donator compartment was necessary.

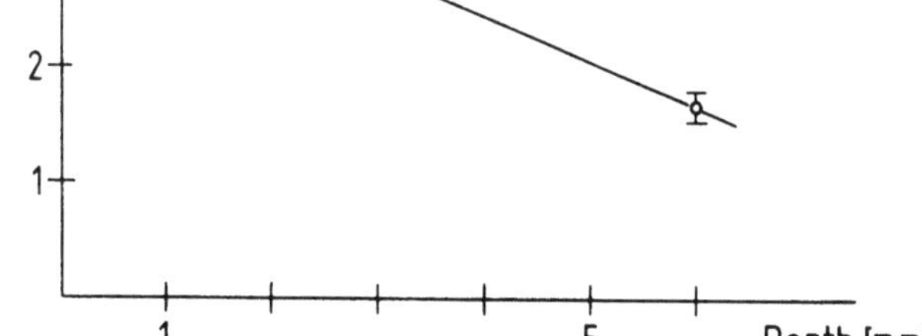

Fig. 3. Effect of the relative position (depth) of the donator compartment on the apparent permeability coefficient P.

From the results (fig. 3) the following conclusions have to be drawn:

1) The hydrodynamic effect is extremely important.
2) The results of the permeabilities are not absolute values as they depend on the hydrodynamic conditions of the measurement equipment.
3) The measurement equipment needs to be defined in detail, similar or better than the standardization in the case of dissolution apparatus, or of other equipment, such as to measure the hardness of materials, e.g. Brinell, Rockwell or Vickers Hardness.

Thus, the relative position of the donator compartment was kept constant at 3 mm immersion depth for all subsequent experiments.

Effect of the rotational speed of the donator compartment: The permeation experiments of salicylic acid with and without the Intestine 407 cell monolayer on the supporting membrane (PAN) were performed at f = 25; 36; 50 and 100 rpm rotational speed of the donator compartment. Only at 100 rpm a detachment of cells from the monolayer could be observed after the experiment. The results are presented in fig. 4 and show the expected linear dependence of the transfer resistances R_{TM} and R_{TZ} according to equation (9):

$$R_{TM}(s/cm) = (1236 \pm 340) + (1588 \pm 280)\ f^{-\frac{1}{2}} \qquad r = 0.970$$

$$R_{TZ}(s/cm) = (2735 \pm 113) + (1404 \pm 85)\ f^{-\frac{1}{2}} \qquad r = 0.998$$

Slope **b** and y - intercept **a** were calculated by linear regression analysis and include results of standard errors. The slopes **b** are expected to be identical **b** = $(b_{TM} + b_{TZ})/2 = 1500\ s^{1/2}/cm$. The difference in the intercepts R_{TZ} and R_{TM} should be equal to R_Z at $f = \infty$. The result at 100 rpm (detachment of cells) was not taken into account (see fig. 4).

Determination of permeability coefficients

Permeation controlled by membrane resistance (supporting membrane and cell monolayer): The results listed in table 1 originate from replicated (n = 5) permeation experiments with salicylic acid, biotin, monolayers of Intestine 407 and MDBK cells. The rotational speed is mentioned as a reference of the source of results.

Permeation controlled by unstirred aqueous boundary layer: The results are listed in table 2 and show only small differences between the transfer resistance of the supporting membrane (PAN) with and without cell monolayer (Intestine 407). Because of the sensitivity of the cells the rotational speed was kept at 36 rpm.

Table 1. Total transfer resistances R_{TZ}, R_{TM} and permeation coefficient P_Z of the cell monolayer (Intestine 407, MDBK)

Compound Membrane/ cell type	R_{TZ} (s/cm)	R_{TM} (s/cm)	P_Z (x 10^{-4} cm/s)	f (rpm)
Salicylic acid:				
PAN/INT 407	4570±836	3460±174	9.0 ±3.1*	36
"	2735±113	1236±340	6.7 ±1.6	∞
Polyamide 66/ MDBK	1168±272	230±15	10.7 ±1.3*	36
Biotin:				
PAN/INT 407	8580±524	6060±187	3.97±0.88	36

*standard error of the mean value

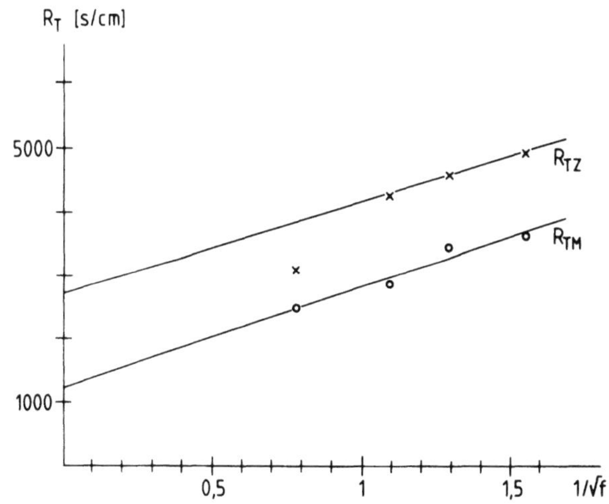

Fig. 4. Results of validation experiment with salicylic acid: effect of the rotational speed of the donator compartment on the total transfer resistances R_{TZ}, R_{TM} of the supporting membrane (PAN) with and without cell monolayer (Intestine 407).

Table 2. Total transfer resistances R_{TZ}, R_{TM} and permeability coefficient P_Z of the cell monolayer (Intestine 407)

Compound	R_{TZ} (s/cm)	R_{TM} (s/cm)	P_Z (10^{-4} cm/s)
Ascorbic acid	4360±1220	3660± 604	--
Butyric acid (Na)	4730± 523	4600± 497	--
Octanoic acid(Na)	5120± 230	5210± 572	--
Testosterone	4280± 508	4230± 287	--
Progesterone	5120± 476	4250±1420	--
Mean value (all substances)	4722± 591	4390± 676	--
	R_T = $(R_{TZ}+R_{TM})/2$ = 4500 s/cm		

Thus it is only possible to give an estimate for the permeability coefficients P_{TZ} and P_{TM} which include the resistance of the supporting membrane. The mean permeability values estimated from the mean resistances (over all substances) are $P_{TZ} = 2.2 \times 10^{-4}$ cm/s, $P_{TM} = 2.3 \times 10^{-4}$cm/s and reflect in fact the resistances of the unstirred aqueous boundary layer. In subsequent experiments the polyamide 66 membrane with the lower resistance and the more vigorous MDBK cells were used for the permeation of progesterone (see fig. 5). The rotational speed was kept at 36 rpm:

R_{TZ} (mean,n = 5) = (5551 ± 441*) s/cm
R_{TM} (mean,n = 4) = (3432 ± 685*) s/cm
R_Z = (2119 ± 815*) s/cm
$P_Z = (4.7 \pm 1.8) \times 10^{-4}$ cm/s

More reliable results for P_Z can be expected for higher rotational speed if the cells remain attached to the supporting membrane.

* = standard error of the mean

Determination of aqueous diffusion coefficients D_w

In case of the more lipophilic drugs the permeability coefficient is determined by the aqueous boundary layers. Thus D_w can be estimated according to equation (9) assuming $R_T \approx 2R_w$ and $D_w^{2/3} \approx 1.286\, \nu^{1/6} f^{-1/2}/R_T$. With $\nu = 10^{-2}$cm^2/s, f = 36 rpm and R_T = 4500 s/cm the aqueous diffusion coefficient for the more lipophilic drugs is approximately $D_w = 2.2 \times 10^{-6}$ cm^2/s

An analogous calculation on the basis of the permeation experiments with progesterone and the polyamide 66 membrane alone (h_M = 0.01 cm, f = 36 rpm, R_{TM} (n = 4) = 3430 ± 685* s/cm, t_{LAG} =(370 ± 50*s) yield D_w (progesterone) = 3.4 x $10^{-6}cm^2/s$. It is evident that if D_w is calculated according to equation (3) and (14) the same result is found.

An evaluation of D_w using the exact equation (12) is only possible if assumptions are made such as $D_w \approx D_M$ and $K_M \approx 1$ for the hydrophilic supporting membrane. Thus a polynomial expression of degree 6 for $D_w^{1/3}$ has to be resolved. For t_{LAG} = 370 s and R_{TM} = 3430 s/cm, D_w becomes 0.82 x 10^{-6} cm^2/s.

The same supporting membrane (polyamide 66, h_M = 0.01 cm, f = 36 rpm) with MDBK cell monolayers and progesterone showed a mean lag time of t_{LAG} (n = 5) = (501 ± 80*)s and a resistance R_{TZ} = (5551 ± 441*)s/cm.

In a first approximation D_w can be calculated assuming $R_{TZ} \approx 2R_w$ as above, i.e. D_w = 1.6 x $10^{-6}cm^2/s$. As the exact value of K_M and D_M are not known and as the assumption of an ideal trilaminate of the supporting membrane with the cell monolayer does not hold, D_w is not evaluated according to equation (12).

In fig. 4 the slope of the linear relationship is equal to b = 1.286 $\nu^{1/6}D_w^{-2/3}$ i.e. equation (9) can be used to estimate $D_w \approx$ 7.9 x 10^{-6} cm^2/s for salicylic acid,

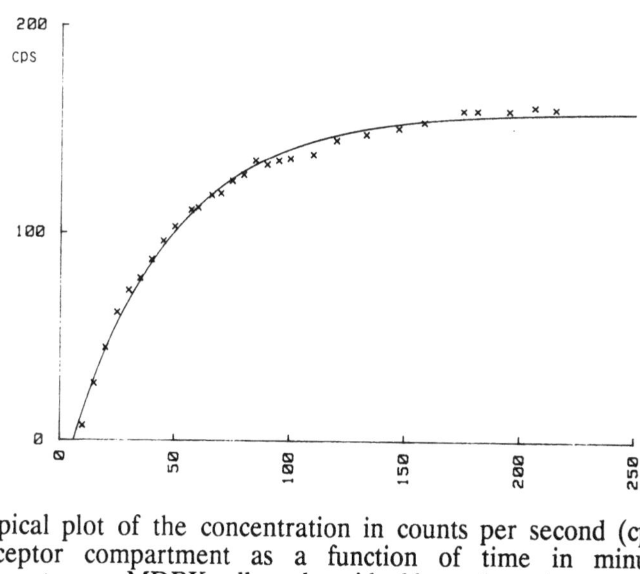

Fig. 5. Typical plot of the concentration in counts per second (cps) in the acceptor compartment as a function of time in minutes (H-3 progesterone, MDBK cells, polyamide 66 membrane)

assuming b = 1500 $s^{1/2}$/cm. Permeation experiments with salicylic acid and the polyamide 66 membrane (h_M = 10^{-2}cm) yield $D_Z = P_Z h_M \approx 1 \times 10^{-5} cm^2/s$.

* = standard error of the mean

Results of permeation experiments with pig intestine

For comparison, permeation experiments were performed with the model substance salicylic acid. Compared to the cell monolayers the patches of the pig intestine show values of higher variability which are most probably due to a swelling process of the tissue during the experiment. In addition to the permeability coefficients, lag times can be determined because of the much larger thickness of this membrane. As expected the permeability coefficient for pig intestine patches is smaller than for cell monolayers. For the evaluation of the permeability coefficient a linear regression analysis permits the calculation of the slope in equation (11) if the final concentration C_A (t =∞) is known (see fig. 6). The results are summarized in table 3.

The linear regression analysis according to equation (13) yields t_{LAG} = 23.5 s + 0.036 R_T with r = 0.995. From the slope a mean thickness h_M = 0.216 cm can be calculated. In table 3 an important variance could be observed for the mean diffusion coefficient $D_M \approx (1.45 \pm 1.0) \times 10^{-5} cm^2/s$.

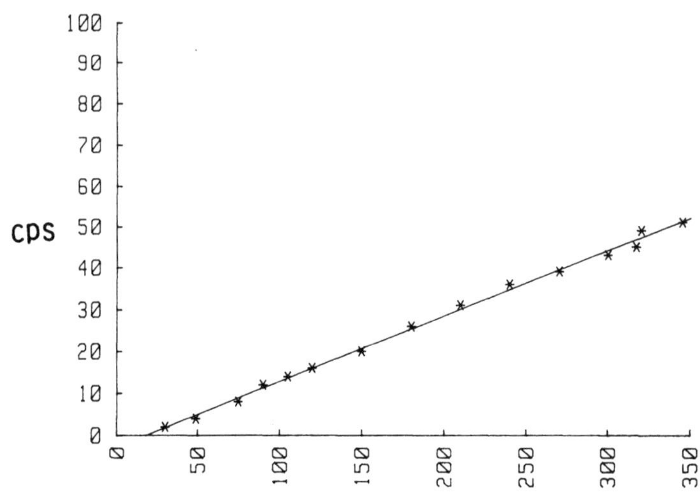

Fig. 6. Typical plot of the concentration in counts per second (cps) in the acceptor compartment as a function of time in minutes (14-C salicylic acid, pig jejunum patch).

Table 3. Lag times and permeability coefficients (salicylic acid, pig jejunum) patches

t_{LAG}(s)	Slope b (cps/min)	r	P_M cm/s 10^{-5}	h_M cm	$D_M=P_M h_M$ cm^2/s 10^{-5}
1526	0.1366	0.9901	2.32	0.212	0.49
346	0.5675	0.9966	9.63	0.200	1.9
385	0.6500	0.9990	11.0	0.255	2.8
1381	0.1650	0.9990	2.8	0.232	0.64
				D_W(mean) =	1.45 ± 1.0

Table 4. Correlation of in vitro permeability coefficients P_{TZ}, P_Z to in vivo results of Ho et. al [10]

Compound	P_{TZ} 10^{-4}cm/s	P_{app}[10] 10^{-4}cm/s	P_Z 10^{-4}cm/s	P_M[10] 10^{-4}cm/s
Salicylic acid	2.2±0.4	2.15	9.0±3.1*	10.2
" (PA66,MDBK)	8.6±2.0	2.15	10.7±1.3*	10.2
Butyric acid	2.1±0.2	1.88	-- **	3.5
Octanoic acid	2.0±0.1	2.95	-- **	38.6
Testosterone	2.3±0.3	2.00	-- **	>20.0
Progesterone	2.0±0.2	2.32	-- **	>23.2

* standard error of the mean
** standard errors too important to calculate P_Z correctly

Correlation with in situ permeability coefficients for the rat jejunum

The apparent permeability coefficients P_{TZ} (supporting membrane: PAN; INT 407 cell monolayer) for salicylic acid , butyric acid, octanoic acid, testosterone and progesterone were compared with the results which were obtained by Ho et al. [10] from in situ jejunum perfusion experiments (modified Doluisio method, 37°C, isosmotic buffer pH6, hydrodynamic condition: high stirring). The results are summarized in table 4 and include the permeation experiment for salicylic acid with polyamide 66 (PA 66) supporting membrane and MDBK cell monolayer.

CONCLUSIONS

The experimental system presented here can be easily used for permeation experiments with different substances under the precondition of strictly-controlled hydrodynamic parameters. As shown, the use of Intestine 407 cell monolayer grown on PAN membrane permits the determination of apparent permeability coefficients which can be correlated satisfactorily to results obtained by Ho et. al [10] in in situ rat intestine perfusion experiments.

When the hydrophilic salicylic acid was tested with monolayers of Intestine 407 (on PAN membrane) and MDBK cells (on polyamide 66) the permeability coefficients P_Z [(9.0 $\pm$3.1) x 10^{-4} cm/s;(10.7 $\pm$ 1.3) x 10^{-4} cm/s] of the cell monolayers were within the limits of the experimental error in both systems, despite the considerable difference in the transfer resistances of both supporting membranes: R_T (PAN) = 0.346 x 10^4 s/cm; R_T (polyamide 66) = 0.023 x 10^4 s/cm. Regarding lipophilic drugs, it has been shown that the unstirred aqueous boundary layers represent the major transfer resistance. Therefore, in such experiments the standard error is naturally higher than with the hydrophilic drugs in which the standard error is below 35%. Experiments performed with higher rotational speed of the donator compartment should lead to more reliable P_Z values with lower standard errors.

The results of our experiments justify further investigation of the additional potentials of the presented absorption model system. It offers the possibility to investigate in detail the effects of absorption enhancers and it will greatly facilitate in vitro studies on transport properties of new drug delivery systems, and new types of drugs such as peptides and proteins.

REFERENCES

[1] US-Patent 4,812, 407; europ. Pat. Appl. 0242326 and other Pat. Appl.

[2] C. Barnes, Physics **5**, 4 (1934) cited by G.L. Flynn, S.H. Yalkowsky and T.J. Roseman J. Pharm. Sci. **63**, 479 (1974).

[3] V.G. Levich, Physicochemical Hydrodynamics; Prentice Hall, Englewood Clifts, N.J., 1962, p. 57-62.

[4] J.A. Barrie, J.D. Levine, A.S. Michaels and P. Wong Trans. Faraday Soc. **59**, 869 (1963).

[5] Q. Henle, F. Deinhardt, J. of Immunology **79**, 54 (1957.

[6] S.H. Madin and N.B. Darby, Proc. Soc. Exp. Biol. Med. **98**, 574 (1958).

[7] B. Schmid, H. Leuenberger, unpublished data.

[8] T. Lindl, J. Bauer, Zell- und Gewebekultur, Gustav Fischer Verlag, Stuttgart, 1987, p. 151.
[9] Ch. Watters, Analytical Biochemistry **88**, 695 (1978).
[10] N. F. H. Ho, J.Y. Park, W. Morozowich and W.I. Higuchi, Physical model approach to the design of drugs with improved intestinal absorption, in Design of pharmaceutical properties through prodrugs and analogs, E.B. Roche, editor, American Pharmaceutical Association, Washington, D.C. 1977, p.136-227.

TRANSFERRIN ENDOCYTOSIS AND FLUID PHASE UPTAKE IN THE DIFFERENTIABLE INTESTINAL CELL LINE HT-29

Christian Huet, Odile Godefroy, Cristina Ibarra*, Alice Dautry-Varsat** and Daniel Louvard

Unité de Biologie des Membranes, Département de Biologie Moléculaire, and **Unité de Biologie Somatique, Département d'Immunologie, Institut Pasteur, 75724-Paris Cedex-15, France. *Département de Biologie, C.E.N. Saclay, 91191 Gif sur Yvette, France

CELL DIFFERENTIATION AND CELL POLARITY

Cells such as epithelial or neuronal cells, when fully differentiated, exhibit a clear functional asymmetry related to morphological asymmetry (Gumbiner & Louvard, 1985).

Intestinal epithelial cells which are covering the intestinal villi, are facing the lumen which corresponds to the external milieu. This monolayer is produced by stem cells located in the crypts. Stem cells can divide and daughter cells migrate along the crypt axis toward the apex of intestinal villi (Leblond & Cheng, 1976). During this migration, cells are differentiating, acquiring an enormous apical surface membrane with numerous microvilli and enzymes such as aminopeptidase and sucrase-isomaltase (Kenny & Maroux, 1982).

In the intestinal epithelial cells, the surface membrane is divided into an apical domain facing the external milieu and a basolateral domain facing the internal milieu. These two membrane domains are delineated by tight junctions that prevent lipid and protein diffusions (Van Meer et al., 1986a, 1986b). The basolateral membrane contains surface markers that can also be found on other cell types (HLA, $Na^{+}K^{+}$ ATPase, transferrin receptor), (Kyte, 1976; Kirby & Parr, 1979; Fuller & Simons, 1986), whereas apical membrane contains the specific intestinal enzymes such as sucrase-isomaltase and aminopeptidase (Louvard, 1980; Reggio et al, 1982; Trugnan et al, 1987).

During differentiation, intestinal cells acquire at their apical pole a brush border which tremendously enhance the surface exchange with nutrients. Concomitantly, cytoskeletal proteins are acquiring a polarized distribution along the axis of the enterocyte. Actin microfilaments are associated with villin, fimbrin, 110 Kd protein, forming the axis and the rootlets of microvilli (Mooseker, 1985).

The terminal differentiation of a daughter cell from the stem cell giving functional enterocytes implies deep morphological functional and biochemical modifications. The func-

Pharmaceutical Applications of Cell and Tissue Culture to Drug Transport
Edited by G. Wilson *et al.*, Plenum Press, New York, 1991

tional polarity of these epithelial cells has to be maintained as they form the border between the external milieu and the organism. *In vitro* studies gave us information on the mechanisms involved when cells acquire and maintain their polarity (Gonzalez-Mariscal et al.,1985; Vega-Salas et al., 1987).

Eukaryotic cells are highly organized. Proteins can circulate in two directions between the intracellular compartments. In polarized cells transport can occur from one pole to the other (transcytosis) or from one pole back to the same membrane domain (diacytosis).

The secreted proteins are known to be synthetized on polysomes bound to the rough endoplasmic reticulum, transferred to the Golgi apparatus and secreted or stocked in secretion vesicules.(for review Meyer, 1982; Kelly, 1985). In a polàrized cell, secretion can well be also polarized: specifically basolateral or apical secretion has been demonstrated (Kondor-Koch al., 1985; Urban et al., 1987; Traber et al., 1987; Tojo et al., 1987).

One of the puzzling question asked of the cellular biologist is how molecules are sorted and how vesicles containing receptors and ligands are addressed to the proper membrane domain.

LIGAND INDUCED ENDOCYTOSIS

Internalization of material and molecules can be achieved by three distinct ways: Phagocytosis which occurs for large structures like bacteria, pinocytosis which is the fluid uptake with dissolved molecules, ligand induced endocytosis which is a very specific process implying the interaction between the endocytosed molecule (ligand) and its membrane receptor. Once the ligand is bound to its specific receptor, the formed complex penetrates the cells within plasma membrane invaginations. In most cases,but not all (Huet et al., 1980; Montesano et al., 1982), these membrane pits and vesicules are coated with clathrin which can be seen in electron microscopy and isolated (Pearse et al.,1976; 1987). Clathrin molecules are formed by associated triskelions built with three 180 kD heavy chains associated with three 30 kD light chains (Pearse et al., 1987; Ungewickel et al., 1981; Zaremba et al., 1983). These triskelions are flexible structures able to quickly reassemble from a soluble cytoplasmic pool (Lubinski et al., 1984; Goud et al., 1985). In cultured cells coated pit formation can be inhibited by potassium depletion (Larkin et al., 1983).

Receptors could be either initially located in the coated pits (LDL, Asialoglycoprotein, transferrin: Anderson et al., 1982; Wall et al., 1981; Willingham et al.,1984) or randomly dispersed on the cell surface (hormones, EGF: Gavin et al., 1974; Beguinot et al., 1984). Endocytotic vesicules fuse to form endosomes. Two kinds of endosomes can be distinguished (Helenius et al.,1983), one is located near the plasma membrane and the other in the Golgi area. Depending on the molecules, the ligand and the receptor can be both degraded,or the receptor is recycled or the ligand also is recycled (Goldstein et al., 1985). The receptor will accumulate in the tubular region of the endosomes, whereas the ligand will concentrate in the vesicular region.(Geuze et al., 1983).

TRANSFERRIN CYCLE

Transferrin is a major serum glycoprotein which can be found either as Iron free apotransferrin or as ferrotransferrin which carries two Fe atoms. The binding of ionic iron occurs to the transferrin with a high affinity ($K_d = 10^{-20}$ M) in the plasma at neutral pH (Aisen et al., 1978).Transferrin is made of a single peptidic chain (MW= 80 kD). Two sites of fixation are involved in Fe binding and another domain is specific for the binding to the membrane receptor. Transferrin receptor is a dimer (MW= 180 kD) with two subunits bound with a disulfide bond. (Dautry-Varsat et al., 1984). It is inserted in the plasma membrane and three domains can be distinguished:a cytoplasmic fragment (MW=5 kD), a transmembrane segment and an extracellular domain (MW=70 kD) which is trypsin sensitive and which binds the transferrin. The cytoplasmic fragment is necessary for the endocytotic process to occur and is also phosphorylated (Iacoppetta et al., 1988).

The transferrin receptor has a better affinity for the ferrotransferrin than for the apotransferrin. After binding, the ligand receptor complex is internalized (Karin et al., 1981) and it will reach the lysosomes where acidic pH will induce the iron to be released from the protein carrier. The ligand receptor complex is recycled to the cell surface where the pH is neutral. That will induce the ligand releasing because of the weakness of the interactions at such a pH (Klausner at al., 1983). In polarized cultured cells(MDCK) and in enterocytes it has been shown that the transferrin receptor is essentially localized on the basolateral membrane (Fuller et al., 1986; Banerjee et al., 1986).

In tissues, the transferrin cycle is adapted to the local conditions of cell growth acquired as cell polarity is exhibited.

DIFFERENTIABLE HUMAN INTESTINAL HT29-18 CELL CLONES

The HT-29 cell line, derived from a human colon adenocarcinoma (Fogh et al., 1975) is able to differentiate if galactose replaces glucose in the culture medium (Pinto et al., 1982). We have isolated a clone (HT29-18) and absorptive and mucus secreting subclones from this cell line which display differentiated properties of the parent cells. HT29-18 cells grown in glucose containing medium form multiple layers of round cells without specific cell-cell contact. In contrast, when grown in galactose containing medium, they form a monolayer with tight junctions and exhibit a well differentiated brush border at their apical membrane which faces the culture medium.

We have also demonstrated that the differentiated cloned population HT29-18/gal is heterogenous: although 90% of the cell population shows morphological characteristics of absorptive cells, about 10% contain mucous granules and are similar to intestinal goblet cells. We have isolated two subclones, HT29-18 C1 and HT29-18 N2 from the differentiated HT29-18/gal cells. HT29-18 C1 cells show morphological characteristics of polarized absorptive cells when grown either in glucose or in galactose containing media. The HT29-18 N2 clone is also polarized in both culture conditions and is similar to goblet cells in vivo. It grows as a monolayer, exhibits tight junctions and contains numerous mucous granules whose exocytosis can be triggered by parasympathomimetic drugs. The clone HT29-18 first isolated appeared as a multipotent cell population out of which we isolated several subclones that differentiate either as absorptive (HT29-18 C1) or as mucous (HT29-18 N2) cells. In contrast to the parent cell line the subclones retain most of their differentiated properties in glucose containing medium (Huet et al., 1987; Godefroy et al., 1988).

CULTURE OF HT29-18 C1 SUBCLONE ON A PERMEABLE SUBSTRATE

In order to investigate the polarization of endocytosis using HT29-C1 subclones, we have developed culture conditions which enable them to form confluent monolayers on permeable substrates. Polycarbonate membranes (porosity= 3µm; Nucleopore Co.) were carbonated by using a carbon evaporator (Edwards, Crawley, England). Mini-Marbrook chambers (4mm diameter) were specially designed. Carbon evaporation on the filter surface and glow discharge greatly favored cell adhesion to this substrate. Cells were seeded at $8x10^5$ cells/filter in 0.8 ml medium inside the chamber. During the growth phase, the mini-chambers were soaked in 10 cm Petri dishes with 30 ml medium in a 10% CO_2/air incubator. During the experimental phase, filter devices were incubated in Costar 24-well dishes which allow the use of 0.8 ml medium on the basolateral side and 0.2 ml in the apical side. Cells were used about nine days after seeding. Tightness of the cell monolayer was controlled by the constant level of medium on the upper compartment of the chamber when a 3-4 mm hydrostatic pressure was applied overnight. Electrical parameters were measured using a modified Ussing chamber. Monolayers were mounted as a flat sheet between two compartments corresponding to the mucosal and the serosal compartments. The potential differences were measured with Ag/AgCl electrodes with 3M KCl bridges. Cells were developing as a monolayer over the entire surface of the filters and the morphological properties were similar to those of cells grown on Petri dishes: They have epithelial characteristics (polarity, junctions, microvilli). In initial conditions, monolayers display an electrical resistance value of 100 W/cm2 and develop an initial transepithelial potential of 0.4 mV. When incubated in Ca^{++} free medium the resistance decreases and no more transepithelial potential can be detected. As already demonstrated, in the absence of calcium, tight junctions are disrupted and no electrical resistance is measured (Godefroy et al., 1988).

IMMUNOCYTOCHEMICAL DISTRIBUTION OF PLASMA MEMBRANE PROTEINS IN DIFFERENTIATED AND UNDIFFERENTIATED HT29-18 CLONES

The polarized distribution of HLA and of transferrin receptors was examined on frozen sections of human colon by immunofluorescence using monoclonal antibodies. These proteins were found to be localized on the basolateral membranes of the columnar cells in the human colon but absent from the apical surfaces. It is worth pointing out that both membrane proteins are facing the internal milieu.

HT29-18/glucose cells, which are not differentiated, showed a uniform staining of their entire surface. These results demonstrate the uniform distribution of these proteins over the entire cell surface and the lack of distinct membrane domains. The distribution of these receptors is similar on nonconfluent and on confluent cultures.

On well confluent HT29-18/galactose cells, immunocytochemistry shows that no staining can be seen either with HLA or with antibodies against transferrin receptors on intact cells. This result demonstrates that no membrane protein is present on the apical membrane. In these experiments, Ca^{++} containing buffers were used and intercellular tight junctions were preserved preventing the antibodies access to the basal pole of the cells. When cells are incubated and fixed under Ca^{++} free conditions, the entire cell surface is then accessible and the binding of the antibodies is allowed. Homogeneous basolateral staining for HLA and transferrin receptors is observed. Under these conditions, very little staining was observed on the apical membranes.

On differentiated HT29-18 C1 cells, receptor localization was performed using a monoclonal antibody on longitudinal sections of monolayers grown on filters. A bright staining of the basolateral membrane is observed while their apical membrane seems to be completely devoid of receptor.

Both differentiated cells grown on coverslips and on porous substrates display asymmetric distribution of the transferrin receptor.

LOCALIZATION OF TRANSFERRIN BINDING SITES: FITC-TRANSFERRIN BINDING AND INTERNALIZATION

Binding of FITC-labelled transferrin (fluorescein isothiocyanate labelled transferrin) was performed to visualize distribution and internalization of transferrin receptor on living cells.Homogeneous membrane staining is seen when living undifferentiated HT29-18/glu cells are labelled for 1 hour at 4°C in Ca^{++} containing medium. In such conditions, no membrane redistribution can take place.

In similar conditions of incubation, FITC-labeled transferrin has no access to the basolateral side of the differentiated monolayer and no staining is observed. When the cells are incubated in Ca^{++} free conditions, the tight junctions are disrupted, FITC-labeled transferrin has access to the basolateral side of the cell and homogeneous basolateral membrane staining is observed. These observations correlate exactly with the staining observed with a monoclonal antibody directed against human transferrin receptors

When cells are incubated at 37°C after receptor ligand binding, a punctate staining is observed on both cell types which corresponds to the presence of transferrin in endocytosed compartments. We observed that transferrin internalization has occurred in differentiated monolayers as well as in undifferentiated cells. This internalization is not prevented by low calcium concentration in the medium. Control experiments have shown that the same staining is observed with or without calcium for FITC-labelled transferrin on undifferentiated HT29-18/glu cells.

QUANTITATIVE BINDING OF ^{125}I-TRANSFERRIN AND ANALYSIS OF mRNA ENCODING FOR TRANSFERRIN RECEPTOR IN BOTH STATES OF DIFFERENTIATION

We used the binding of ^{125}I-labeled transferrin to assay the cell surface distribution of the transferrin receptor. Differentiated HT29-18 cells were incubated with ^{125}I-labeled transferrin for one hour on ice and the amount of transferrin bound was determined. Apical binding was measured by incubation in Ca^{++} containing medium that maintains tight junctions; it was compared with the total binding obtained in Ca^{++} free medium that gives access to the basolateral membrane of the cells. Non specific binding was estimated by the use of an excess of cold transferrin. Ca^{++} free medium has no effect on binding of transferrin to undifferentiated cells, which indicates that the lack of Ca^{++} does not interfere with transferrin binding. These experiments indicate that 97% of the transferrin receptor is found on the basolateral membrane of differentiated HT29-18 cells.

Differentiated HT29-18 C1 subclone monolayers were grown on plastic dishes and assayed for transferrin binding either in calcium containing medium , or in calcium free medium. Binding in the presence of calcium represents apical binding while in the absence of calcium observed values represent apical and basolateral values, e.g. total binding. Apical binding is 3-5% of total binding at concentrations ranging from 1 nM to 80 nM ^{125}I-labeled transferrin. This value is interpreted as non-specific binding because no saturation can be observed; it was subtracted from binding measurements on entire cell surface membrane (obtained in Ca^{++} free medium) to give an estimation of the amount bound on the basolateral membrane. Scatchard plot analysis of basolateral binding on differentiated HT29-18 C1 cells gave us about 100000 high affinity binding sites per cell (Kd = 2.5 nM). No striking difference between differentiated and undifferentiated cells was observed.

As it was shown for the receptor itself, localized with a monoclonal antibody during differentiation of HT29-18 clone (Godefroy et al., 1988), transferrin receptors display a basolateral distribution during acquisition of polarity in the HT29-18 C1 subclones used in this study. Transferrin receptor affinity is comparable to those obtained for other cell lines (Bleil and Bretscher, 1981; Fuller and Simons, 1986).

Northern blot analysis of total RNA of undifferentiated HT29-18/glu cells and of differentiated HT29-18/gal cells showed no significiant differences in size and level of transcripts encoding for transferrin receptor accumulated in these cells. Size of transcripts (4 kb) are identical on polarized and on non polarized cells (Kühn et al., 1984).

Since the level of RNA and the membrane receptor density are equivalent in both states of differentiation of HT29-18 subclones, we conclude that no significant changes occur in rates of synthesis during differentiation. The striking difference between the two cell types consists in the restriction of transferrin receptor to the basolateral domain of HT29 differentiated cells.

QUANTITATIVE ^{125}I-TRANSFERRIN INTERNALIZATION AND RECYCLING

Internalization and recycling has been studied quantitatively using ^{125}I-labeled transferrin. ^{125}I-labeled transferrin was bound to undifferentiated cells in the same conditions as FITC-transferrin. The cells were washed and either solubilized or incubated for different times at 37°C. Supernatants and cell extracts were then counted. Results show that after 20 minutes half of transferrin bound has recycled into the medium.

We took advantage of the ability of transferrin to dissociate from its receptor at low pH (pH = 2.8) to demonstrate that transferrin is effectively internalized at 37°C and not only clustered at the cell surface. An acidic wash was performed after binding of iodinated transferrin at 4°C and before or after warming up at 37°C. Eighty-two % of bound transferrin can be released before warming up while only 14 % is recovered after 37°C incubation. These results show that iodinated transferrin is internalized inside the cell and no is more present on the cell surface.

Binding and recycling studies were also performed on the differentiated HT29-18-C1/glu subclone. As monolayers grown on filter devices are very sensitive to temperature variations, binding of ^{125}I-labeled transferrin during one hour was obtained at room temperature after a 30 minute transferrin depletion in serum-free media. The binding values obtained consist of the cell surface binding plus the uptake by internalization which is slowred down at 20°C. Nevertheless, polarity of binding is still observed, while after one hour at 20°C ten times as much ^{125}I-labeled transferrin is taken up by the basolateral side of the monolayer as by the apical side. After transferrin binding to the basolateral membrane, filters are carefully washed and reincubated for several times at 37°C. Supernatants from apical and basolateral sides are collected and counted as well as the corresponding cell extracts. Half of ^{125}I-labelled transferrin is recycled in about 40 minutes. This is longer than for undifferentiated cells, but time of diffusion across the filter must then be taken into account. Transferrin recycling as well as binding is then strictly polarized and restricted to the basolateral domain of differentiated HT29-18-C1/ glu subclone.

As a control,the same experiments were performed after apical transferrin uptake; they showed that the remaining 5% of transferrin bound to the apical domain at 20°C recycles into the basolateral media and into the apical media equivalently. This transport was considered as background level for this method with this cell line, resulting from fluid-phase uptake. No evidence of a specific pathway was demonstrated at the apical side of the cells.

EFFECT OF AMMONIUM CHLORIDE ON THE RECYCLING OF ^{125}I-TRANSFERRIN

NH_4Cl-induced pathway modifications on transferrin recycling during differentiation of HT29 cells was investigated. We repeated previous recycling experiments in both states of differentiation of HT29 cells, using 20 mM NH_4Cl in binding, washing and reincubation solutions. In these conditions, transferrin recycling was slower in undifferentiated than in differentiated cultures. After two hours, 30% of the radioactivity remained bound to the cells. Recycling polarity is still efficient in differentiated HT29-18-C1/glu subclone since after 3 hours, about 70 % of total bound radioactivity is recovered in the basolateral supernatant while only 5 % is found in the apical supernatant. It has been shown in HepG2 cells that NH_4Cl blocks dissociation of iron from transferrin but not the recycling of ferrotransferrin to the surface (Ciechanover et al., 1983). No effect of NH_4Cl on the sorting pathway of transferrin recycling was observed here on polarized HT29 cultures.

FLUID PHASE UPTAKE IN POLARIZED HT29 CELLS

Extracellular proteins can enter cells after being trapped in fluid droplets that are internalized through invaginations of plasma membrane. This process is called fluid-phase endocytosis. In order to evaluate the reliability of our filter system and a possible disymetric fluid-phase uptake, we measured HRP transport through the monolayer of HT29 cells.

First, "passive" transcytosis was measured: we incubated filter grown monolayers with HRP on one side for two hours. HRP activity was then measured on the other side of the layers. Results show that uptake is two times greater from the basolateral side than from the apical side. The same incubations at 4°C showed an inhibition of this transport. "Leaky" monolayers were easily detected by this method as measured HRP transcytosed was 5 to 20 times higher than the control values.

The system we have developed, gives the opportunity to compare HRP uptake to transferrin uptake by the two distinct membrane domains of the polarized HT29 cells. After one hour chase in transferrin-free media, filters were incubated for two hours with both HRP and transferrin on one side at 20°C. They were then rinsed and the release of HRP and transferrin during the next hour at 37°C was measured on both sides of the filter. After a period of uptake at 20°C by a given membrane domain, we then investigated to which membrane domain the internalized vesicles were targeted. Remaining intracellular HRP and transferrin were determined at the end of this experiment and values of released HRP and transferrin were given in percentage of initial total uptake. In this experiment, known cycle of transferrin is considered as a double-label which monitors the integrity of the monolayer. Results show that 90% of apical uptake returns to the apical side and 90% of basolateral uptake returns to the basolateral side. No accumulation of peroxidase is observed, as after one hour 90% of internalized peroxidase is released in the medium. As already described, transferrin has only a polarized basolateral recycling when bound on the basolateral membrane which expresses its receptor. The same experiment done at 20°C shows that the peroxidase release time is longer, but no significiant change in the polarity was observed.

Control experiments were done at 4°C and demonstrated that HRP release was reduced by 50% during the first twenty minutes. These experiments, the first described on such a polarized system, demonstrate that fluid phase endocytotic vesicles will return preferentially to the membrane domain from which they come from.

DISCUSSION

HT29 cell line differentiation can be modified by culture conditions and represents a unique in vitro model of polarity acquisition in an epithelial cell line. Establishment of polarity of a HT29-18 clone occurs within change of culture conditions: it consists of an adaptation of the cell line to grow in defined conditions. This adaptation is accompanied by cell death. So far, the intermediate stages of this polarity acquisition cannot be investigated step by step on the same cell culture. Nevertheless, the use of well-defined subclones with undifferentiated or differentiated characteristics enabled us to study modification of endocytosis during polarization.

The establishment of cell polarity in epithelial cells is characterized by the acquisition of two membrane domains delineated by tight junctions which separate two distinct membrane surfaces, with specific lipid and protein composition (Gumbiner and Louvard, 1985). We took advantage of the properties of HT29 subclones to demonstrate that transferrin receptor localization can switch from an uniform membrane distribution on undifferentiated cells to a strictly polarized basolateral distribution on differentiated cells.

This work led us to develop differentiated HT29-18 C1/glu cultures on permeable substrates. This kind of approach has been developed on the MDCK cell line (Cereijido et al., 1978), and is now extended to numerous types of epithelia (Grasset et al., 1984; Garcia et al., 1986). HT29-18-C1/glu monolayers develop a transepithelial potential of 0.4 mV and display an electrical resistance of 100 $\Omega.cm_2$. Such electrical parameters are characteristic of "leaky" epithelia.

Permeable substrate culture of differentiated HT29-18 C1/glu monolayers enabled us to demonstrate that recycling of transferrin was efficient and occurred only at the basolateral membrane domain. In this cell line, as phenotypical differentiation appears, functional differentiation occurs : sorting signals for basolateral insertion of the transferrin receptor appears as well as targeting signals for basolateral recycling of endocytic vesicles containing apotransferrin and its receptor. In vivo localization of transferrin receptor has shown that it is found on the basolateral membrane of enterocytes (Banerjee et al., 1986; Godefroy et al., 1988). Localization of transferrin receptor in vivo is related to the presence of blood transferrin in the internal milieu. For cells grown on impermeable substrates, transferrin is present in the culture medium, mostly on the apical side of the monolayer. For cells grown on permeable substrates, transferrin is present on both sides of the monolayer. Nevertheless, our results show that transferrin receptor expression and recycling is independent of external transferrin localization. The transferrin receptor in HT29 cell line has kept an internal sorting signal, which corresponds to the tissue from which the cell line was isolated. A similar observation was made on MDCK cells, a kidney cell line which has also kept a basolateral localization and recycling of its receptor (Fuller et al., 1986). Polarity efficiency seems to be controlled by endogenous factors and is independent of external milieu.

It has been shown that, after internalization of ferrotransferrin bound to its receptor, endocytic vesicles undergo acidification (Dautry-Varsat et al., 1983), and this low pH in endocytic vesicles is essential for the dissociation of iron from transferrin (Ciechanover et al., 1983). No depolarization effect on transferrin recycling was observed by use of ammonium chloride in HT29 cells. We can postulate that maintenance of low pH, and so far iron-transferrin dissociation, is not required for the polarized targeting of transferrin recycling endocytic vesicles. In MDCK cells, the sorting process which mediates the polarized secretion of basement membrane components is inhibited by the weak base NH_4Cl (Caplan et al., 1987), an agent which raises the pH of intracellular acidic compartments (Maxfield, 1982). The ammonium chloride effect on this secretion shows a loss of polarity . On endothelial cells grown on permeable substrates, it was shown that ammonium chloride did not disrupt the polarity of the regulated secretory pathway of von Willebrand factor (Sporn et al., 1989). These authors were studying secretion in the same way as we were studying recycling of an endocytosed vesicle. The same compartments may not be implicated in this transport.

Horseradish peroxidase has been used as a fluid-phase marker to analyze fluid movements across the cell. We have described transcytosis in differentiated HT29-18-C1 monolayers : two times more HRP crosses the cell from the basolateral to the apical medium than from the apical to the basolateral medium. To estimate the effective value of HRP uptake polarity, diffusion across the filter should have been taken into account. An exhaustive study has been made on transcytosis in the MDCK cell. In this work, each plasma membrane was evaluated showing that fluid uptake was directly related to the surface of each membrane domain (Von Bonsdorff et al., 1985). No similar study has been made on differentiated HT29 cells at the present time, that could show the same properties.

We then studied diacytosis of HRP during transferrin uptake and recycling on differentiated HT29 monolayers. The goal of this experiment was to describe to which cell surface domain endocytosed HRP will preferentially recycle after its uptake. This represents the first study done for this purpose on a polarized epithelium. Taking advantage of our precedent results, transferrin was used as double-labelled control of the normal behavior of the monolayer. We were able to demonstrate that in absorptive cells, there was no accumulation compartment and that after one hour, 90 % of peroxidase was recycled to the cell surface. After internalization, 90 % of vesicles return to the same membrane domain. This shows that fluid-phase is strongly polarized in differentiated HT29-18-C1 clones and that transferrin recycling is only one of these recycling phenomena. We have noticed that in differentiated HT29 clones, no accumulation can be observed, while after one hour 90% of endocytosed HRP is recycled to the medium. The analysis of endocytic vesicles in non polarized cells such as macrophages, fibroblasts or BHK cells has shown an accumulation of fluid phase markers inside the cells (Steiman et al., 1976; Marsh and Helenius, 1980). These authors have described a linear accumulation of fluid phase markers inside the cells at 37°C after several hours of incubation. Our procedure was different as we have studied HRP release after one hour uptake by one membrane domain at 20°C, a temperature known to allow endocytosis but to inhibit several steps in intracellular vesicle traffic. These conditions were determined to be identical to these for transferrin binding studies. We have not demonstrated that after this period of time we have obtain a saturation step, but only that the majority of HRP uptake during one hour at 20°C is released after one hour incubation at 37°C. This demonstrates that endocytosed HRP is rapidly recycled, 80% to the membrane domain from which it has been taken up, 10% to the opposite membrane domain; only 10% remain inside the cell.

This study, which compares fluid-phase to receptor-mediated endocytosis in HT29 subclones, demonstrates that polarity acquisition is not only the fate of specific membrane movements but seems to be an intrinsic change in all cell physiology.

REFERENCES

Aisen P, Leibman A, & Zweier J (1978) Stoichiometric and site characteristics of the binding of iron to human transferrin. J. Biol. Chem. 256: 1930- 1937.

Anderson R G W, Brown M S, Beisigel U, & Goldstein J L (1982). Surface distribution and recycling of the low density lipoprotein receptor as visualized with anti-receptor antibodies. J. Cell. Biol. 93: 523-531.

Banerjee D, Flanagan P R, Cluett J, & Valberg L S. (1986) Transferrin receptors in the human gastrointestinal tract. Relationship to body iron stores. Gastroenterology 91: 861-869.

Beguinot L, Lyall R M, Willingham M C, & Pastan I (1984). Down regulation of the epidermal growth factor receptor in KB cells is due to receptor internalization and subsequent degradation in lysosomes. Proc. Natl. Acad. Sci. USA 81: 2384-2388.

Bleil D, & Bretscher M S (1982) Transferrin receptor and its recycling in Hela cells. EMBO J.1: 351: 355.

Caplan M J, Stow J L, Newman A P, Madri J, Anderson H C, Farquard M G, Palade G E, & Jamieson J D. (1987) Dependence on pH of polarized sorting of secreted proteins. Nature 329: 632: 635.

Cereijido M, Robbins E S, Dolan W J, Rotundo C A, & Sabatini D D. (1978) Polarized monolayers formed by epithelial cells on a permeable and translucent support. 77: 853- 880.

Ciechanover A, Schwartz A L, Dautry-Varsat A, & Lodish H F (1983) Kinetics of internalization and recycling of transferrin and the transferrin receptor in a human hepatoma cell line J. Biol. Chem. 258: 9681: 9689.

Dautry-Varsat A, Ciechanover A, & Lodish H (1983) pH and the recycling of transferrin during the receptor mediated endocytosis. Proc. Natl. Acad. Sci. USA, 80: 2258- 2262.

Fogh J, & Trempe G (1975) New tumor cell line. In: Human Tumor cells in vitro. (J. Fogh, ed), New York, Plenum Press. p115-141.

Fuller S D, & Simons K (1986). Transferrin receptor polarity and recycling accuracy in "tight" and "leaky" strains of Madin-Darby Canine Kidney Cells. J.Cell. Biol. 103:1767-1779.

Garcia J G N, Silfinger-Birnboim A, Del Vecchio P J, Fenton J W, & Malik A B. (1986) Thrombin induces increases in albumin transport across cultured endothelial monolayers. J. Cell. Physiol. 128: 96- 104.

Gavin J R, Roth J, Neville D M, De Meyts P, & Buell D N (1974). Insulin dependent regulation of insulin receptor concentra-

tion: a direct demonstration in cell culture. Proc. Natl. Acad. Sci. USA, 71: 84-88.
Geuze H J, Slot J W, & Strous G J (1983). Intracellular site of asialoglycoprotein receptor-ligand uncoupling: double-label immunoelectron microscopy during receptor mediated endocytosis. Cell 32: 277-287.
Godefroy O, Huet C, Blair L, Sahuquillo-Merino C, & Louvard D (1988) Differentiation of a clone isolated from the HT29 cell line: polarized distribution of histocompatibility antigens (HLA) and of transferrin receptors. Biol. Cell. 63: 41-55.
Goldstein J L,Brown M S, Anderson R G W , Russel D W, & Scneider W J (1985). Receptor-mediated endocytosis: concepts emerging from the LDL receptor system. Ann. Rev. Cell Biol. 1: 1-39.
Gonzalez-Mariscal L, Chavez de Ramirez B, & Cereijido M (1985). Tight junction formation in cultured epithelial cells (MDCK). J. Membr. Biol. 86:113-125.
Goud B, Huet C, & Louvard D (1985). Assembled and unassembled pools of clathrin: a quantitative study using an enzyme immunoassay. J. Cell Biol. 100: 521-527.
Grasset E, Pinto M, Dussaulx E, Zweibaum A, & Desjeux J F (1984) Epithelial properties of a human colonic carcinoma cell line Caco-2 : electrical parameters. Am. J. Physiol. 247: C260- C267.
Gumbiner B, & Louvard D (1985). Localized barriers in the plasma membrane: a common way to form domains. Trends Biochem. Sci. 10: 435-438.
Helenius A, Mellman A, Wall D, & Hubbard A (1983) Endosomes. Trends in Biochem. Sci. 8: 245-250
Huet C, Ash J F, & Singer S J (1980). The antibody-induced clustering and endocytosis of HLA antigens on cultured fibroblasts . Cell, 21: 429-438.
Huet C, Sahuquillo-Merino C, Coudrier E, & Louvard D (1987) Absorptive and mucus secreting subclones isolated from a multipotent intestinal cell line (HT-29) provide new models for cell polarity and terminal differentiation. J. Cell. Biol. 105: 345-357.
Iacoppetta B J, Rothenberger S,& Kühn L C (1988) A role for the cytoplasmic domain in transferrin receptor sorting and coated pit formation during endocytosis. Cell, 54: 485- 489.
Karin M & Mintz B (1981) Receptor mediated endocytosis of transferrin in developmentally totipotent mouse teratocarcinoma stem cells. J. Biol. Chem. 256: 3245- 3252.
Kelly R B (1985) Pathways of protein secretion in eukaryotes. Science, 230: 25-32.
Kenny A J, & Maroux S (1982). Topology of microvillar membrane hydrolases of kidney and intestine. Physiol. Rev., 62: 91-128.
Kirby W N, & Parr E L (1979). The occurrence and distribution of H2 antigens on mouse intestinal epithelial cells. J. Histochem. Cytochem. 27: 746-750.
Klausner R D, Ashwell G, Van Renswoude J V,Harford J B, & Bridges K R (1983a). Binding of apotransferrin to K562 cells: explanation of the transferrin cycle. Proc. Natl. Acad. Sci. USA 80: 2263- 2266.
Kondor-Koch C , Bravo R , Fuller S D , Cutler D & Garoff, H (1985). Exocytic pathways exist to both the apical and the basolateral cell surface of the polarized epithelial cell MDCK. Cell 43: 297-306.

Kühn L C, Mc Clelland A,& Ruddle F H, (1984) Gene transfer, expression and molecular cloning of the human transferrin gene receptor. Cell 37: 95- 103.
Kyte J (1976). Immunoferritin determination of the distribution of (Na-K) ATPase over the plasma membranes of renal convolulated tubules. II Proximal segment. J. Cell. Biol. 68: 304-318.
Larkin J M.,Sztul E S., & Palade G E (1986). Phosphorylation of the rat hepatic polymeric IgA receptor. Proc. Natl. Acad. Sci. USA, 83: 4759-4763.
Leblond C P, & Cheng H (1976). Identification of stem cells in the small intestine of the mouse. in "Stem cells of renewing populations. "Cairnic A.B.,Lala P.K. &D.G. Osmond eds. Academic Press, pp 7-31.
Louvard D (1980). Apical membrane aminopeptidase appears at site of cell-cell contact in cultured kidney epithelial cells. Proc. Natl. Acad. Sci. USA 77: 4132-4136.
Lubinski I, & Huet C (1984). Independent variation in the number of coated pits and of coated vesicles in cultured fibroblasts. Bio. Cell. 52: 119-128.
Marsh M, & Helenius A (1980) Adsorptive endocytosis of Semliki Forest Virus. J. Mol.Biol. 142: 439- 454.
Maxfield F R (1982) Weak bases and ionophores rapidly and reversibly raise the pH of endocytotic vesicles in cultured mouse fibroblasts. J. Cell. Biol. 95: 676- 681.
Meyer D I (1982). The signal hypothesis: a working model. Trends in Biochem. Sci. 7: 320-321.
Montesano R, Roth J, Robert A & Orci L (1982). Non coated membrane invaginations are involved in binding and internalization of cholera and tetanus toxins. Nature 296: 651-653.
Mooseker M (1985). Organization, chemistry and assembly of the cytosqueletal apparatus of the intestinal brush borders. Ann. Rev. Cell Biol. 1: 261-293.
Pearse B M F (1976). Clathrin: a unique protein associated with intracellular transfer of membrane coated vesicles. Proc. Nat. Acad. Sci. USA 73: 1225-1259.
Pearse B M F (1987). Clathrin and coated vesicles. EMBO J. 6: 2507-2512.
Pinto M, Appay M D, Simon-Assmann P, Chevalier G, Dracopoli N, Fogh J & Zweibaum A (1982) Enterocytic differentiation of cultured human colon cancer cells by replacement of glucose by galactose in the medium. Biol. Cell. 47: 323-330.
Reggio H,Coudrier E, & Louvard D (1982). Surface and cytoplasmic domains in polarized epithelial cells . in Membranes in Growth and Development, Alan Liss, Inc. 150 Fith avenue, New York, NY 10011, pp 89-105.
Steinman R M, Mellman I S, Muller W A, & Cohn Z A (1983). Endocytosis and the recycling of plasma membrane. J. Cell Biol. 104: 1261- 1268.
Sporn A, Marder V J, & Wagner D D (1989) Differing polarity of the constitutive and regulated secretory pathways for von Willebrand factor in endothelial cells. J. Cell. Biol. 108: 1283- 1289.
Tojo S J, Germeraad S, King D S, & Fristrom J W, (1987). Polarized secretion of an ectopic protein in Drosophila salivary gland in vivo. EMBO J. 6: 2249-2254.
Traber M G, Kayden H J, & Rindler M J (1987). Polarized secretion of newly synthesized lipoproteins by the Caco 2 human intestinal cell line. J. Lipid Res., 28: 1350- 1363.

Trugnan G, Rousset M, Chantret I, Barbat A, & Zweibaum A (1987). The posttranslational processing of sucrase isomaltase in HT 29 cells is a function of their state of enterocytic differentiation. J. Cell. Biol. 104: 1199-1205.
Ungewickel E, & Branton D (1981). Asssembly units of clathrin coats. Nature 289: 420-422.
Urban J, Parczyck K, Leutz A, Kayne M, & Kondor-Koch C (1987). Constitutive apical secretion of an 80 kD sulfated glycoprotein complex in the polarized epithelial Madin-Darby canine cell line. J. Cell. Biol. 105: 2735-2743.
Van Meer G, Gumbiner B, & Simons K (1986). The tight junction does not allow lipid molecules to diffuse from one epithelial cell to the next. Nature 232: 639-641.
Van Meer G, & Simons K (1986). The function of tight junctions in maintaining differences in lipid composition between the apical and the basolateral cell surface domains of MDCK cells. EMBO J. 5: 1455-1464.
Vega-Salas D E, Salas P J I, Gundersen D & Rodriguez-Boulan E (1987). Formation of the apical pole of epithelial (Madin-Darby Canine Kidney) cells: polarity of an apical protein is independent of tight junctions while segregation of a basolateral marker requires cell-cell interactions. J. Cell. Biol. 104: 905-916.
Von Bonsdorff C H, Fuller S D,& Simons K (1985).Apical and basolateral endocytosis in Madin-Darby-Canine Kidney (MDCK) cells grown on nitro cellulose filters. EMBO J. 4: 2781-2792.
Wall D A, & Hubbard A L (1981). Galactose specific recognition systems of mammalian liver: receptor distribution on the hepatocyte cell surface. J. Cell. Biol. 90: 687-696.
Willingham M C, Hanover J A.,Dickson R B. & Pastan I (1984). Morphologic characterization of the pathway of transferrin endocytosis and recycling in human KB cells. Proc. Natl. Acad. Sci. USA 81: 175-179.
Zaremba S. & Keen J M. (1983). Assembly polypeptides from coated vesicles mediate reassembly of unique clathrin coats. J. Cell. Biol.97: 1339-1347.

CULTURE OF HUMAN GOBLET CELLS

Christian L. Laboisse

U239 INSERM

Faculté Bichat

16, rue Huchard

75018 Paris

France

INTRODUCTION

Mucins comprise a heterogeneous set of high molecular weight glycoproteins (HMGs). All mucins share several characteristics: they are rich in serine and threonine and contain several hundreds of oligosaccharide chains in O-glycosidic linkages to these aminoacids. The gel-forming properties of mucins are essential for the protection of the epithelial lining of the gastrointestinal tract. The mucus gel was first thought of as being a physical barrier whose main function was to shield against harmful agents. As a consequence, the lesions of the gastrointestinal mucosa were assumed to be largely due to qualitative or quantitative defects of this gel. Now several recent studies suggest that intestinal mucins should be viewed as an essential component of a complex ecosystem which comprises in the distal part of the intestine, the colonic microflora, the luminal content, the epithelial lining of the mucosa and its secretory products.

The intestinal mucus gel is secreted by specialized epithelial cells, the so-called goblet cells. These cells are structurally organized for the synthesis, processing, glycosylation, storage and vectorial discharge of mucins. The intracellular topography of their synthetic steps has been inferred from several approaches: (i) immunocytochemistry with antibodies prepared against mature or deglycosylated mucins (apomucins)[1] (ii) labeling of tissue sections with lectin-gold complexes[1] (iii) subcellular fractionation used in combination with specific glycosyltransferase assays[2], (iiii) immunolocalization of glycosyltransferases by immunoelectron microscopy with polypeptide epitope-purified antibodies[3]. Altogether, the results obtained are consistent with the view that glycosylation of the polypeptide core of mucins commences in the cis region of the Golgi apparatus and that glycosylation is completed during the passage from the cis to the trans region of the golgi apparatus, before they become concentrated in forming mucous granules[1].

The organization of mucin granule content is not known with precision. The mucin glycoproteins are highly concentrated in the granule and rapidly hydrate and expand after secretion[4].

Regional differences in the composition of mucins have been described using biochemical and immunohistochemical methods[5]. On the basis of these findings, it is likely that the variations in the composition of mucins represent an adaptation to various physiological functions. Well controlled experiments relating the molecular characteristics of mucins to specific functions are still needed, however. Several lines of investigation suggest that the composition of mucins vary not only in separate intestinal regions, but also during cell maturation. Lectins and monoclonal antibodies have revealed differences between mucins accumulated in the crypt and those of the surface epithelium[6]. This suggests that as goblet cells mature during their migration from the bottom of the crypts to the surface epithelium, they synthesize different types of mucins.

Pharmaceutical Applications of Cell and Tissue Culture to Drug Transport
Edited by G. Wilson *et al.*, Plenum Press, New York, 1991

Given the pivotal role played by mucins in the physiology of the GI tract, it is likely that mucus secretion "in vivo" is affected by inputs from endocrine, paracrine, neurocrine and immune pathways. It can also be assumed that the effects of these four pathways on mucous cells are interdependent. However, despite their major physiological importance, very little is known about the chemical transmitters that regulate the secretion of mucus. In the same way their receptors on mucous cells and their intracellular transduction pathways are still poorly understood and are largely inferred from studies on other cell types.

There are several reasons for the failure to break new ground in this field of research. It is our inability to isolate a homogeneous preparation of goblet cells from the intestinal mucosa that is generally considered as the key problem. To further complicate this situation, the specificity and sensitivity of many of the currently used methods for quantifying the secretion of mucus are often questionable. For example, assays based solely on the incorporation of a radiolabeled glycoprotein precursor into acid precipitable material in incubation media have been subject to various criticisms[4]. In fact non-mucin components are rapidly labeled by radioactive precursors such as (^{35}S)sulfate, and (^{3}H)- or (^{14}C)monosaccharides. In comparison, goblet cell mucins are poorly labeled since the formation of new mucin granules is a slow process. Finally, as pointed out by Roomi et al.[7], as well as by Neutra and Forstner[4], the release of total radiolabeled glycoproteins is an unreliable index of mucin secretion. The lack of standardized methodology to quantify mucins often makes it difficult to compare the results obtained by various groups working on the regulation of mucin secretion.

While many ingenious experimental techniques have been devised to study the specialized functions of the intestinal cells "in vivo" and "ex vivo", there are obvious limitations to such procedures. Therefore several investigators have turned their attention towards cell culture systems which may offer an approach to circumvent some of the limitations of "in vivo" studies. The purpose of this chapter is twofold: (i) to provide a brief overview of some culture systems that are potentially useful for studying the secretion of mucins from the intestinal mucosa and more specifically from the colon and (ii) to discuss the advantages and the limitations of these systems.

"IN VITRO" SYSTEMS FOR STUDYING THE SECRETION OF MUCINS

Organ culture methods

Organ culture methods have been widely employed over the past years to study the secretion of glycoproteins by the intestinal mucosa[8,9,10,11]. To culture human intestinal tissue by this technique, mucosal samples are obtained during routine diagnostic colonoscopy with standard biopsy forceps. As soon as they are obtained the explants are placed, cut surface down , on a wire-mesh screen in organ culture plates. The central well of the dish is filled with an appropriate medium. For studies of mucosal synthetic and secretory functions, radioactive precursors such as (^{3}H)Leucine, (^{3}H)Glucosamine, or (^{14}C)Glucosamine can easily be added to the medium. In recent studies utilizing explants of human colonic mucosa in organ culture, Smith and Podolsky[11] examined the incorporation of precursors in high molecular weight glycoproteins and their subsequent appearance in media. The incorporation of (^{14}C)Serine and (^{3}H)Glucosamine into tissue mucin glycoprotein was found to remain linear throughout the 24-h incubation. Endogenously labeled mucin glycoproteins remained low in media for the 1st 12h of maintenance in vitro, but increased in a linear manner during further incubation[11].

In efforts to identify agents that could potentially evoke a mucin secretory response, Specian and Neutra[12], and Neutra et al.[13] have extensively used a morphological/autoradiographic assay on rabbit mucosal biopsies in organ culture. They found that acetylcholine and cholinomimetic drugs function as goblet cell secretagogues in normal intestinal tissue. The ultrastructural analysis of the goblet cells stimulated by acetylcholine showed that the rapid secretion of the stored granules was accomplished by sequential fusion of secretory granule membranes with the plasma membrane and with each other, a process referred to as "compound exocytosis"[12].

However the interpretation of the mechanisms involved in the secretory response of goblet cells to secretagogues is hampered by the cellular complexity and heterogeneity of the mucosal explants. To further complicate interpretation of organ culture studies, some transmitters may be present in the mucosa, closely associated with the final target, the mucous cell.

Isolated intestinal cells

The rationale for the use of isolated cells and epithelial sheets for studying the secretion of mucins is that certain fundamental problems, particularly the elucidation of the site of action of secretagogues, have been difficult to solve in studies based on organ culture techniques.

Several methods have been devised for isolating either epithelial sheets or monodispersed populations of rat[14,15] or rabbit[16] colonocytes. There is considerable similarity in the steps involved in these procedures. Ahnen et al.[15] emphasized the importance of combining EDTA chelation and mechanical dissociation of everted colon segments. Functional activity of the dispersed cells as assessed by linear rates of incorporation of (^{3}H)leucine and (^{3}H)fucose was found to be maintained for 2.5-3h[15].

However colonic mucosa represents such a complex mixture of cell types that enrichment of various subpopulations becomes essential for certain studies. Although the isolation of populations of mature and immature rat colonocytes has been achieved by using a sequential dissociation technique[15], a method for separating highly enriched populations of goblet cells is still needed.

Cell culture systems

The ideal model for studying the secretion of mucins from goblet cells would be a monolayer culture of goblet cells that have retained in vitro the regulatory mechanisms operative in the same cells "in vivo". Such a model would have several advantages: (1) the mucins would be produced in aseptic conditions, (2) they would be less contaminated than those produced "in vivo" or in organ culture, (3) a continuous monolayer culture would provide mucins with stable characteristics allowing reproducible experiments. Thus much effort has been devoted in recent years to the culture of epithelial cells from the human colonic mucosa. Attempts at culturing disaggregated tissue have generally been unsuccessful. Recently Buset et al.[17] have described an improved method for the attachment and growth of normal human colonic cells from minute biopsies. The best substrate for cell attachment was found to be a mixture of ungelled collagen I and bovine serum albumin. Chopra et al.[18] have found that serial passaging of cultures from the normal human fetal colonic mucosa required supplementing the medium with insulin, transferrin, epidermal growth factor and cholera toxin. Unfortunately the authors did not evaluate the percentage of mucus-secreting cells in their cultures.

Interestingly, Whitehead et al.[19] have hypothesized that the three dimensional organization of the crypt is probably important for maintaining the replication and differentiation of intestinal cells. Furthemore they devised a nonenzymatic crypt isolation method which was found to yield viable intact crypts from human colonic mucosa. In culture the epithelial cells survived for extended periods only when the crypts were cultured in collagen gels over a feeder layer of bovine aortic endothelial cells. Cells containing mucus were present in the cultured crypts at all stages of culture. This study demonstrated the presence of a population of cells capable of synthesizing DNA after 7 days in culture.

More recently Gibson et al.[20] have developed a method for maintaining colonic epithelial cells in short term culture. Colonic epithelial sheets and whole crypts were isolated by digestion of mucosal scrapings with dispase and collagenase and then separated from mucus and undigested fragments of lamina propria by passage through a small needle followed by a coarse stainless steel mesh. The structural integrity of the freshly isolated cells was found to be very good. The cells were placed in culture and monolayer sheets of cells were observed to increase in size for three days. However it is still not known whether it will be possible to exploit these cultures for studying mucus secretion.

Taken together these studies point to the fact that it is still very difficult to establish monolayer cultures of epithelial cells from normal human colon.

Another approach to develop monolayer cultures from the intestine has been to culture pre-malignant epithelium from the human colon. Some of the long term cultures were found to retain morphological differentiation characteristics of colonic cells, in particular the ability to synthesize and secrete mucin[21,22,23]. Finally, it has been possible to transform colon cells "in vitro" in order to generate altered cell lines with considerably extended "in vitro" growth capacities[24,25,26]. However the full characterization of these transformed lines is still under way.

THE Cl.16E CELL LINE: A MODEL FOR MUCUS SECRETION

Due to the difficulty of culturing normal intestinal cells, some human colonic cancer cell lines have attracted a great deal of research interest[27,28].

As early as 1984, we reported the isolation and the characterization of stably differentiated clones[29] from the human colonic adenocarcinoma cell line HT-29[30]. The development of these clonal cell lines was made possible by the discovery in our laboratory that the treatment of HT-29 cells with sodium butyrate was able to cause the emergence of stably differentiated sub-populations from the undifferentiated parental cells in culture[29]. This discovery and the ability to grow these subpopulations as independent clonal cell lines made possible the development of "in vitro" models to explore the regulation of some intestinal specialized functions.

In particular, the establishment in our laboratory of a goblet cell line[29] (Cl.16E) prompted us to test whether it could be a useful model for the "in vitro" study of mucins.

We have approached the study of (i) the structure of mucins and (ii) the regulatory mechanisms of mucus secretion using the Cl.16E cell line that is a stably differentiated clonal derivative of the HT29 colonic adenocarcinoma cell line.

Under standard culture conditions the Cl.16E cells differentiate at confluency into typical goblet cells, which are arranged in a monolayer[29,31]. Nearly 90% of the cells are differentiated into goblet cells in late postconfluent cultures, as assessed by alcian blue staining of filter-grown cells. These cells are highly polarized, and they secrete a visco-elastic gel at the cell surface.These characteristics of differentiation are highly stable throughout long-term culture.

Purification and partial characterization of the secreted mucins

In order to elucidate the nature of the gel-forming glycoproteins secreted by the Cl.16E cells, the gel was solubilized by repeatidly pipetting the conditioned medium over the cell layer and submitted to SDS-Polyacrylamide gel electrophoresis in 3-16% gels. Then the HMGs blocked in the stacking gels were eluted[32]. The demonstration that the HMGs prepared in this way were pure mucins was based on the demonstration that (i) they were devoid of mannose, (ii) the oligosaccharide side chains were linked to the peptide core by O-glycosidic bonds[32]. Interestingly these studies suggested that the size heterogeneity of oligosaccharide side chains was considerably less in Cl.16E mucins than in normal colonic mucins.

When the mucins from Cl.16E were subjected to SDS-PAGE under reducing conditions, they entered the gels and gave several bands: a periodate/Schiff-stained band which had barely entered the gel, corresponding to the glycosylated domains of the mucins, and several Coomassie Blue-stained bands. In particular our results showed that the purified mucins contained a 120Kd "link" peptide[32]. This finding is consistent with other recent evidence showing that a "link" glycopeptide of molecular mass 118Kd is common to all gastrointestinal mucins[33].

Cl.16E cell line as a model for studying the regulation of mucin secretion

Polyclonal antibodies were raised against the mucins from Cl.16E cells. Using immunohistochemistry, they were found to react with the mucus granules of the normal human intestinal goblet cells as well as with the superficial mucous cells of the gastric mucosa[32,34]. These antibodies did not react with cellular structures other than mucus granules.

These antibodies have been instrumental in developing an immunofiltration assay for mucins secreted in culture[35]. The development of this assay made it possible to quantify the mucins secreted either in basal conditions or under stimulation. Recent investigations using filter cultures which give improved access of the basolateral surface of cells to secretagogues, have shown that the cholinergic agent carbachol was able to stimulate the secretion of mucins from Cl.16E[34]. Carbachol was found to cause a significant increase in mucus secretion with a maximal stimulation occuring after 30 min. of incubation[34]. A dose-dependent effect was found, which was inhibited by atropine. These results indicate that the effects of carbachol are mediated by muscarinic receptors present on the Cl.16E cells.

Our results prompted us to extend our studies one step further and to test the regulatory effects of neuropeptides on the cultured cells. In a recent study in collaboration with Laburthe's group we have characterized the effects of the Vasoactive Intestinal Peptide (VIP) on Cl.16E cells[36]. Several important conclusions were drawn from this study. First, the Cl.16E cells expressed receptors for VIP with properties very similar to those previously reported for epithelial cells isolated from normal human colon, including the affinity, the ability to discriminate between several VIP-related peptides, and the Mr (63,000) as estimated by affinity cross-linking experiments. Second, VIP strongly potentiated carbachol-induced mucin secretion. This potentiating effect of VIP was mimicked by forskolin and by DbcAMP. Finally, VIP elicited a dramatic increase of intracellular cAMP levels in filter-grown Cl.16E cells with a dose-response curve (ED50, 4 nM) very similar to that observed for the modulation of mucin secretion[36]. Collectively these findings support the view that the Cl.16E cell line is an excellent model for studying the effects of neuroendocrine secretagogues as well as their interdependent actions on goblet cells.

Finally the recent demonstration that cholera toxin strongly stimulates the secretion of mucins from Cl.16E cells suggests that this cell line may contribute important information with regard to the action of toxins on mucous cells[34].

CONCLUSION

From our studies, it is apparent that the Cl.16E cell line provides a powerful reductionist approach to elucidating the regulatory mechanisms of mucus secretion. In this context, this clone will be instrumental in unravelling the complex interactions of secretagogues acting through different signalling pathways, i.e. the cAMP pathway and the phosphoinositide/calcium pathway.

However, with a reductionist approach, using a homogeneous cell line, we lose the opportunity to maintain "in vitro" the spatial relationship between the various cell types that exist in the normal intestinal epithelium. The recent isolation by our group of a clonal cell line which differentiates into two cell types i.e. enterocytic cells and mucous goblet cells, provides a model recreating "in vitro" a spatial relationship between different cell types, close to that existing "in vivo"[37]. The existence of this new cell line (Cl.27H) supports the view that the various epithelial cell types of the intestinal crypt originate from the same precursor[37]. Finally this new model provides a unique approach to understand which factors, capable of affecting the secretory functions of goblet cells, are contributed by epithelial non-mucous cells.

Despite many intensive studies, very little is known about the factors that regulate the replication and maturation of normal colonic epithelial cells. In this context, recent work emphasizing the role played by the three dimensional organization of the colonic crypt for maintaining the potential of the cells to replicate and differentiate, is probably the starting point of promising studies aimed at elucidating the mechanisms that control the renewal of colonocytes. This work complements the approach which consists in examining the dependence of colonic epithelial cells upon components of the extracellular matrix[38]. The knowledge gained from these studies will probably be the key to culturing successfully goblet cells from the normal human colonic mucosa.

REFERENCES

1. M. Deschuyteneer, A.E. Eckhardt, J. Roth, and R.L. Hill, The subcellular localization of apomucin and nonreducing terminal N-Acetylgalactosamine in porcine submaxillary glands, J. Biol. Chem. 263:2452 (1988).
2. H. Schachter, and D. Williams, Biosynthesis of mucus glycoproteins, in: "Mucus in Health and Disease", Part II , Advances in experimental Medicine and Biology vol.144, E.N. Chantler, J.B. Elder, and M. Elstein eds., Plenum Press, New York (1982).
3. D.J. Taatjes, J. Roth, J. Weinstein, and J.C. Paulson, Post-golgi apparatus localization and regional expression of rat intestinal sialyltransferase detected by immunoelectron microscopy with polypeptide epitope-purified antibody. J. Biol. Chem. 263:6302 (1988).

4. M.R. Neutra, and J.F. Forstner, Gastrointestinal mucus: synthesis, secretion, and function, in: "Physiology of the gastrointestinal tract" L.R. Johnson, J. Christensen, M.J. Jackson, E.D. Jacobson, and J.H. Walsh eds., Raven Press, New York (1987).
5. J. Bara, F. Loisillier, and P. Burtin, Antigens of gastric and intestinal mucous cells in human colonic tumours, Br. J. Cancer 41:209 (1980).
6. L. Chambraud, A. Bernadac, J.P. Gorvel, and S. Maroux, Renewal of goblet cell mucus granules during the cell migration along the crypt-villus axis in rabbit jejunum: an immunolabeling study, Biol. Cell 65:151 (1989).
7. N. Roomi, M. Laburthe, N. Fleming , R. Crowther, and J. Forstner, Cholera-induced mucin secretion of rat intestine: lack of effect of cAMP, cycloheximide, VIP, and colchicine, Am. J. Physiol. (Gastrointest. liver physiol. 10) 247:G140 (1984).
8. J.S. Trier, Organ-culture methods in the study of gastrointestinal-mucosal function and development, New Engl. J. Med. 295:150 (1976).
9. J.T. LaMont, and A.S. Ventola, Synthesis and secretion of colonic glycoproteins, Biochim. Biophys. Acta. 629:553 (1980).
10. M.R. Neutra, R.J. Grand, and J.S. Trier, Glycoprotein synthesis, transport, and secretion by epithelial cells of human rectal mucosa, Lab. Invest. 26:535 (1977).
11. A.C. Smith, and D.K. Podolsky, Biosynthesis and secretion of human colonic mucin glycoproteins, J. Clin. Invest. 80:300 (1987).
12. R.D. Specian, and M.R. Neutra, Regulation of intestinal mucus secretion.I. Role of parasympathetic stimulation. Am. J. Physiol. 242 (Gastrointest. Liver Physiol. 5):G370 (1982).
13. M.R. Neutra, L.J. O'Malley, and R.D. Specian, Regulation of intestinal goblet cell secretion.II. A survey of potential secretagogues. Am. J. Physiol. 242 (Gastrointest. Liver Physiol. 5):G380 (1982).
14. T.A. Brasitus, Isolation of proliferative epithelial cells from the rat cecum and proximal colon. Anal. Biochem. 123:364 (1982).
15. D.J. Ahnen, T.A. Reed, and J.M. Bozdech, Isolation and characterization of populations of mature and immature rat colonocytes, Am. J. Physiol. 254 (Gastrointest. Liver Physiol. 17):G610 (1988).
16. J.D. Kaunitz, Preparation and characterization of viable epithelial cells from rabbit distal colon, Am. J. Physiol.254 (Gastrointest. Liver Physiol. 17):G502 (1988).
17. M. Buset, S. Winawer, and E. Friedman, Defining conditions to promote the attachment of adult human colonic epithelial cells, In Vitro Cell. Devel. Biol. 23:403 (1987).
18. D.P. Chopra, K.M. Siddiqui, and R.A. Cooney, Effects of insulin, transferrin, cholera toxin, and epidermal growth factor on growth and morphology of human fetal normal colon epithelial cells, Gastroenterology 92:891 (1987).
19. R.H. Whitehead, A. Brown, and P.S. Bhathal, A method for the isolation and culture of human colonic crypts in collagen gels,In Vitro Cell. Devel. Biol. 23:436 (1987).
20. P.R. Gibson, E. van de Pol, L.E. Maxwell, A. Gabriel, and W.F. Doe, Isolation of colonic crypts that maintain structural and metabolic viability in vitro, Gastroenterology, 96:283 (1989).
21. C. Paraskeva, B.G. Buckle, D. Sheer, and C.B. Wigley , The isolation and characterization of colorectal epithelial cell lines at different stages in malignant transformation from familial polyposis coli patients, Int. J. Cancer, 34:49 (1984).
22. J.K.V. Willson, G.N. Bittner, T.D. Oberley, L.F. Meisner, and J.L. Weese, Cell culture of human colon adenomas and carcinomas, Cancer Res. 47: 2704 (1987).
23. C. Paraskeva, S. Finerty, and S. Powell, Immortalization of a human colorectal adenoma cell line by continuous "in vitro" passage: possible involvement of chromosome 1 in tumour progression, Int. J. Cancer 41:908 (1988).
24. M.P. Moyer, and J.B. Aust, Human colon cells: culture and in vitro transformation, Science224:1445 (1984).
25. R.D. Berry, S.C. Powell, and C. Paraskeva, "In vitro" culture of human foetal colonic epithelial cells and their transformation with origin minus SV40 DNA, Br. J. Cancer 57:287 (1988).
26. S. Emami, L. Mir, C. Gespach, and G. Rosselin, Transfection of fetal rat intestinal epithelial cells by viral oncogenes: establishment and characterization of the E1A-immortalized SLC-11 cell lines. Proc. Natl. Acad. Sci. USA 86:3194 (1989).
27. C.L. Laboisse, Differentiation of colon cells in culture, in: "Cell and molecular biology of colon cancer", L.H. Augenlicht, ed., CRC Press, Boca Raton (1989).
28. C.L. Laboisse, The structure of gastrointestinal mucins: searching for the Rosetta stone, Biochimie 68:611 (1986).
29. C. Augeron, and C.L. Laboisse, Emergence of permanently differentiated cell clones in a human colonic cancer cell line in culture after treatment with sodium butyrate, Cancer Res. 44: 3961 (1984).

30. J. Fogh , and G. Trempe, New human tumor cell lines, in "Human Tumor Cells "in vitro"", J. Fogh, ed., Plenum Press, New York, (1975).
31. C. Augeron, J.J. Maoret, C.L. Laboisse, and E. Grasset, Permanently differentiated cell clones isolated from the human colonic adenocarcinoma cell line HT29: possible models for the study of ion transport and mucus production, in "Ion gradient-coupled transport", F. Alvarado and C.H. Van Os, eds., INSERM symposium N°26, Elsevier (1986).
32. J.J. Maoret, J. Font, C. Augeron, P. Codogno, C. Bauvy, M. Aubery, and C.L. Laboisse, A mucus-secreting human colonic cancer cell line - purification and partial characterization of the secreted mucins, Biochem. J. 258:793 (1989).
33. A.M. Roberton, M. Mantle, R.E.F. Fahim, R.D. Specian, A. Bennick, S. Kawagishi, P. Sherman, and J.F. Forstner, The putative "link" glycopeptide associated with mucus glycoproteins - composition and properties of preparations from the gastrointestinal tracts of several mammals, Biochem. J. 261:637 (1989).
34. I. Roumagnac, and C. Laboisse, A mucus-secreting human colonic epithelial cell line responsive to cholinergic stimulation, Biol. Cell 61:65 (1987).
35. I. Roumagnac, and C.L. Laboisse, A simple immunofiltration assay for mucins secreted by a human colonic epithelial cell line, J. Immunol. Methods 122:265 (1989).
36. M. Laburthe, C. Augeron, C. Rouyer-Fessard, I. Roumagnac, J.J. Maoret, E. Grasset, and C. Laboisse, Functional VIP receptors in the human mucus-secreting colonic epithelial cell line Cl.16E, Am. J. Physiol. 256 (Gastrointest. Liver Physiol. 19):G443 (1989).
37. C.L. Laboisse, J.J. Maoret, N. Triadou, and C. Augeron, Restoration by polyethylene glycol of characteristics of intestinal differentiation in subpopulations of the human colonic adenocarcinoma cell line HT29. Cancer Res.48:2498 (1988).
38. M. Kedinger, K. Haffen, and P. Simon-Assmann, Intestinal tissue and cell cultures, Differentiation 36:71 (1987).

ENDOTHELIAL BARRIERS

Kenneth L. Audus
Department of Pharmaceutical Chemistry
The University of Kansas
School of Pharmacy
Lawrence, Kansas

INTRODUCTION

The endothelial cells lining the vasculature comprise an interface elaborated for controlling the distribution of substances between the blood and the interstitial fluid. Additionally, this interface performs synthetic, degradative, and receptor functions that vary greatly, depending on the location and nature of the surrounding tissue. Consequently, endothelia are currently understood to be a dynamic, heterogenous population of cells. Many of the functions of the endothelial cell have direct relevance to the pharmaceutical chemist's goal of providing selective, efficient, and effective delivery of therapeutic substances to tissue target sites. The purpose of this review is to briefly detail the cell biology of vascular endothelium and properties of these cells that one desires in a cell culture system appropriate for addressing problems in pharmaceutical chemistry.

GENERAL CHARACTERISTICS AND FUNCTIONS OF ENDOTHELIAL BARRIERS

Exchange of nutrients, waste products, and therapeutic substances between the blood and the tissues occurs at the level of the smallest blood vessels, the capillaries. A capillary is about 20-100 nm in length and has an inside diameter of approximately 5-10 μm. Exchange of materials between the blood and the tissues is facilitated by the close proximity, within 40-80 μm, of capillaries to tissues cells (Simionescu and Simionescu, 1983; Smith and Kampine, 1984). Capillary endothelium is defined as a single, thin (<0.5 μm thick) layer of endothelial cells, its basement membrane, and an occasional pericyte. The basement membrane is comprised of glycoproteins and collagen, and contributes to the capillary structure by providing a growth support for the endothelium (Simionescu and Simionescu, 1983). The precise role of the pericyte is unknown, however, emerging evidence from *in vitro* studies points toward important contributions in attenuating endothelial cell growth (Orlidge and D'Amore, 1987; Antonelli-Orlidge et al., 1989). Though, largely unexplored, potential tissue cell:endothelial cell interactions may also be important in influencing capillary endothelial cell functions. For instance, in the central nervous system astrocytes have been observed to control marker enzyme and "tight" tight

Pharmaceutical Applications of Cell and Tissue Culture to Drug Transport
Edited by G. Wilson *et al.*, Plenum Press, New York, 1991

intercellular junction expression (DeBault and Cancilla, 1980; Tao-Cheng et al., 1988).

The ultrastructural characterizations by Palade in 1953 are credited as the first to demonstrate the considerable variety and complexity of endothelia. Based on morphology, endothelial barriers may be classified into three general types. Table 1 contains a list of the endothelial cell types and some characteristics for each type. Despite exciting findings by Palade (1953), a few decades passed before capillary endothelia were recognized as anything but a passive lining. Currently, in addition to ultrafiltration, the endothelial cell is known to play several important physiological roles. Basic functions performed by endothelial barriers include those summarized in Table 2.

The filtration/and or permeability properties of endothelia are generally of greatest interest to the pharmaceutical chemist. For the most part, a good relationship exists between morphology and the water and electrolyte filtration characteristics of endothelial barriers in different tissues. For example, water, electrolytes, and serum proteins may readily cross both open fenestrated and discontinuous endothelial barriers through either large intercellular and transcellular openings, or pinocytic processes. On the other hand, the nonfenestrated continuous endothelial barrier, with tight intercellular junctions, is adapted in some tissue locations (e.g., blood-brain barrier) to greatly restrict transcellular movement of water and ions (Renkin, 1988). Several lines of evidence indicate that endothelia are asymmetric, have a complex glycocalyx, and several receptor systems for mediating either endocytosis or transcytosis of macromolecules (Simionescu, 1988). In addition, endothelial cells retain several enzyme systems for degrading neurotransmitters and hormones (Shepro and Dunham, 1986). Expression of any or all of these biochemical properties is not universal for endothelia. Clearly then, all of these properties affect the distribution of substances across endothelia and, for these reasons, extrapolations based on purely endothelial cell morphology will not apply.

DESIRABLE CHARACTERISTICS FOR ENDOTHELIAL CELL CULTURE SYSTEMS

A substantial portion of the morphological and biochemical knowledge accumulated on endothelia has resulted from the development of tissue culture systems. The first endothelial cell culture system was established in the early 1970's, by Jaffe et al. (1973). Subsequently numerous endothelial cell culture systems have been described. To simply list the different systems is beyond the scope of this paper, however, such a list would indicate that endothelia from many body tissues and from a variety of species has been attempted with generally successful results. As a consequence of the diversity of endothelia, selection of the appropriate capillary endothelial population for establishment of a tissue culture system for pharmaceutical applications should be dictated by the tissue site targeted for drug delivery. To discuss desirable characteristics of an endothelial cell culture system, reference will be made to the development of primary cultures of brain capillary endothelial cells. Specifically, morphological, biochemical, and selected permeability properties will be outlined that demonstrate the appropriateness of an *in vitro* model to investigate drug distribution across the endothelial barrier comprising the BBB. Later chapters will address properties of tissue culture systems derived from peripheral endothelia.

Table 1. Some Characteristics of The Three Types of Endothelial Barriers *

Type	*Intercellular or Transcellular Openings (Å)*	*Pinocytosis*	*Representative Tissue Locations*
Continuous (common)	40-90	+	muscle, exocrine pancreas, gonads, skin, connective tissue
Continuous	<20	attenuated	blood-brain barrier
Fenestrated	800-1000	+	nasal cavity, gastrointestinal tract, kidney
Discontinuous	>1000	+	liver, spleen, bone marrow

* Simionescu and Simionescu, 1983; Smith and Kampine, 1984; Renkin, 1988.

The Blood-Brain Barrier *In Vitro*

Brain capillary endothelial cells are a continuous type of endothelium and comprise the so-called blood-brain barrier (BBB). By morphology (see Table 1) and function, endothelia of the BBB can be contrasted to endothelia of other body tissues. A primary function of brain capillary endothelia is to assist in the maintenance of a constant environment in the central nervous system. To perform this function, mammalian brain capillary endothelia have evolved systems to tightly control the movement of water and water-soluble substances between the blood and the brain. Regulation of the transendothelial movement of materials at the BBB is achieved in large part through tight intercellular junctions, the existence of few pinocytic vesicles, and the absence of fenestrations (Reese and Karnovsky, 1967; Brightman and Reese, 1969). There are at least eight specific transendothelial carrier mechanisms that facilitate the delivery of required blood-borne nutrients (e.g., amino acids, glucose) across the BBB. BBB endothelia also express numerous enzymatic systems, a complex glycocalyx, and a predominant electronegative charge on the luminal surface (Vorbrodt, 1988). Although often overlooked, these biochemical properties contribute significantly to the brain capillary endothelial cell's ability to regulate transendothelial movement of substances (Cornford, 1985). BBB endothelia have about three times more mitochondria than their peripheral counterparts. This feature suggests BBB endothelia retain sufficient energy resources to support their various transport and metabolic activities (Cornford, 1985).

The unique endothelia of the BBB have been isolated and grown successfully in both primary and passaged culture systems. In our laboratories, homogenous populations of bovine brain capillary endothelial cells have been isolated by enzymatic and centrifugation techniques (Audus and Borchardt, 1986a; 1987). Once isolated the

Table 2. Functions of Endothelial Barriers *

Function	*Example(s)*
Filtration	Actively and passively regulates the distribution of nutrients and waste products between the blood and the tissues.
Biosynthesis	Synthesis of Factor VIII, platelet derived growth factor, hormones, adhesion factors, heparin, plasminogen activator, and collagen.
Metabolism	Enzymes for degradation of neurotransmitters, peptides, various esters.
Receptor	On stimulation by circulating hormones may secrete smooth muscle effectors such as endothelium derived relaxing factor.
Angiogenesis	Vascularization of tumors and wound repair.
Container	Entire endothelium of 70 kg man covers about 700 m^2 and is comprised of approximately 6 x 10^{13} cells. Container for the blood.

* Simionescu and Simionescu, 1983; Hammersen and Hammersen, 1985.

endothelial cells can be conveniently frozen at -70°C until needed. Following thawing and washing, suspended cells can be seeded onto a fibronectin-treated surface that is precoated with rat-tail collagen. Generally, ten to fourteen days later the endothelial cells form confluent monolayers. The cells continue to exist in monolayers for up to 20 days in culture, indicating contact inhibition is retained on confluence (Audus and Borchardt, 1986a; 1987). The following discussion summarizes characterization of these primary cultures of bovine brain microvessel endothelial cell monolayers as an appropriate *in vitro* model for drug transport and metabolism studies. Passaged culture systems for BBB studies have been described (DeBault and Cancilla, 1980) and a succeeding chapter here will address specific properties of passaged BBB endothelial cell cultures.

Morphology: Tight intercellular junctions, few pinocytic vesicles, and the absence of fenestra are considered morphological markers for BBB endothelia. In primary culture, bovine brain capillary endothelial cell monolayers retain tight intercellular junctions (Audus and Borchardt, 1986a; 1987). These junctions have been shown to be typical tight intercellular junctions but are not identical, either in configuration or extent, to those at the BBB *in vivo*. As indicated above, several researchers are now focusing on the astrocyte as providing some factor which may modulate tight intercellular junctions of the BBB (DeBault and Cancilla, 1980; Tao-Cheng et al., 1988; Trammel and Borchardt, 1989; Raub et al., 1989). In this culture system, an attenuated pinocytic activity is observed both qualitatively with electron microscopy and quantitatively by assaying the uptake of the fluid-phase endocytosis marker, Lucifer Yellow (Guillot et al., 1989).

Table 3. Calculated Total Volume Turnover of the Endocytic Compartment for Selected Cell Types

	Volume/hour
Bovine Brain Microvessel Endothelial Cell Monolayer[a]	1.2%
L-cell Fibroblast[b]	9-15%
Mouse Macrophage[b]	26-34%

a From Guillot et al., 1989.
b From Besterman et al., 1981.

Qualitatively, the appearance of Lucifer Yellow uptake in the monolayers of this culture system has been shown to be consistent with the presence of few pinocytic vesicles observed in the early electron microscopic studies of Reese and Karnovsky (1967) and Brightman and Reese (1969). Unfortunately, quantitative information for fluid-phase endocytosis at the BBB *in vivo* is not available, thus comparisons has only been made with other cell types. Listed in Table 3, for comparison, are examples of the calculated total volume turnover of the endocytic compartment based on uptake of fluid-phase markers for selected cell types. As is apparent from data in Table 3, the endocytic activity of BBB endothelia in primary culture was much lower than other cell types. Consistent with the BBB *in vivo*, an abundance of mitochondria and **no** fenestra have been observed in the *in vitro* model (Audus and Borchardt, 1986a; 1987).

Biochemical Features: Table 4 contains a current list of some of the biochemical features of primary cultures of brain microvessel endothelium. Biochemical markers for the BBB include alkaline phosphatase and γ-glutamyl transpeptidase. Both enzyme systems have been biochemically and histochemically demonstrated in the primary cultures of brain capillary endothelial cells (Audus and Borchardt, 1986a; Baranczyk-Kuzma et al, 1986). As summarized in Table 5, for both enzymes, cultured cells were enriched relative to specific activity of the enzymes in crude cerebral gray matter homogenates (Baranczyk-Kuzma et al., 1986). The endothelial origin of the cultured cells has been confirmed by the retention of angiotensin converting enzyme (Baranzyk-Kuzma et al., 1986), Factor VIII antigen (Audus and Borchardt, 1986a), and with less specific markers, vimentin and the uptake of acetylated-low density lipoprotein (Guillot et al., 1989). Collectively, the enzyme studies with primary cultures of brain microvessel endothelial cells suggest that expression of typical BBB biochemistry has been retained *in vitro*. Thus, the *in vitro* system can be used to address drug metabolism at the BBB. As detailed in later chapters, the model has been used to investigate the potential fate of various catecholamine esters (Scriba and Borchardt, 1989) and peptides (van Bree et al., 1989) at the BBB.

Polarity: Cellular polarity is particularly critical in the realistic characterization of transcellular drug transport mechanisms in *in vitro* tissue culture models. Asymmetry of the endothelial lining has been suggested by several studies. In peripheral endothelia, the luminal presence of glycoproteins, podocalyxin and podoendin, indicates asymmetry. The location of several constituents (e.g., the abluminal location of Na,K^+-ATPase, and the predominant luminal alkaline

Table 4. Biochemical Features of Primary Cultures of Bovine Brain Microvessel Endothelial Cell Monolayers

Enzyme or Marker	*Reference*
Acid Hydrolases	a
Sulfatase	
Acid Phosphatase	
β-Galactosidase	
Adenylate Cyclase	b
Aminopeptidases	c
Blood-Brain Barrier Enzymes	d,e
γ-Glutamyl Transpeptidase	
Alkaline Phosphatase	
Catecholamine-Degrading Enzymes	d,f
Monoamine Oxidase A & B	
Catechol O-Methyltransferase	
Phenol Sulfotransferase	
Cholinergic Enzymes	g
Acetylcholinesterase	
Butyrylcholinesterase	
Endothelial Cell Markers	d,e,h
Acetylated-LDL uptake	
Angiotensin Converting Enzyme	
Factor VIII Antigen	
Vimentin	

a Baranczyk-Kuzma et al., 1989b.
b A.Baranczyk-Kuzma, K.L.Audus, F.L.Guillot, and R.T. Borchardt, unpublished observations.
c Baranczyk-Kuzma and Audus, 1987.
d Baranczyk-Kuzma et al., 1986.
e Audus and Borchardt, 1986.
f Baranczyk-Kuzma et al., 1989a.
g A.Trammel and R.T. Borchardt, unpublished observations.
h Guillot et al., 1989.

phosphatase and anionic charges due to sulfated glycoproteins) indicate polarity of the BBB (Vorbrodt, 1988). *In vitro* models have played a significant role in the characterization of the polarity of the BBB (Joo, 1985). Consequently, tissue culture models can be expected to play a role in further defining the role of endothelial cell polarity in determining the distribution of drugs across this barrier.

In our laboratories, polarity of the *in vitro* BBB has been observed in recent receptor and permeability studies. Exposure of the apical side of brain microvessel endothelial cell monolayers to nanomolar concentrations of angiotensin II typically reduces the permeability to fluorescein conjugated dextran or fluorescein by 80%. Similar exposure of the basolateral side of the monolayers to angiotensin II does not result in permeability changes (Guillot and Audus, 1989a). These experiments suggest that the angiotensin II

Table 5. Specific Activities of Blood-Brain Barrier Marker Enzymes in Crude Cerebral Cortex Homogenates and Confluent Monolayers of Brain Microvessel Endothelial Cells *

Enzyme	*Crude Cerebral Cortex Homogenates*	*Confluent Brain Endothelial Cell Monolayers*
γ-Glutamyl Transpeptidase	6.0 ± 0.5 nmol/mg/min	11.5 ± 0.6 nmol/mg/min
Alkaline Phosphatase	3.3 ± 0.2 μmol/mg/min	35.3 ± 3.2 μmol/mg/min

* Baranczyk-Kuzma et al., 1986.

receptor controlling permeability pathways across the monolayers resides on the apical side of cultured cells. Studies on the efflux of dextran and transferrin loaded monolayers has also indicated that materials were preferentially returned to extracellular medium on the apical side of the cells (T.J. Raub, unpublished observations). These observations were consistent with the apparent vectorial movement of macromolecules from brain to blood observed *in vivo* (Broadwell et al., 1988; Vorbrodt, 1988). In contrast, the V_{max} for the transport of choline across brain microvessel endothelial cells in the apical to basolateral direction has been shown to be greater than for the opposite direction (Trammel and Borchardt, 1987).

Receptors and Permeability: Regulation of endothelial cell permeability can be mediated through receptor-mediated mechanisms. Receptors may trigger transendothelial permeability changes through either transcytotic pathways or paracellularly. Several receptors have been identified on primary cultures of brain microvessel endothelium as indicated in Table 6. Both angiotensin II and atrial natriuretic factor receptors control at least water and electrolyte passage across endothelial barriers. Recent evidence from our laboratories established that stimulation of angiotensin II receptors enhances fluid-phase endocytosis yet, attenuates substantially, the overall permeability of monolayers of brain microvessel endothelial cells. Accumulating evidence suggests that angiotensin II stimulates the apical membrane uptake of materials, the material in intracellular compartments, and inhibits other pathways for transcytotic passage (Guillot and Audus, 1989a; 1989b). A widely accepted pharmacological approach to controlling water and electrolyte permeability in certain pathophysiological conditions (e.g., vasogenic edema) has not been identified. Based on observed responsiveness to vasoactive substances, the *in vitro* model holds promise as a system to further elucidate receptor-mediated mechanisms for the purpose of identifying pharmacological approaches to regulating vascular permeability under certain conditions.

A potential technique for specific delivery of substances across the BBB involves development of chimeric peptides and proteins that may undergo receptor-mediated transcytosis. The presence of various receptors on brain microvessel endothelial cells, listed in Table 6, holds promise for developing chimeric drug delivery schemes. The existence of these receptor systems, however, provides a strong indication that the *in vitro* model can be used to explore the

Table 6. Peptide and Protein Receptors on Primary Cultures of Brain Microvessel Endothelial Cells

Receptors Controlling Permeability Characteristics	
Atrial Natriuretic Factor	Smith et al., 1989
Angiotensin II	Guillot and Audus, 1989a; 1989b
Potential Receptors for Transcytotic Transport Mechanisms	
Ricin	Raub and Audus, 1987
Transferrin	Newton and Raub, 1989
Modified Albumins	Smith and Borchardt, 1989
Insulin,IGF-1,IGF-2	Rosenfeld et al., 1987; Keller et al., 1989

pharmaceutical applications of chimeric systems. Subsequent chapters here will discuss further the relevance of permeability to transferrin and lipoprotein receptors.

Permeation of substances across brain microvessel endothelial cell monolayers has been shown to be related to both lipophilicity and molecular weight in a manner consistent the BBB *in vivo* (Rim et al., 1986; Shah et al., 1989; Raeissi and Audus, 1989). Retention of absolute permeability characteristics of the *in vivo* BBB in an *in vitro* system has been a challenge (Audus and Borchardt, 1986a; 1987). Transendothelial electrical resistance has been estimated at 160 ± 18 ohms/cm^2 (T.J. Raub, unpublished results). As indicated above, endothelial cell:astrocyte interactions appear important in tight junction development and several investigators have begun attempts to identify the nature of these interactions. The current *in vitro* BBB model can be used in drug transport studies, however. Markers for extracellular leakage, to normalize for background leakiness of the monolayers, have been employed successfully to allow characterization of transendothelial transport mechanisms (Audus and Borchardt, 1986b). Transcellular transport of leucine, for example, was shown to be saturable, competitive with other amino acids, and energy-independent in this *in vitro* system (Audus and Borchardt, 1986b). In addition, the kinetic parameters *in vitro* appear in good agreement with true kinetic parameters *in vivo* (Greenwood et al., 1989). The transport parameters of several amino acid drugs including, baclofen (van Bree et al., 1988), α-methyldopa (Chastain and Borchardt, 1989), and acivicin (Chastain and Borchardt, 1987), have also been explored with this system. More recently, the mechanisms of passage of small peptides, vasopressin, leu-enkephalin, and delta sleep-inducing peptide have been investigated with this model system (Raeissi and Audus, 1989; Reardon and Audus, 1989; Thompson and Audus, 1989).

SUMMARY

Morphological, permeability, and biochemical properties are not universal among endothelial barriers. Specific endothelial barriers are comprised of dynamic, heterogenous populations of cells whose physiological roles generally depend of the location and nature of the surrounding tissue. The establishment of endothelial barriers *in vitro* can be accomplished with both primary and passaged culture systems. As outlined here for the blood-brain barrier, desirable properties of endothelial barriers for applications in pharmaceutical studies *in vitro* include endothelial cell morphology, biochemical features, polarity, receptors, and permeability properties representative of the barrier in the target tissue. Current tissue culture methods permit establishment of *in vitro* models that are similar in many respects (e.g., morphology, biochemistry, polarity, receptors, transcellular carrier systems) to the *in vivo* endothelial barriers. Deficiencies of the *in vitro* endothelial barriers include an absolute permeability or "leakiness," that is greater than observed *in vivo*. Permeability properties of *in vitro* endothelial barrier models of the future will likely be closer to *in vivo* once a better understanding of the role of both circulating and tissue derived factors in modulating endothelial cell functions are incorporated. Despite the inability of the *in vitro* endothelial barriers to fully mimic absolute permeability characteristics observed *in vivo*, current tissue culture systems have been shown to be, and will continue to be, valuable tools for the pharmaceutical chemist in defining mechanisms controlling distribution of drugs between the blood and interstitial fluids.

ACKNOWLEDGEMENTS

This work was supported by The American Heart Association - Kansas Affiliate, a Life Sciences Contacts Award from Eli Lilly & Co., and The Upjohn Company.

REFERENCES

Antonelli-Orlidge, A., Saunders, K.B., Smith, S.R., and D'Amore, P.A., 1989, An activated form of transforming growth factor beta is produced by cocultures of endothelial cells and pericytes, Proc. Natl. Acad. Sci. USA, 86:4544-4548.

Audus, K.L. and Borchardt, R.T., 1986a, Characterization of an *in vitro* blood-brain barrier model system for studying drug transport and metabolism, Pharm. Res., 3:81-87.

Audus, K.L. and Borchardt, R.T., 1986b Characteristics of the large neutral amino acid transport system of bovine brain microvessel endothelial cell monolayers, J. Neurochem., 47:484-488.

Audus, K.L. and Borchardt, R.T., 1987, Bovine brain microvessel endothelial cell monolayers as a model system for the blood-brain barrier, Ann. N.Y. Acad. Sci., 507:9-18.

Baranczyk-Kuzma, A. and Audus, K.L., 1987, Characteristics of aminopeptidase activity from brain microvessel endothelium, J. Cerebr. Blood Flow Metab., 7:801-805.

Baranczyk-Kuzma, A., Audus, K.L., and Borchardt, R.T., 1986, Catecholamine-metabolizing enzymes of bovine brain microvessel endothelial cell monolayers, J. Neurochem., 46:1956-1960.

Baranczyk-Kuzma, A., Audus, K.L., and Borchardt, R.T., 1989a, Substrate specificity of phenol sulfotransferase from primary cultures of bovine brain microvessel endothelium, Neurochem. Res., 14:689-691.

Baranczyk-Kuzma, A., Raub, T.J., and Audus, K.L., 1989b, Demonstration of acid hydrolase activity in primary cultures of bovine brain microvessel endothelium, J. Cerebr. Blood Flow Metab., 9:280-289.

Besterman, J.M., Aihart, J.A., Woodworth, R.C., and Low, R.B., 1981, Exocytosis of pinocytosed fluid in cultured cells: Kinetic evidence for rapid turnover and compartmentation, J. Cell Biol., 91:716-727.

Brightman, M.W. and Reese, T.S., 1969, Junctions between intimately apposed cell membranes in the vertebrate brain, J. Cell Biol., 40:649-677.

Broadwell, R.D., Balin, B.J., and Salcman, M., 1988, Transcytotic pathway for blood-borne protein through the blood-brain barrier, Proc. Natl. Acad. Sci. USA, 85:7820-7824.

Chastain, J.E. Jr. and Borchardt, R.T., 1989, L-α-Methyldopa transport across bovine brain microvessel endothelial cell monolayers, a model of the blood-brain barrier, Neurosci. Res. Commun., 4:147-152.

Chastain, J.E. Jr. and Borchardt, R.T., 1987, Potential substrates for the large neutral amino acid transport system of bovine brain microvessel endothelial cell monolayers, J. Cell Biol., 105:328a.

Cornford, E.M., 1985, The blood-brain barrier, a dynamic regulatory interface, Mol. Physiol., 7:219-260.

DeBault, L.E. and Cancilla, P.A., 1980, γ-Glutamyl transpeptidase in isolated brain endothelial cells: Induction by glial cells *in vitro*, Science, 207:653-655.

Greenwood, J., Hazell, A.S., and Pratt, O.E., 1989, The transport of leucine and aminocyclopentanecarboxylate across the intact, energy-depleted rat blood-brain barrier, J. Cerebr. Blood Flow Metab., 9:226-234.

Guillot, F.L. and Audus, K.L., 1989a, Vasoactive peptide regulation of blood-brain barrier permeability, Pharm. Res., 6:S211.

Guillot, F.L. and Audus, K.L., 1989b, Biochemistry of angiotensin peptide-mediated effects on fluid-phase endocytosis in brain microvessel endothelial cell monolayers, J. Cell Biol., 107:809a.

Guillot, F.L., Audus, K.L., and Raub, T.J., 1989, Fluid-phase endocytosis by primary cultures of bovine brain microvessel endothelial cell monolayers, Microvasc. Res., in press.

Hammersen, F. and Hammersen, E., 1985, Some structural and functional aspects of endothelial cells, Basic Res. Cardiol., 80:491-501.

Jaffe, E.A., Nachman, R.L., and Becker, C.G., 1973, Culture of human endothelial cells derived from umbilical veins. Identification by morphologic and immunologic criteria, J. Clin. Invest., 52:2745-2764.

Joo, F., 1985, The blood-brain barrier *in vitro*: Ten years of research on microvessels isolated from the brain, Neurochem. Int., 7:1-25.

Keller, B.T., Smith, K.R., and Borchardt, R.T., 1988, Transport barriers to absorption of peptides, Pharmacetisch Weekblad, 10:38-39.

Newton, C.R. and Raub, T.J., 1989, Characterization of the transferrin receptor in primary cultures of bovine brain capillary endothelial cells, J. Cell Biol., 107:770a.

Orlidge, A. and D'Amore, P.A., 1987, Inhibition of capillary endothelial cell growth by pericytes and smooth muscle cells, J. Cell Biol., 105:1455-1462.

Palade, G.G., 1953, Fine structure of blood capillaries, J. Appl. Physiol., 24:1424-1433.

Raeissi, S. and Audus, K.L., 1989, Brain microvessel endothelial cell monolayer permeability to delta sleep-inducing peptide, J. Pharm. Pharmacol., 41:in press.

Raub, T.J. and Audus, K.L., 1987, Adsorptive endocytosis by bovine brain capillary endothelial cells *in vitro*, J. Cell Biol., 105:312a.

Raub, T.J., Kuentzel, S.L., and Sawada, G.A., 1989, Characteristics of primary bovine cerebral microvessel endothelial cell monolayers on membrane filters: Permeability and glial induced changes, J. Cell Biol., 109:315a.

Reardon, P.M. and Audus, K.L., 1989, Arginine-vasopressin distribution across the *in vitro* blood-brain barrier, Pharm. Res., 6:S88.

Reese, T.S. and Karnovsky, M.J., 1967, Fine structural localization of a blood-brain barrier to exogenous peroxidase, J. Cell Biol., 34:207-217.

Renkin, E.M., Transport pathways and processes, in: "Endothelial Cell Biology in Health and Disease," N. Simionescu and M. Simionescu, eds., Plenum Press, New York (1988), pp. 51-68.

Rim, S., Audus, K.L., and Borchardt, R.T., 1986, Relationship of octanol/buffer and octanol/water partition coefficients to transcellular diffusion across brain microvessel endothelial cell monolayers, Int. J. Pharm., 32:79-84.

Rosenfeld, R.G., Pham, H., Keller, B., Borchardt, R.T., and Pardridge, W.M., 1987, Structural comparison of receptors for insulin, insulin-like growth factor-I and -II (IGF-I and IGF-II) in brain and blood-brain barrier, Biochem. Biophys. Res. Commun., 149:159-166.

Scriba, G.K.E. and Borchardt, R.T., 1989, Metabolism of catecholamine esters by cultured bovine brain microvessel endothelial cells, J. Neurochem., 53:610-615.

Shah, M.V., Audus, K.L., and Borchardt, R.T., 1989, The application of bovine brain microvessel endothelial cell monolayers grown onto polycarbonate membranes *in vitro* to estimate the potential permeability of solutes throught the blood-brain barrier, Pharm. Res., 6:624-627.

Shepro, D. and Bernadette, D., 1986, Endothelial cell metabolism of biogenic amines, Ann. Rev. Physiol., 48:335-345.

Simionescu, M., Receptor-mediated transcytosis of plasma molecules by vascular endothelium, in: "Endothelial Cell Biology in Health and Disease," N. Simionescu and M. Simionescu, eds., Plenum Press, New York (1988), pp. 69-104.

Simionescu, N. and Simionescu, M., The cardiovascular system, in: "Histology," L. Weiss, ed., Elsevier Biomedical, New York (1983), pp. 371-433.

Smith, K.R. and Borchardt, R.T., 1989, Permeability and mechanism of albumin, cationized albumin and glycosylated albumin transcellular transport across monolayers of cultured brain capillary endothelial cells, Pharm. Res., 6:466-473.

Smith, K.R., Kato, A., and Borchardt, R.T., 1988, Characterization of specific receptors for atrial natriuretic factor on cultured bovine brain microvessel endothelial cells, Biochem. Biophys. Res. Commun., 157:308-314.

Smith, J.J. and Kampine, J.P., 1984, "Circulatory Physiology, Williams and Wilkins, Baltimore, pp 127-137.

Tao-Cheng, J.H., Nagy, Z., and Brightman, M.W., 1988, Tight junctions of cerebral endothelium *in vitro* are enhanced by astroglia, J. Neurosci., 7:3293-3299.

Thompson, S.E. and Audus, K.L., 1989, Aspects of leu-enkephalin transport and metabolism at the blood-brain barrier, Pharm. Res., 6:S175.

Trammel, A.T. and Borchardt, R.T., 1987, Choline transport in cultured brain microvessel endothelial cells, Pharm. Res., 4:S41.

Trammel, A.T. and Borchardt, R.T., 1989, The effects of astrocytes and glioma cells on the permeability of cultured brain microvessel endothelial cells, Pharm. Res., 6:S88.

van Bree, J., Audus, K.L., and Borchardt, R.T., 1988, Carrier-mediated transport of baclofen across monolayers of bovine brain endothelial cells in primary culture, Pharm. Res., 5:369-371.

van Bree, J.B.M.M., De Boer, A.G., Verhoef, J.C., Danhof, M., and Breimer, D.D., 1989, Transport of vasopressin fragments across the blood-brain barrier: *In vitro* studies using monolayer cultures of bovine brain endothelial cells, J. Pharmacol. Exp. Ther., 249:901-905.

Vorbrodt, A.W., 1988, Ultrastructural cyto-chemistry of blood-brain barrier endothelia, Prog. Histochem., 18:1-96.

TRANSPORT OF DRUGS ACROSS THE BLOOD-BRAIN BARRIER: IN VITRO AND IN VIVO STRATEGIES

Joost B.M.M. van Bree, Albertus G. de Boer,
Meindert Danhof and Douwe D. Breimer

Division of Pharmacology, Center for Bio-Pharmaceutical Sciences
University of Leiden
Leiden, The Netherlands

INTRODUCTION

The transport rate across the BBB is an essential kinetic parameter for drugs acting on the central nervous system, because it determines the time to onset and the intensity of drug action.[1,2] For the study of BBB-passage in vivo, various models have been described, however most techniques require the use of many experimental animals and offer only limited information.[3-5] Moreover, in the in vivo situation the estimated transport parameters are inevitably influenced by physiological factors which cannot be kept under control during the experiment (e.g. cerebral blood flow, hormone levels, stress levels).[6] Therefore comparison of BBB transport parameters obtained in different experimental settings is hazardous.

A brief overview will be given of the different in vitro and in vivo models used to assess drug transport into the central nervous system, together with their specific applicability and the problems involved.

IN VITRO METHODOLOGIES

Since the anatomical features of the blood-brain barrier are localized in the brain microvessels and more specifically in the endothelial cells, efforts have been made to obtain in vitro preparations of the microvessels. Later on, the emphasis was put on the endothelial cells which can be cultured in monolayers, thereby representing the blood-brain barrier in vitro.

In the late sixties and early seventies, the first reports were published which described procedures for isolating brain microvessels. Siakotos et al. described a large scale procedure in which more than 1 kg of fresh brain material was needed,[7] however, the scene for further development was set by the work of Brendel and colleagues who developed a method based on mechanical dispersion of the brain tissue.[8] This technique was further developed in

Pharmaceutical Applications of Cell and Tissue Culture to Drug Transport
Edited by G. Wilson *et al.*, Plenum Press, New York, 1991

subsequent years, with ultimately homogeneous cultures of brain microvessel endothelial cells on porous carriers. These allow the design of relatively sophisticated transport studies.

In general, to obtain microvessels or endothelial cells from brain tissue, a procedure is required in which the tissue is minced or digested in order to set the microvessels free. Subsequently, the free vessels have to be isolated from the neuronal debris. After further digestion and/or purification, endothelial cells can be obtained. It is very important that the product of the isolation is well characterized before it is used for further studies. In many instances microscopic assessment of the morphological features can be used, but in order to get information about homogeneity and purification level of the isolated product, histochemical or immunochemical tests are indispensable. A number of enzymes are present in the microvessel endothelial cells, which are absent in the surrounding brain tissue: alkaline phosphatase (AP),[9] and gamma-glutamyl transpeptidase (GT).[10] Endothelial markers in general are often used: angiotensin converting enzyme (ACE),[11] Factor VIII/Von Willebrandt factor,[12] and acetyl-LDL uptake.[13] Assessment of the expression and activity level of a combination of these markers can be used for positive and affirmative identification of isolated microvessels and endothelial cells in culture.

A review of the isolation and culture techniques as described in the literature is presented in Table I. Mechanical dispersion was the most wide-spread procedure at the beginning of the development of the <u>in vitro</u> models. Subsequent to the dispersion step, the brain tissue was further homogenized by passing it through nylon mesh sieves. For separation of the microvessels from the neuronal tissue the density difference between these two was used. The neuronal tissue has a high fat content and thus a lower density. By means of suspending the homogenate in a concentrated albumin solution (25%) with centrifugation (albumin flotation technique),[14] or by means of density gradient centrifugation,[15] the separation can be established. Further purification can be achieved by filtering over a glass bead column to which the microvessel fragments selectively adhere. The microvessels obtained after mechanical isolation are metabolically active,[16] and when plated into tissue culture show an overall cell viability of 50%.[17] This figure was contradicted by the results of Williams et al., who compared the various mechanical dispersion techniques with an enzymatic digestion scheme with collagenase that he himself had designed: microvessels originating from mechanical dispersion techniques showed a viability of only 5%, while cultures from microvessels isolated according to the enzymatic procedure showed a viability of 84%. Furthermore he demonstrated, by measuring the enzyme 5-nucleotidase, which is specific for astroglia, that preparations from mechanical origin were much more contaminated with astrocytes and pericytes.[18]

Table I. Overview of the development of in vitro methodologies to study the blood-brain barrier and transport processes across the blood-brain barrier

Author	Source	Method	Goal	Charact.	Remarks
Goldstein 1975	Rat cerebral cortex	Mech.Dispersion. Albumin flotation. Glass bead Filtr.	Isolated brain capillaries	A.P. + yGT +	Isolated capillaries are metabolically active.
Panula 1978	Rat whole brain	Mech. Dispersion. Gradient Centrifug. Direct plating of capillary pellet.	Brain capillary cultures	A.P. +	Heterogenous culture with some endothelial cells.
Hjelle 1978	Bovine cortex	Mech. Dispersion. Purification by mesh filtration.	Isolated brain capillaries	yGT +	Carrier-mediated mechanism for amino acid uptakeis still functional in vitro.
De Bault 1979	Mouse whole brain	Mech. Dispersion. Mesh Filtration. Direct plating of capillary fragm.	Endothelial cell culture	Morphology	Plating efficiency 50 %. Heterogenous culture.
Phillips 1979	Bovine/Rat white matter	Trypsin digestion. and subsequent plating.	Endothelial cell culture	FVIII + A.P. ±	Culture largely contaminated by A.P. negative cells.
Wiliams 1980	Rat cerebral cortex	Collagenase digest. Albumin flotation. 2nd Collagenase digest. Glass bead filtration.	Endothelial cell culture	yGT + A.P. + Na/K ATP-ase	84% viability versus 3% viability for mech. disp.techniques.
Spatz 1980	Rat brain	Mech. Dispersion. Trypsin-collagenase digestion. Trypsin passage upon confluence.	Establishment of endothelial cell line	A.P. ± yGT ±	Cell line looses histochemical markers No formation of tight junctions.
Bowman 1981	Rat cerebral cortex	Mech. Dispersion. Dextran Flotation. glass bead filtr. collagenase/dispase digestion of capillaries. Percoll density gradient.	Endothelial cell culture.	FVIII + ACE +	Homogenous culture. Approx. 90 % plating efficiency. Formation of tight junctions in vitro.
Bowman 1983	Bovine cortex	Isolation procedure as above. Culture on collagen-coated nylon mesh.	endothelial monolayer cultures on porous support, allowing "transport across" studies.	As Above	Barrier formation in vitro.

The use of collagenase probably gives better results in cell viability, since the endothelial cell membrane is left undamaged (in contrast to the situation when mechanically induced shear-stress occurs). Homogeneity is improved because the collagenase specifically degrades the basal membrane (primarily consisting of collagen), thereby effectively removing adjacent pericytes and astroglia from the microvessels. Cerebral cortex is normally used for the isolation procedures since it is very rich in capillaries.

Phillips et al. used white matter for their isolation in order to prevent contamination with endothelial cells originating from larger blood vessels.[19] In the resulting cultures however only 61% of cells showed a moderately positive staining for AP leaving reasons to doubt the endothelial origin of the cells.

The first well defined cell-line of cerebrovascular endothelial origin was established by Spatz et al. Although cells homogeneously stained positive for AP and GT, electron microscopic examination failed to demonstrate the existence of tight junctions. Only gap junctional complexes could be detected.[20] This is an important finding, because it shows the potential lack of correlation between histochemical and morphological criteria for positive identification. The cultured cells are undoubtedly of endothelial origin, but if one of the important functional characteristics i.e. the presence of tight junctions is missing, the cell-line is of little use for the study of the barrier function.

The isolation and culture techniques were further improved by the use of continuous density gradient centrifugation subsequent to a collagenase/dispase digestion of rat cerebral microvessels.[21] Primary cultures were free of contaminating glia and pericytes. The resulting culture was positive for Factor VIII and ACE and electron microscopy revealed the existence of tight junctions between the cells. Occasionally, a second cell type was seen in the cultures which was not positive for Factor VIII. Probably this was a contamination with fibroblasts, but in a later study it was shown that this problem can be avoided by the use of plasma derived serum.[22] The main disadvantage of this procedure as described by Bowman et al was the need for approximately 40 rat brains per isolation, yielding approximately 10 to 20 million cells. Therefore, the procedure was adapted for isolation from bovine brains, 2 brains yielding approximately 200 million cells.

With the isolated microvessel technique, uptake of substrates into the vessels could be studied, but the main disadvantage was that the substrate mainly approached the microvessels from the abluminal side, which is literally opposite to the in vivo situation. With the development of cell culture techniques this disadvantage disappeared, however again only endothelial uptake could be assessed. When culture procedure was modified in such a way that cells could be cultured on collagen coated nylon mesh transcellular transport studies could be performed.[23] Currently this technique is to be preferred over the use of isolated microvessels.

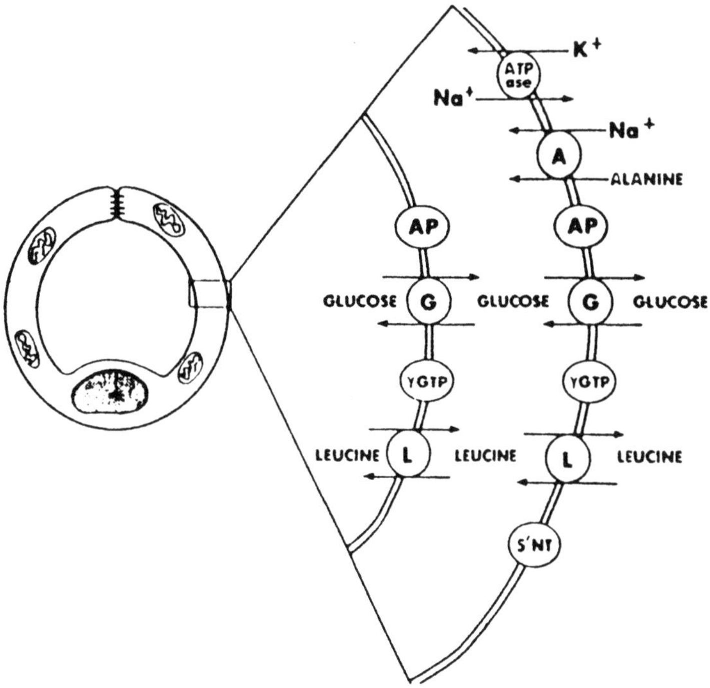

Fig. 1. Schematic representation of the luminal and abluminal distribution of enzyme and carrier systems of brain endothelial cells. Most of these results were obtained by the use of isolated capillaries (from ref. 30).

ISOLATED BRAIN MICROVESSELS

With the development of the in vitro blood-brain barrier a new technique came available enabling experiments which were not possible in vivo. Two basic fields of interest are covered with the in vitro model. Improved knowledge of the physiology and biochemistry of the blood-brain barrier was obtained from the in vitro research in terms of the specific characteristics of the blood-brain barrier. The isolated microvessels are well suited for studying uptake of nutrients and for the localization or identification of specific enzymes and substrate binding sites. Several specialized uptake and enzyme systems have been identified or further characterized.[24-28] Some of the systems proved to be specifically localized on the luminal or abluminal side, implying that the blood-brain barrier expresses polarity.[29-31] Figure 1 gives an overview of the luminal and abluminal occurrence of a number of transport and enzyme systems. The second main application of the isolated brain microvessel technique is in drug transport and disposition studies. However, drug transport cannot be studied using isolated microvessels. The technique is more suitable to study drug binding and metabolism, as illustrated by a few examples: Enkephalin was shown to be rapidly degraded by the isolated microvessels,[32] and recent work demonstrated the presence of the drug metabolizing enzymes cytochrome P450 and P-450 reductase in rat brain microvessels.[33] Furthermore, specific binding sites for atrial natriuretic factor were detected in brain microvessels;[34] the isolated brain capillaries technique has a number of drawbacks for application in drug transport studies. It inevitably is a heterogenous system: by using immunohistochemical techniques, two enzymes, glial fibrillar acidic protein and glutamine synthetase, occurring in astroglia, could be demonstrated on brain microvessels. This indicates that the vast majority of the isolated microvessels contained more than just endothelial cells and the surrounding basal membrane.[35] Study of drug transport using isolated brain capillaries is restricted to measuring uptake into the isolated microvessels,[36,24,16] whereas it is not possible to study transport across the microvessel wall.

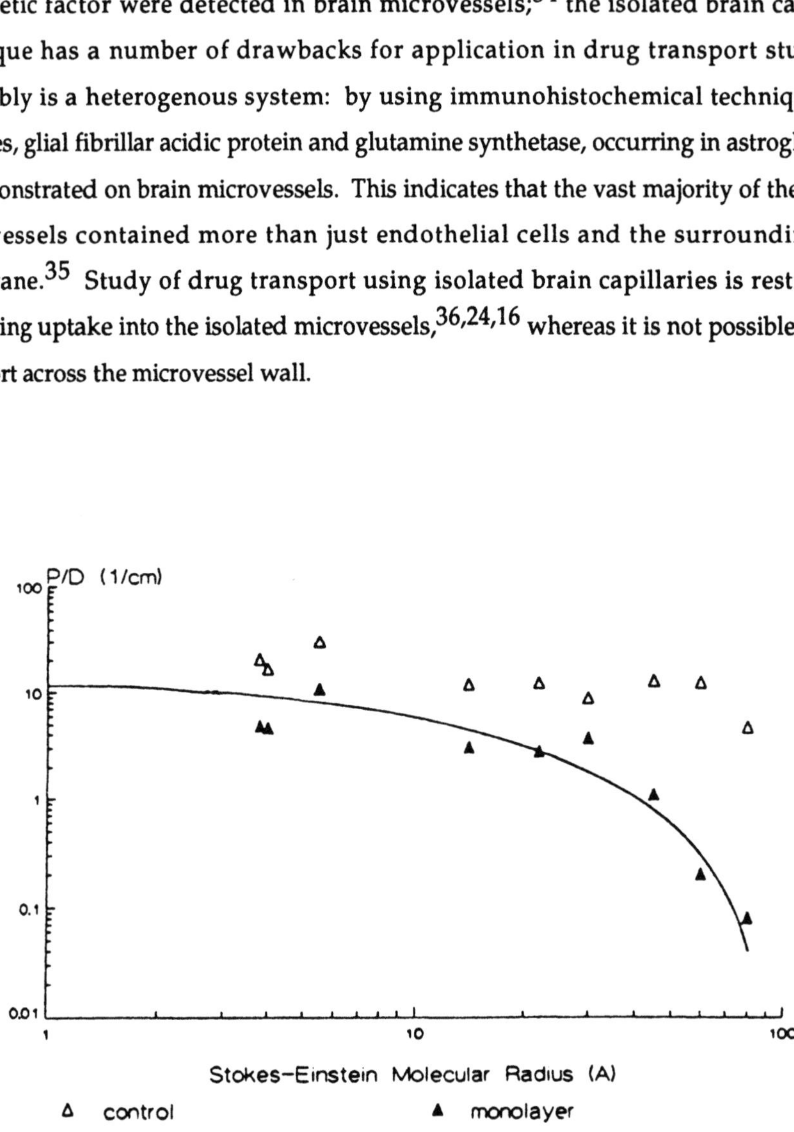

Fig. 2. The effect of molecular size on permeability of para-cellularly transported FITC-labelled dextrans. The line represents the best fit to a hydrodynamic model from which the pore characteristics were calculated (from ref. 44).

Recently, techniques have been developed to isolate endothelial cells from brain microvessels and bring them into culture with retention of specific histochemical and morphological properties.[21,37,38] Cells are cultured on porous membranes until they are completely covered with an endothelial cell monolayer. This model enables the assessment of transport profiles and the study of factors which influence drug transport across the endothelial monolayer. The cerebrovascular endothelial monolayer expresses a barrier function against polar substances,[39] and important factors such as molecular size, lipophilicity and protein binding were shown to influence in vitro transport.[40] Metabolizing enzymes and enzyme systems, known to be present in vivo, were also shown to be present in the monolayer cultures. The in vivo blood-brain barrier possesses a number of carrier systems for nutrient transport into the central nervous system.[41] The large neutral amino acid (LNAA) system remains functional in the in vitro situation with comparable characteristics.[42] In a study using the in vitro system it could be demonstrated that the central nervous system muscle relaxing agent baclofen is transported across the blood-brain barrier by means of the LNAA system[43], another functional and characteristic feature of complexes between the endothelial cells. In the in vitro setup, a filtering effect of the monolayer was observed (Figure 2) and application of an hydrodynamic model enabled the calculation of pore size and abundance. Comparison of the in vitro pore data with in vivo tight junction characteristics lead to the conclusion that the in vitro pores have similar characteristics as the in vivo tight junctions.[44] With this in vitro model the problems and complexities of the in vivo situation can be partially circumvented and it is potentially a valuable tool for the study of drug transport into the central nervous system.

A specific example of the application of the in vitro model in the study of blood-brain barrier transport of drugs is related to vasopressin and its fragments. The metabolic stability of the AVP1-8 analogues towards the endothelial monolayer was assessed (figure 3).

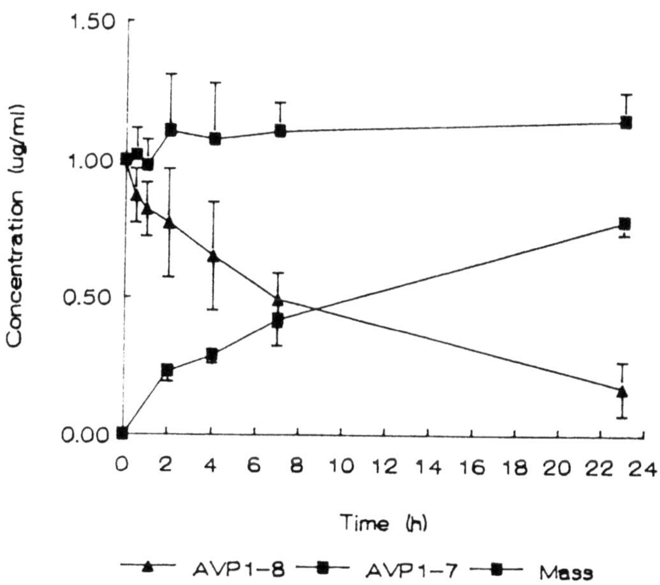

Fig. 3. Metabolic stability of the AVP1-8 peptide when incubated with a 60 cm^2 confluent cerebrovascular endothelial monolayer (from ref. 45).

All compounds appeared to be transported across the cellular layer (figure 4). The extent of the transport is in the same order of magnitude as the corresponding dextrans used to assess paracellular transport. Paracellular transport through aqueous pores or channels is dependent on molecular size.

In case of peptide and proteins, molecular size may not be directly related to molecular weight.[45] Figure 4 clearly illustrates that there is no straightforward relationship between BBB-permeability of AVP and its fragments and molecular weight: Dependent on the hydrophilic characteristics of the outer structure of the peptide or protein molecule, varying degrees of H-bond formation and water-surrounding may occur, resulting in different "effective" molecular sizes of molecules with comparable molecular weight. Therefore, in correlating paracellular transport to molecular characteristics, the use of molecular size should be preferred over molecular weight.

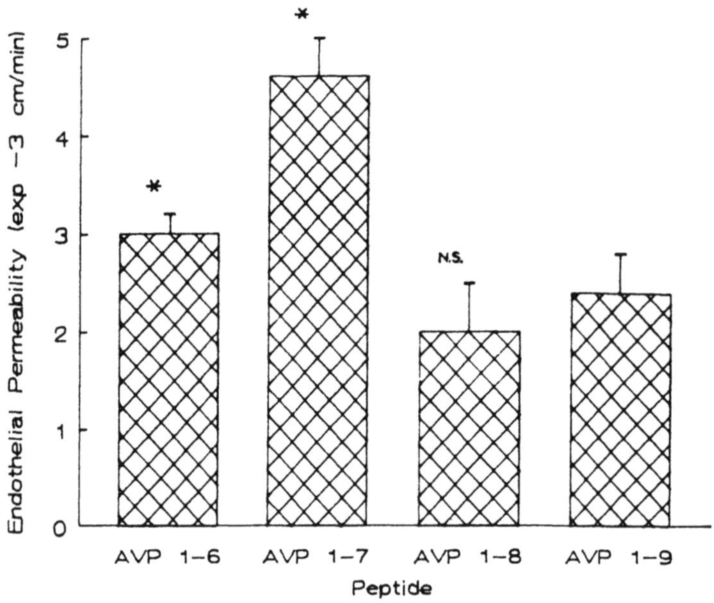

Fig. 4. Endothelial permeability of AVP1-9 and different AVP-fragments (from ref. 45).

IN VIVO METHODOLOGIES

Several experimental methods have been used to determine the CNS uptake of peripherally administered drugs and generally these can be divided into single-passage and multi-passage techniques.

The "indicator dilution method" was the first single passage technique reported and could be used to study uptake into several organs or tissues including the CNS: drug is administered into the arterial blood flow and the venous outflow concentration of the organ or tissue is measured.[3]

The brain uptake index, based on measuring extravasation of the drug in brain tissue after carotid injection and single passage through the brain, was used to quantify blood-brain barrier passage using a reference substance (often tritiated water) and a non-diffusible vascular space marker (14-carbon-labelled sucrose or inulin) as internal standard.[4,46] The single-passage techniques have a limited applicability, since their use is restricted by a number of assumptions:[47,48] there should only be unidirectional drug transport, the diffusion pattern of the drug studied and other solutes serving as intravascular or diffusion reference marker substance have to be equivalent, and the concentrations of the drug and the marker substances at the site of diffusion are assumed to be equal to the concentrations of these compounds in the injected solutions, assuming that no mixing of the injected fluid and blood occurs after injection.[49,50] Furthermore, drugs with extremely slow uptake, such as peptides, cannot be studied accurately due to the short exposure time when using single-passage techniques. Hence, this methodology is often considered as inadequate for application to hydrophilic drugs, mainly because of its lack of sensitivity.[51-54]

Drugs with unfavourable physico-chemical profiles for BBB transport should preferentially be studied with multi-passage techniques, since the exposure time is much longer and thus measurable concentrations in brain tissue can be obtained. These techniques were originally designed as steady-state experiments during which plasma levels were kept constant by means of intravenous infusion,[55] but later on mathematical models were developed which enabled calculation of permeability-surface area products after intravenous bolus administration.[56] Generally, this was done by determining the amount of drug in brain tissue after sacrifice at a certain time after administration. The amount of drug in the CNS was divided by the integral of the plasma drug concentration and time (AUC) thus, in pharmacokinetic terms, the clearance of drug from blood or plasma into the CNS was assessed. However, in many instances brain concentration is measured after a certain time period following drug administration and transport into the CNS is quantified as percentage of the dose administered or as a concentration relative to the plasma concentration. For an accurate estimation of this parameter for drugs with moderate to slow CNS uptake, it is essential that the brain concentration obtained is corrected for the amount of intravascular drug, requiring quantitative knowledge of the regional blood volumes of the brain.[57] Since sampling of brain tissue can only be done after sacrifice of the animal (single point method) it is not possible to study the time profile of BBB transport in an individual animal.

A recently developed multi-passage technique based on serial CSF sampling and numerical deconvolution enables the determination of a complete transport profile in a single living animal. It in principle also allows one to obtain information on the major mechanism involved for the transport of the given compound into the central nervous system: e.g. passive diffusion versus active transport. The time course of cumulative cleared plasma volume due to blood-brain barrier transport is calculated. When a linear relationship is obtained, the transport processes involved are first order and slope of the line is the blood-brain barrier clearance, a model-independent pharmacokinetic parameter. When a non-linear relationship exists, due to carrier-mediated mechanisms, the kinetic constants of the carrier can be determined. The general applicability of the method was established using three drugs with different lipophilicity: atenolol, acetaminophen and antipyrine.[58] The resulting transport curves are depicted in figure 5. All compounds were transported by means of a linear process and, as expected, blood-brain barrier clearance increased with increasing lipophilicity.

When this method was applied to the AVP1-8 peptide, similar results were obtained. Figure 6 shows the mean transport curve. Appreciable linearity was obtained for each individual rat with blood-brain barrier clearance values ranging between 0.20-1.96 l/min. When AVP1-8 is, for instance, compared with atenolol, a hydrophilic ß-blocking agent which is said not to penetrate the central nervous system, blood-brain barrier clearance is yet a factor 7-20 lower.

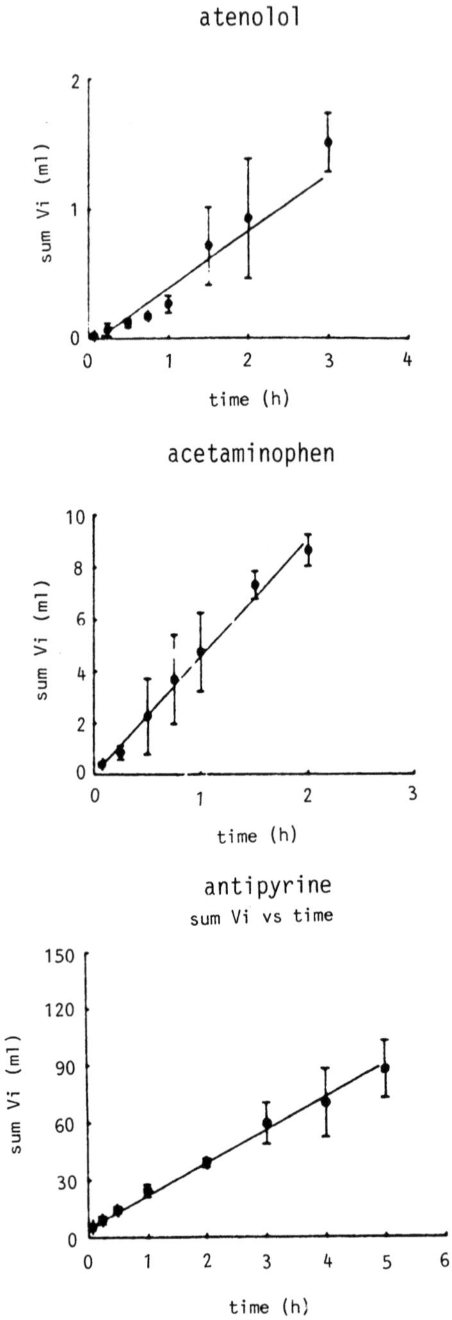

Fig. 5. Mean in vivo transport curves (n=6) for atenolol, acetaminophen and antipyrine as assessed by means of the unit impulse response methodology (from ref. 58).

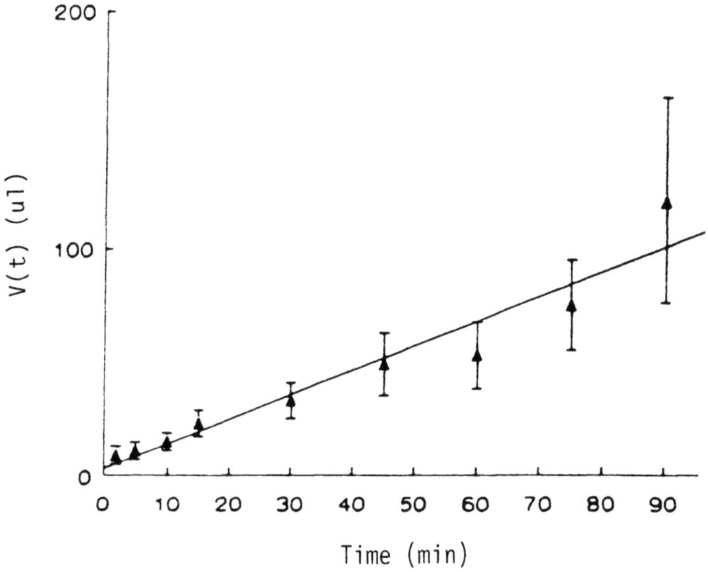

Fig. 6. Mean in vivo transport curve (n=6) of AVP1-8 as assessed by means of the unit impulse response methodology.

This finding demonstrates that, as predicted by the in vitro model, AVP1-8 is able to cross the blood-brain barrier, but as seen from the in vivo data, transport only occurs to a very minor extent.

CONCLUSIONS

The availability of in vitro models for the study of drug transport across the blood-brain barrier offers a valuable experimental tool, allowing more detailed research of events and factors involved in the transport processes. It should be noted however, that no in vitro model will ever be able to exactly mimic the in vivo properties of the cellular barrier. Therefore, it is important that an in vitro method is used in combination with an in vivo approach. In such a fashion, the in vitro model can offer a detailed insight into the events at the cellular level, while the in vivo experiments monitor the overall transport, in the integrated intact physiological system forming the blood-brain barrier.

A combination of in vitro and in vivo methods represents the optimal strategy in studying drug transport across the blood-brain barrier. The specific advantages and features of each method can be combined and exploited to find answers to complex questions like those related to peptides and protein delivery into the central nervous system.

REFERENCES

1. Goldstein, A., Aronow, L. and Kalman, S.M.: (1974) Principles of drug action. pp 193-198, Wiley and Sons, New York.

2. Danhof, M. and Levy, G.: (1984) Kinetics of drug action in disease states. Int. J. Pharmacol. Exp. Ther. 229:44-50.

3. Crone, C.: (1963) The permeability of capillaries in various organs as determined by use of the indicator dilution technique. Acta Physiol. Scand. 58:292-305.

4. Oldendorf, W.H.: (1970) Measurement of brain uptake of radiolabeled substances using a tritiated water internal standard. Brain Res. 24:372-376.

5. Rapoport, S.I., Ohno, K. and Pettigrew, K.D.: (1979) Drug entry into the brain. Brain Res. 172:354-359.

6. Hertz, M.M. and Paulsen, O.B.: (1980) Heterogeneity of cerebral capillary flow in man and its consequences for estimation of blood-brain barrier permeability. J. Clin. Invest. 65:1145-1151.

7. Siakotos, A.N., Rouser, G. and Fleischer, S.: (1969). Isolation of highly purified human and bovine brain endothelial cells and nuclei and their phospholipid composition. Lipids 4:234-239.

8. Brendel, K., Meezan, E. and Carlson, E.C.: (1974) Isolated brain microvessels: a purified metabolically active preparation from bovine cerebral cortex. Science 185:953-955.

9. Djuricic, B.M. and Mrsulja, B.B.: (1977) Enzymatic activity of the brain: microvessels vs total forebrain homogenate. Brain Res. 138:561-564.

10. Orlowski, M., Sessa, G. and Green, J.P.: (1974) Gamma-glutamyl transpeptidase in brain capillaries: possible site of a blood-brain barrier for amino acids. Science 184:66-68.

11. Caldwell, P.R.B., Segal, B.C., Hsu, K.C., Das, M. and Soffer, K.L.: (1978) Angiotensin converting enzyme: accumulation in medium from cultured endothelial cells. Biochem. Biophys. Res. Commun. 82:1147-1153.

12. Jaffe, E.A., Hoyer, L.W. and Nachman, R.L.: (1973) Synthesis of antihemophilic factor antigen by cultured human endothelial cells. J. Clin. Invest. 52:2757-2764.

13. Voyta, J.C., Via, D.P., Butterfield, C.E.; Zetter, B.R.: (1984) Identification and isolation of endothelial cells based on their increased uptake of acetylated-low density lipoprotein. J. Cell Biol. 99:286-297.

14. Goldstein, G.W., Wolinski, J.S., Csejtey, J. and Diamond, I.: (1975) Isolation of metabolically active capillaries from rat brain. J. Neurochem. 25:715-717.

15. Panula, P., Joo, F. and Rechardt, L.: (1978) Evidence for the presence of viable endothelial cells in cultures derived from dissociated rat brain. Experientia 34:95-97.

16. Hjelle, J.T., Baird-Lambert, J., Cardinale, G., Spector, S. and Udenfried, S.: (1978) Isolated microvessels the blood-brain barrier in vitro. Proc. Natl. Acad. Sci. U.S.A. 75:4544-4548.

17. DeBault, L.E., Kahn, L.E., Frommes, S.P. and Cancilla, P.A.: (1979) Cerebral microvessels and derived cells in tissue culture: isolation and preliminary characterisation. In Vitro 15:473-487.

18. Williams, S.K., Gillis, J.F., Matthews, M.A., Wagner, R.C. and Bitenski, M.W.: (1980) Isolation and characterization of brain endothelial cells: morphology and enzyme activity. J. Neurochem. 35:374-381.

19. Phillips, P., Kumar, P., Kumar, S., and Waghe, M.: (1979) Isolation and

characterization of endothelial cells from rat and cow brain white matter. J. Anat. 129:261-272.

20. Spatz, M., Bembry, J., Dodson, R.F., Hervonen, H. and Murray, M.R.: (1980) Endothelial cell cultures derived from isolated cerebral microvessels. Brain Res. 191:577-582 (1980).

21. Bowman, P.D., Betz, A.L., Ar, D., Wolinsky, J.S., Penney, J.B., Shivers, R.R. and Goldstein, G.W.: (1981) Primary culture of capillary endothelium from rat brain. In Vitro 17:353-362.

22. Bowman, P.D., Betz, A.L. and Goldstein, G.W. (1982) Primary culture of microvascular endothelial cells from bovine retina: selective growth using fibronectin coated substrate and plasma derived serum. In Vitro 18:626-632.

23. Bowman, P.D., Ennis, S.R., Rarey, K.E., Betz, A.L. and Goldstein, G.W.: (1983) Brain microvessel endothelial cells in tissue culture: a model for study of blood-brain barrier permeability. Ann. Neurol. 14:396-402.

24. Betz, A.L., Csejtey, J. and Goldstein, G.W.: (1979) Hexose transport and phosphorylation by capillaries isolated from rat brain. Am. J. Physiol. 236:c96-c102.

25. Cangiano, C., Cardelli, P., James, J.H., Rossi, F., Patriziz, M.A., Brackett, K.A., Strom, R. and Fischer, J.E.: (1983) Brain microvessels take up large neutral amino acids in exchange for glutamine. J. Biol. Chem. 258:8949-8954.

26. Goldstein, G.W.: (1979) Relation of potassium transport to oxidative metabolism in isolated brain capillaries. J. Physiol. (Lond.) 286:185-195.

27. Betz, A.L. and Goldstein, G.W.: (1981) Development changes in metabolism and transport properties isolated from rat brain. J. Physiol. (Lond.) 312:365-376.

28. Choi, T.B. and Pardridge, W.M.: (1986) Phenylalanine transport at the human blood-brain barrier: studies with isolated human brain capillaries. J. Biol. Chem. 261:6536-6541.

29. Betz, A.L. and Goldstein, G.W.: (1978) Polarity of the blood-brain barrier: neutral amino acid transport into isolated brain capillaries. Science 202:225-227.

30. Betz, A.L., Firth, J.A. and Goldstein, G.W.: (1980) Polarity of the blood-brain barrier: distribution of enzymes between the luminal and the abluminal membranes of brain capillary endothelial cells. Brain Res. 192:17-28.

31. Betz, A.L.: (1983) Sodium transport in capillaries isolated from rat brain. J. Neurochem. 41:1150-1157.

32. Pardridge, W.M. and Mietus, L.J.: (1981) Enkephalin and the blood-brain barrier: studies of binding and degradation in isolated brain capillaries. Endocrinology 109:1138-1142.

33. Ghersi-Egea, J.F., Minn, A. and Siest, G.: (1988) A new aspect of the protective function of the blood-brain barrier: activities of four drug-metabolizing enzymes in isolated rat brain. Life Sci. 42:2515-2523.

34. Chabrier, P.E., Roubert, P., Plas, P. and Braquet, P.: (1988) Blood-brain barrier and atrial natriuretic factor. Can. J. Physiol. Pharmacol. 66:276-279.

35. White, F.P., Dutton, G.R. and Norenberg, M.D.: (1981) Microvessels isolated from rat brain: localization of astrocyte processes by immunohistochemical techniques. J. Neurochem. 36:328-332.

36. Goldstein, G.W., Wolinski, J.S. and Csejtey, J.: (1975) Isolation of metabolically active capillaries from rat brain. J. Neurochem. 25:715-717.

37. Diglio, C.A., Grammas, P., Giacomelli, F. and Wiener, J.: (1982) Primary culture of rat cerebral microvascular endothelial cells: isolation, growth and characterization. Lab. Invest. 46:554-563.

38. Folkman, J., Haudenschild, C.C. and Zetter, B.R.: (1979) Long-term culture of capillary endothelial cells. Proc. Natl. Acad. Sci. U.S.A. 76:5217-5221.

39. Audus, K.L. and Borchardt, R.T.: (1986) Characterization of an in vitro model for studying drug transport and metabolism. Pharm. Res. 3:81-87.

40. Rim, S., Audus, K.L. and Borchardt, R.T.: (1986) Relationship of octanol/water and octanol/buffer partition coefficients to transcellular diffusion across brain microvessel endothelial cells. Int. J. Pharm. 32:79-84.

41. Baranczyk, A., Audus, K.L. and Borchardt, R.T.: (1985) Catecholamine metabolizing bovine brain microvessel endothelial cell monolayers. J. Neurochem. 46:1956-1969.

42. Audus, K.L. and Borchardt, R.T.: (1986) Characterization of the large neutral amino acid transport system of bovine brain microvessel endothelial cell monolayers. J. Neurochem. 47:484-488.

43. Van Bree, J.B.M.M., Audus, K.L. and Borchardt, R.T.: (1988) Carrier-mediated transport of baclofen across monolayers of bovine brain endothelial cells in primary culture. Pharm. Res. 5:369-371.

44. Van Bree, J.B.M.M., DeBoer, A.G., Danhof, M., Ginsel, L.A. and Breimer, D.D.: (1988) Characterization of an in vitro blood-brain barrier: effects of molecular size and lipophilicity on cerebrovascular endothelial transport rates of drugs. J. Pharmacol. Exp. Ther. 247:1233-1239.

45. Van Bree, J.B.M.M., DeBoer, A.G., Verhoef, J.C., Danhof, M. and Breimer, D.D.: (1989) Transport of vasopressin fragments across the blood-brain barrier: in vitro studies using monolayer cultures of bovine brain endothelial cells. J. Pharmacol. Exp. Ther. 249:901-905.

46. Bradbury, M.W.B., Patlak, C.S. and Oldendorf, W.H.: (1975) Analysis of brain uptake and loss of radiotracers after intracarotid injection. Am. J. Physiol. 229:1110-1115.

47. Fenstermacher, J.D., Blasberg, R.G. and Patlak, C.S.: (1981) Methods for quantifying the transport of drugs across the blood-brain systems. Pharmac. Ther. 14:217-248.

48. Oldendorf, W.H.: (1974) Blood-brain barrier permeability to drugs. Ann. Rev. Pharmacol. 14:239-248.

49. Yudelevich, D.L. and Derose, N.: (1971) Blood-brain transfer of glucose and other molecules measured by rapid indicator dilution. Am. J. Physiol. 220:841-846.

50. Pardridge, W.M., Landaw, E.M., Miller, L.P., Braun, L.D. and Oldendorf, W.H.: (1983) Carotid injection technique: bounds for bolus mixing by plasma and brain. J. Cerebral Blood Flow Metabol. 5:576-583.

51. Kastin, A.J., Zadina, J.E., Banks, W.A. and Graf, M.V.: (1984) Misleading concepts in the field of brain peptides. Peptides 5:249-253.

52. Fawcett Wilson, J.: (1988) Low permeability of the blood-brain barrier to nanomolar concentrations of immunoreactive alpha-melanotropin. Psychopharmacol. 96:262-266.

53. Zlokovic, B., Lipovac, M.N., Begley, D.J., Davson, H. and Rakic, L.: (1988) Slow penetration of thyrotropin-releasing hormone across the blood-brain barrier of an in situ perfused guinea pig brain. J. Neurochem. 51:252-257.

54. Banks, W.A. and Kastin, A.J.: (1985) Permeability of the blood-brain barrier to neuropeptides: the case for penetration. Psychoneuroendocrinology 10:385-399.

55. Brodie, B.B., Kurz, H. and Shanker, L.S.: (1960) The importance of dissociation constant and lipid solubility in influencing the passage of drugs into CSF. J. Pharmacol. Exp. Ther. 130:519-528.

56. Ohno, K., Pettigrew, K.D. and Rapoport, S.I.: (1978) Lower limits of cerebrovascular permeability to non-electrolytes in the conscious rat. Am. J. Physiol. 235:H299-H307.

57. Preston, E. and Haas, N.: (1986) Defining the lower limits for blood-brain barrier permeability: factors affecting the magnitude and interpretation of permeability-area products. J. Neurosci. Res. 16: 709-719.

58. Van Bree, J.B.M.M., Baljet, A.V., Van Geyt, A., DeBoer, A.G., Danhof, M. and Breimer, D.D.: (1989) The unit impulse response procedure for the pharmacokinetic evaluation of drug entry into the central nervous system. J. Pharmacokin. Biopharm. in press.

METABOLISM OF CATECHOLAMINE PRODRUGS BY CULTURED BOVINE BRAIN MICROVESSEL ENDOTHELIAL CELLS

Gerhard K. E. Scriba[1] and Ronald T. Borchardt[2]

[1] School of Pharmacy, University of Münster, 4400 Münster, F.R.G. and [2] School of Pharmacy, University of Kansas, Lawrence, KS 66045, U.S.A.

INTRODUCTION

The endothelium of the brain capillaries, the anatomical site of the blood-brain barrier, consists of a single layer of cells. In contrast to the peripheral microvasculature, it is characterized by tight intracellular junctions, the absence of fenestrations and few pinocytotic vesicles (Brightman, 1977). Moreover, high enzymatic activity including metabolizing enzymes such as monoamine oxidase (MAO; EC 1.4.3.4) (Lai *et al.*, 1975; Hardebo *et al.*, 1980), cholinesterases (Joo and Csillik, 1966; Duricic and Mrsulja, 1977), aminopeptidases (Brecher *et al.*, 1978) and cytochrome P-450 (Ghersi-Egea *et al.*, 1988) have been reported in the microvessel wall. Both, anatomical and enzymatic properties of the endothelium, are responsible for the restricted movement of solutes from the blood into the brain.

Strict regulation of the movement of catecholamines and catecholamine precursors across the blood-brain barrier is required for normal brain function. High activity of aromatic amino acid decarboxylase (AADC; EC 4.1.1.28), MAO and catechol-O-methyltransferase (COMT; EC 2.1.1.6) are found in brain capillaries (Lai *et al.*, 1975; Hardebo *et al.*, 1980) and form an efficient enzymatic barrier for monoamines and their precursors (Fig. 1). As a result only negligible amounts of biogenic amines penetrate the blood-brain barrier (Oldendorf, 1971; Cornford *et al.*, 1982).

However, lipophilic derivatives of the biogenic amines norepinephrine and dopamine have been shown to penetrate the blood-brain barrier leading to significant cerebral tissue levels and subsequent pharmacological effects (Creveling *et al.*, 1969; Borgman *et al.*, 1973). The recent introduction of the positive inotropic drug ibopamine (IBO), the diisobutyryl ester of the catecholamine epinine (EPN) (N-methyldopamine) (De Angelis, 1985), led us to investigate the potential that catecholamine diesters can deliver biogenic amines to the brain.

Pharmaceutical Applications of Cell and Tissue Culture to Drug Transport
Edited by G. Wilson *et al.*, Plenum Press, New York, 1991

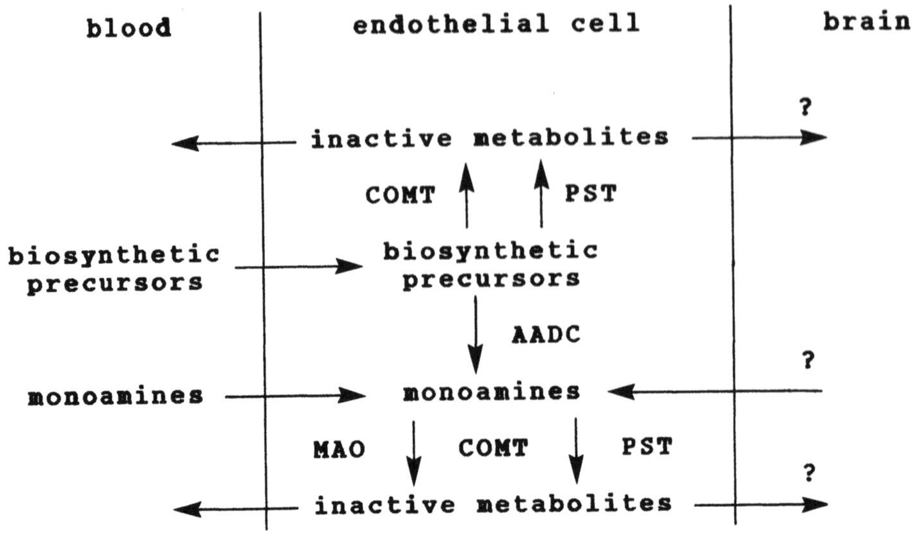

Fig. 1. Enzymatic barrier of brain microvessels to monoamines and monoamine precursors. AADC, aromatic amino acid decarboxylase; COMT, catechol-O-methyltransferase; MAO, monoamine oxidase; PST, phenol sulfotransferase.

Primary cultures of endothelial microvessels have been developed as an *in vitro* model for the blood-brain barrier (Bowman *et al.*, 1983, Audus and Borchardt, 1986a). In our laboratories, a culture system derived from bovine brain microvessels has been shown to retain many anatomical, enzymatic and permeability characteristics of the mammalian blood-brain barrier (Audus and Borchardt, 1986a, b, 1987; Rim *et al.*, 1986; Baranczyk-Kuzma *et al.*, 1986), and, thus, represents a good working model to study the various aspects of drug penetration and/or metabolism at the blood-brain interface. In contrast to other *in vitro* models such as isolated cerebral microvessels, the present cell culture system can be used to study transcellular passage as well as cellular uptake and metabolism.

METABOLISM OF CATECHOLAMINE ESTER PRODRUGS BY CULTURED BOVINE BRAIN MICROVESSEL ENDOTHELIAL CELLS

The mammalian blood-brain barrier displays high activity of neurotransmitter-metabolizing enzymes. With regard to the degradation of catecholamines, COMT and MAO are present (Lai *et al.*, 1975; Hardebo *et al.*, 1980). Moreover, butrylcholinesterase (EC 3.1.1.8) and acetylcholinesterase (EC 3.1.1.7) have been found (Joo and Csillik, 1966; Duricic and Mrsulja, 1977). The activities of these four enzymes have been characterized in the bovine derived cell culture model of the blood-brain barrier (Fig. 2) (Baranczyk-Kuzma *et al.*, 1986; Trammel and Borchardt, 1987). In addition, phenol sulfotransferase (PST; EC 2.8.2.1) has been found in cultured microvessel endothelium (Baranczyk-Kuzma *et al.*, 1986, 1989). PST mediates the sulfate conjugation of catecholamines and other phenolic compounds. The enzyme is present in the brain but its cellular localization has not yet been determined (Young *et al.*, 1984). In contrast to the *in vivo* situation, AADC activity could not be demonstrated in the cell cultures (Chastain and Borchardt, 1989).

Most enzymatic activites peak at day 10 in culture when the monolayers reach confluency (Fig. 2). In contrast to whole brain homogenate, MAO-A is predominant in the cell cultures. The blood-brain barrier activity and the ratio MAO-A / MAO-B is known to be species-dependent (Kalaria and Harik, 1987). In contrast to the bovine microvessel cell cultures, in most species including humans the activity of MAO-B was found to be equal or higher than the activity of MAO-A. The COMT level of cultured endothelium is about 5 fold lower than that of brain homogenate and about 100 fold lower than MAO. A similar ratio has been described for freshly isolated rat brain microvessels (Hardebo *et al.*, 1980). Between the esterases characterized butrylcholinesterase activity exceeds acetylcholinesterase activity (Fig. 2). This also corresponds to the *in vivo* situation (Duricic and Mrsulja, 1977). Due to its enrichment in brain capillaries butrylcholinesterase is considered a marker enzyme for this tissue (Landers et al., 1962).

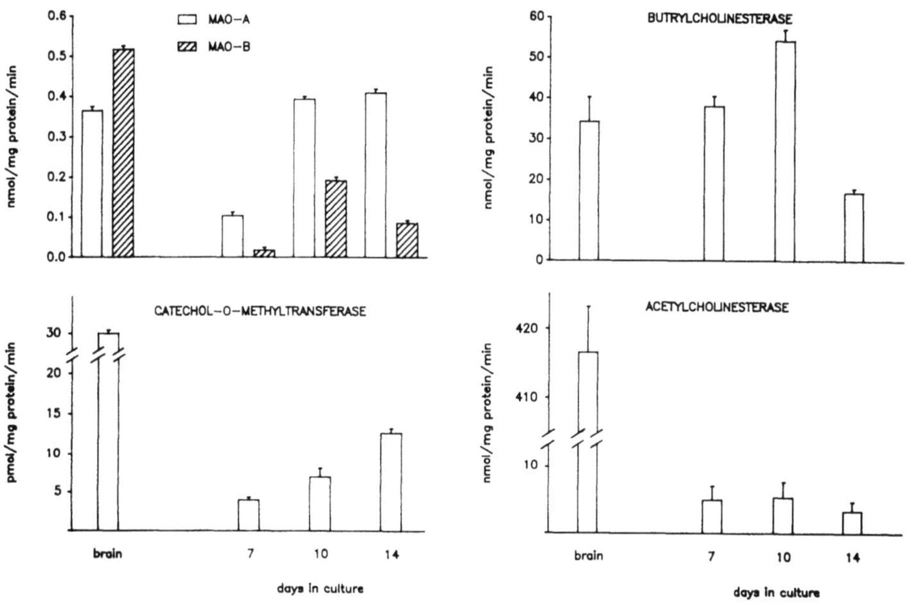

Fig. 2. Activities of MAO, COMT, acetylcholinesterase and butrylcholinesterase in cultured bovine brain microvessel endothelial cell monolayers (Baranczyk-Kuzma *et al.*, 1986; Trammel and Borchardt, 1987).

EPN is the active metabolite of the positive inotropic drug IBO. Moreover, EPN has been previously shown to alter the catecholamine content in hypothalamus and brainstem *in vitro* (Scriba *et al.*, 1988). Therefore, EPN was used as the parent compound in the present study. The structures of the ester prodrugs of EPN, including IBO, are summarized in Table 1.

Table 1. Structures, partition coefficients and permeability coefficients through bovine brain microvessel endothelial cell monolayers of EPN and EPN diester prodrugs.

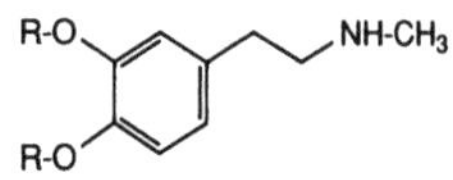

Compound	R	Log Partition Coefficients	Permeability Coefficients[a] X 10^4 cm/min
epinine (EPN)	H	- 1.16	37.8
3,4-O-diacetylepinine (DA-EPN)	CH_3—CO	- 0.48	82.8
3,4-O-diisobutyrylepinine (ibopamine, IBO)	$(CH_3)_2$CH—CO	0.43	93.6
3,4-O-dipivaloylepinine (DP-EPN)	CH_3-C(CH_3)(CH_3)—CO	1.16	138.6
3,4-O-dibenzoylepinine (DB-EPN)	C_6H_5—CO	1.63	129.0

[a] Permeability coefficients through bovine brain microvessel endothelial cell monolayers grown on polycarbonate membranes, 12 μm pore size (Scriba and Borchardt, 1989).

When 10 to 12 day old monolayers of bovine brain microvessel endothelial cells (Audus and Borchardt, 1986a) were incubated with EPN or EPN esters, 3,4-dihydroyphenylacetic acid (DOPAC), the product of MAO degradation, was the only metabolite detectable by HPLC with electrochemical detection (Scriba and Borchardt, 1989). EPN is known to be a substrate for MAO *in vivo* (Wiseman-Distler *et al.*, 1965). Degradation by MAO is also the major route of IBO metabolism after oral administration in rats (Pocchiari *et al.*, 1986). The lack of the formation of O-methylated derivatives by the cultured endothelium may be due to the low activity of COMT in the cell cultures (Fig. 2) (Baranczyk-Kuzma *et al.*, 1986) and/or a low affinity and/or V_{max} for EPN as a substrate for the enzyme. O-Sulfated metabolites were not observed either, although sulfate conjugation of catecholamines in the brain appears to be a major metabolic pathway *in vivo* (Young *et al.*, 1984). There are at least two forms of PST. The thermolabile type of PST sulfates micromolar concentrations of catecholamines such as dopamine and millimolar concentrations of phenol and p-nitrophenol. The thermostable type of PST catalyzes the sulfation of micromolar concentrations of phenol and p-nitrophenol and millimolar concentrations of catecholamines (Young et al., 1984). However, the thermostable form of PST that is present in the endothelial cell cultures did not sulfate catecholamines, such as dopamine, epinephrine and norepinephrine, nor the oxidative metabolite DOPAC (Baranczyk-Kuzma *et al.*, 1986, 1989).

The time course of the formation of DOPAC from EPN and IBO upon incubation of monolayers is illustrated in Fig. 3. Generally, lower levels of DOPAC were detected in the assay medium and within endothelial cells of IBO-treated cells than of EPN-treated cells. In contrast, intracellular IBO levels were about 30 - 40 times higher than EPN levles in EPN-treated cells. Compared to the slow increase observed for intracellular metabolite levels, concentrations of DOPAC in the medium rose rapidly after a lag time of about 20 minutes. As there is no evidence for an active transport mechanism for metabolites such as DOPAC, the levels in the incubation medium can be regarded as the result of passive diffusion out of the endothelial cells.

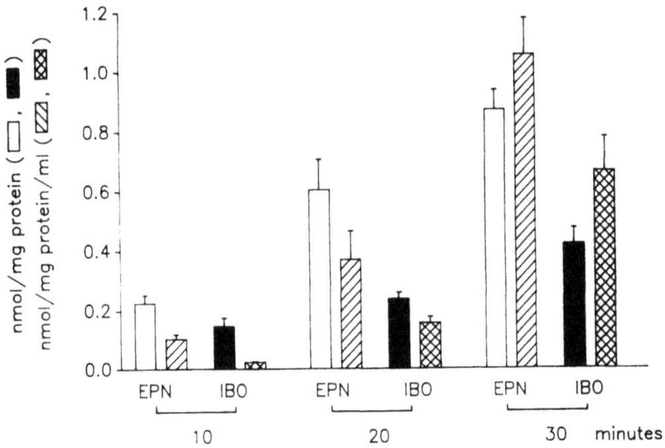

Fig. 3. Time-dependent formation of DOPAC from 100 µM EPN and 100 µM IBO by cultured bovine cerebral microvessel cell monolayers. (□, ■), DOPAC levels of the endothelial cells; (▨, ▩), DOPAC content of the assay buffer. Each value is the mean ± SEM (n = 4 - 6).

Increasing the lipophilicity of EPN by esterification of the phenolic hydroxyl groups greatly increased the concentrations of the esters in the endothelial cell monolayers compared to EPN (Fig. 4). This can be attributed to the faciliation of diffusion through the cell membranes caused by the increased lipophilicity of the prodrugs. Moreover, the high intracellular levels may reflect the partitioning of lipophilic compounds from a hydrophilic compartment (the buffer) into a lipophilic compartment (the cells). DA-EPN led to extremely high intracellular levels. This may be due to the rapid diffusion of the small molecule. Moreover, the diacetyl ester is quickly hydrolyzed. Free EPN might not be able to easily diffuse back into the buffer because of its polarity. This increases the intracellular levels relative to the other esters. Enzymatic hydrolysis of the prodrugs decreased in the order DA-EPN > IBO > DB-EPN ≥ DP-EPN.

The metabolism of EPN and EPN diesters upon incubation with endothelial cell monolayers for 30 minutes is shown in Fig. 5. With the exception of DA-EPN less DOPAC was formed from the ester prodrugs compared to the parent catecholamine. Rapid hydrolysis of the diacetyl ester combined with elevated drug tissue levels of DA-EPN compared to the other EPN esters led to extremely high levels of DOPAC in DA-EPN-treated cells and in the incubation buffer. Although incubation of endothelial cells with DP-EPN and DB-EPN produced prodrug levels similar to those observed in IBO-treated cells (Fig. 4) only negligible metabolism of the dibenzoyl and dipivaloyl esters to DOPAC was found (Fig. 5).

High activity of butrylcholinesterase and low activity of acetylcholinesterase have been found in rat brain microvessels (Joo and Csillik, 1966; Duricic and Mrsulja, 1977). Both enzymes are also present in the bovine endothelial cell cultures (Fig. 2) (Trammel and Borchardt, 1987). They are effectively inhibited by organophosphates. Therefore, it seems likely that these two enzymes are primarily responsible for the hydrolysis of the EPN esters. The slow hydrolysis of the dipivaloyl and the dibenzoyl esters is consistent with these esters being poor substrates for butrylcholinesterase and actetylcholinesterase because of the bulky acyl groups.

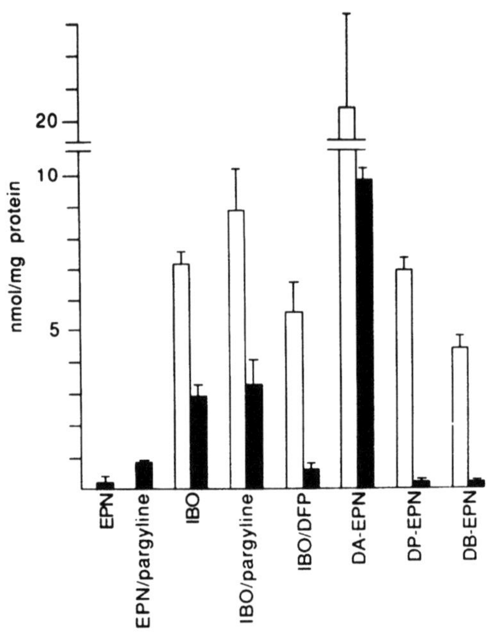

Fig. 4. EPN and EPN diester tissue levels of bovine cerebral microvessel endothelial cell monolayers. The open bars represent the sum of ester and free EPN formed by hydrolysis of the ester, the filled bars EPN or EPN hydrolyzed from the esters, respectively. DFP, diisopropyl fluorophosphate. Each value is the mean ± SEM (n = 4 - 6).

Inhibition of serine-dependent esterases with diisopropyl fluorophosphate attenuated DOPAC levels (Fig. 5). It is not known if esters of catecholamines are substrates or even inhibitors of MAO. The present results suggest, however, that hydrolysis of the ester groups appears to be necessary in order for the molecules to be metabolized by MAO in the brain endothelium.

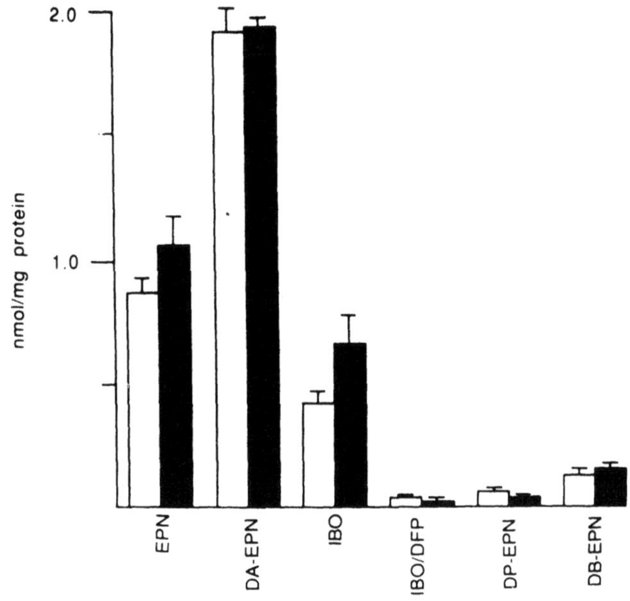

Fig. 5. Metabolism of EPN and EPN diesters by bovine cerebral microvessel endothelial cell monolayers. Intracellular DOPAC levels are represented by the open bars, buffer concentrations by the filled bars. DFP, diisopropyl fluorophosphate. Each value is the mean ± SEM (n = 4 - 6).

Preincubation with pargyline, an inhibitor of MAO, completely suppressed the formation of DOPAC from EPN and the EPN esters confirming the MAO-mediated degradation of the compounds. While no significant changes were observed for the ester prodrugs, EPN content of the cells tripled (Fig. 5) emphasizing the importance of MAO at the blood-brain interface. Pargyline did not effect the hydrolysis of the esters (Fig. 5, IBO/pargyline). The MAO-mediated metabolism of EPN was more sensitive towards inhibition by the MAO-A inhibitor clorgyline than the MAO-B inhibitor deprenyl (Table 2). EPN is known to be a substrate for MAO *in vivo* (Wiseman-Distler et al., 1965) but, to our knowledge, has not been investigated as a substrate for the different types of MAO. However, related monoamines such as dopamine, phenylethylamine (at high concentrations) and N-methylphenylethylamine are substrates for both forms of the enzyme (Suzuki *et al.*, 1980 and 1982; Fowler and Tipton, 1983). As N-methylation does not change the substrate specificity of

MAO (Suzuki *et al.*, 1982) it can be predicted that EPN would be a substrate for both MAO-A and MAO-B. The results further confirm earlier observations that the bovine derived model of the blood-brain barrier exhibits higher activity of MAO-A than MAO-B (Fig. 2) (Baranczyk-Kuzma *et al.*, 1986).

DIFFUSION OF CATECHOLAMINE ESTER PRODRUGS THROUGH CULTURED BOVINE BRAIN MICROVESSEL ENDOTHELIAL CELLS

While there are specific transport systems for water-soluble nutrients such as hexoses, amino acids or nucleosides (Pardridge and Oldendorf, 1977) most solutes enter the brain by passively diffusing through the blood-brain barrier. The rate of passage of water-soluble compounds from the blood into the brain is a function of the lipid solubility of the molecules (Oldendorf, 1974). Using a variety of compounds with

Table 2. IC_{50} values for the inhibition of MAO in cultured bovine brain microvessel endothelial cell monolayers by clorgyline and deprenyl

inhibitor	specificity MAO-A	MAO-B	IC_{50}[a]
deprenyl	-	+	$1.9 \cdot 10^{-7}$M
clorgyline	+	-	$9.1 \cdot 10^{-10}$M

[a] IC_{50} values were derived from the nonlinear inhibition curves of the MAO inhibitors using 100 μM EPN as substrate.

different structures, it has been demonstrated that the brain uptake of solutes *in vivo* can be related directly to the octanol/water or the octanol/buffer partition coefficient and inversely to the molecular weight (Cornford *et al.*, 1982; Levin, 1980). A similar correlation was also established for the bovine derived cell culture model of the blood-brain barrier. Thus, a linear relationship between the permeability coefficient and the log octanol/buffer partition coefficient/(molecular weight)$^{1/2}$ has been described (Rim *et al.*, 1986, Shah *et al.*, 1988). A good correlation between these parameters was also found for this "homologous" series of catecholamine esters (Scriba and Borchardt, 1989). The octanol/buffer partition coefficients of the compounds ranged over two orders of magnitude from the most hydrophilic (EPN) to the most lipophilic (DB-EPN) (Table 1). The flux of the ester prodrugs through cerebral microvessel endothelial cell monolayers grown on porous supports, expressed as permeability coefficients, paralleled the increase of the lipophilicity of the compounds (Table 1). In this study, a high concentration of the compounds (100 μM) was used in order to saturate the enzymes so that the diffusion is only dependent on the lipophilicity of the prodrugs.

The low permeability of the cerebral microvessel monolayers to EPN is comparable to *in vivo* literature data for other catecholamines (Oldendorf, 1971 and 1974; Cornford *et al.*, 1982). The high tissue levels of DA-EPN (Fig. 4) did not correspond to a high permeability coefficient. As this compound does not have structural requirements for any of the known carrier systems, active transport is very unlikely. Therefore, the permeability can be attributed to the lipid solubility and the molecular weight of the ester. Furthermore, this might reflect a difference of diffusion through and partitioning into cell layers.

CONCLUSIONS

Primary cultures derived from bovine cerebral microvessels retain enzymatic and permeability characteristics of the mammalian blood-brain barrier *in vivo*. The movement of catecholamines across the endothelial cells is restricted due to the low permeability of the cell membranes to these compounds and the high activity of MAO within the endothelium. Ester prodrugs of catecholamines penetrate the cell monolayers according to their lipophilicity. Esterase hydrolysis appears to be necessary for further metabolism by MAO. As a function of the nature of the acyl groups either hydrolysis or MAO-mediated oxidation appears to be the rate-limiting step of the metabolic degradation of the ester derivatives of EPN in the microvessels. The brain uptake of catecholamine ester prodrugs should, therefore, be dependent on their lipid solubility and hydrolytic stability towards plasma and endothelium esterases.

ACKNOWLEDGMENT

This work was supported by grant HL24093 from the U.S. Public Health Service and a grant from The Upjohn Co. (Kalamazoo, MI, U.S.A.).

REFERENCES

Audus, K. L., and Borchardt, R. T., 1986a, Characterization of an *in vitro* blood-brain barrier model system for studying transport and metabolism, Pharm. Res., 3:81-87.

Audus, K. L., and Borchardt, R. T., 1986b, Characteristics of the large neutral amino acid transport system of bovine brain microvessel endothelial cell monolayers, J. Neurochem., 47:484-488.

Audus, K. L., and Borchardt, R. T., 1987, Bovine brain microvessel endothelial cell monolayers as a model system for the blood-brain barrier, Ann. N. Y. Acad. Sci., 507:9-18.

Baranczyk-Kuzma, A., Audus, K. L., and Borchardt, R. T., 1986, Catecholamine-metabolizing enzymes of bovine brain microvessel endothelial cell monolayers, J. Neurochem., 46:1956-1960.

Baranczyk-Kuzma, A., Audus, K. L., and Borchardt, R. T., 1989, Substrate specificity of phenol sulfotransferase from primary cultures of bovine brain microvessel endothelium, Neurochem. Res., 7:689-691.

Borgman, R. J., McPhillips, J. J. Stitzel, R. E., and Goodman, I. J., 1973, Synthesis and pharmacology of centrally acting dopamine derivatives and analogs in relation to Parkinson's disease, J. Med. Chem., 16:630-633.
Bowman, P. D., Ennis, S. R., Rarey, K. E., Betz, A. L., and Goldstein, G. W., 1983, Brain microvessel endothelial cells in tissue culture: a model for the study of blood-brain barrier permeability, Ann. Neurol., 14:396-402.
Brecher, P., Tercyak, A., Gavras, H., and Chobanian, A. V., 1978, Peptidyl dipeptidase in rabbit brain microvessels, Biochem. Biophys. Acta, 526:537-546.
Brightman, M. W., 1977, Morphology of blood-brain surfaces, Exp. Eye Res., 25:1-25.
Chastain, J. E. Jr., and Borchardt, R. T., 1989, L-α-methyldopa transport across bovine brain microvessel endothelial cell monolayers: a model of the blood-brain barrier, Neurosci. Res. Commun., 4:147-152.
Cornford, E. M., Braun, L. D., Oldendorf, W. H., and Hill, M. A., 1982, Comparison of lipid-mediated blood-brain barrier penetrability in neonates and adults, Am. J. Physiol., 243:C161-C168.
Creveling, C. R., Daly, J. W., Tokuyama, T., and Witkop, B., 1969, Labile lipophilic derivatives of norepinephrine capable of crossing the blood-brain barrier, Experientia, 25:26-27.
De Angelis, L., 1985, Ibopamine hydrochloride, Med. Actual., 21:67-69.
Duricic, B. M., and Mrsulja, B. B., 1977, Enzymatic activity of the brain: microvessels vs. total brain homogenate, Brain Res., 138:561-564.
Fowler, C. J., and Tipton, K. F., 1983, On the substrate specificities of the two forms of monoamine oxidase, J. Pharm. Pharmacol., 36:111-115.
Ghersi-Egea, J.-F., Minn, A., and Siest, G, 1988, A new aspect of the protective functions of the blood-brain barrier: activities of four drug-metabolizing enzymes in isolated rat brain microvessels, Life Sci., 42:2515-2523.
Hardebo, J. E., Emson P. C., Falck, B., Owman, C., and Rosengren, E., 1980, Enzymes related to monoamine transmitter metabolism in brain microvessels, J. Neurochem., 35:1388-1393.
Joo, F., and Csillik, B., 1966, Topographic correlation between hematoencephalic barrier and the cholinesterase activity of brain capillaries, Exp. Brain Res., 1:147-151.
Kalaria, R. N., and Harik, S. I., 1987, Blood-brain barrier monoamine oxidase: enzyme characterization in cerebral microvessels and other tissues from six mammalian species including human, J. Neurochem., 49:856-864.
Lai, F.M., Udenfriend, S., and Spector, S., 1975, Presence of norepinephrine and related enzymes in isolated brain microvessels, Proc. Nat. Acad. Sci. USA, 72:4622-4625.
Landers, J. W., Chason, J. L., Gonzales, J. E., and Palutke, W., 1962, Morphology and enzymatic activity of rat cerebral capillaries, Lab. Invest., 11:1253-1259.
Levin, V. A., 1980, Relationship of octanol/water partition coefficient and molecular weight to rat brain capillary permeability, J. Med. Chem., 23:682-684.
Oldendorf, W. H., 1971, Brain uptake of radiolabeled amino acids, amines and hexoses after arterial injection. Am. J. Physiol., 221:1629-1639.

Oldendorf, W. H., 1974, Blood-brain barrier permeability to drugs, Ann. Rev. Pharmacol., 14:239-248.
Pardridge, W. M., and Oldendorf, W. H., 1977, Transport of metabolic substrates through the blood-brain barrier, J. Neurochem., 28:5-12.
Pocchiari, F., Pataccini, R., Castelnovo, P., Longo, A., and Casagrande, C., 1986, Ibopamine, an orally active dopamine-like drug: metabolism and pharmacokinetics in rats, Arzneim.-Forsch. (Drug Res.), 36:334-340.
Rim, S., Audus, K. L., and Borchardt, R. T., 1986, Relationship of octanol/buffer and octanol/water partition coefficients to transcellular diffusion across brain microvessel endothelial cell monolayers, Int. J. Pharm., 32:79-84.
Scriba, G. K. E., Hower, J. A., Liang, N. Y., Fennessey, P. V., and Borchardt, R. T., 1988, Effect of ibopamine and the active metabolite epinine on the catecholamine content of rat hypothalamus and brainstem *in vitro*, J. Pharmacol. Exp. Ther., 247:609-616.
Scriba, G. K. E., and Borchardt, R. T., 1989, Metabolism of catecholamine esters by cultured bovine brain microvessel endothelial cells, J. Neurochem., 53:610-615.
Shah, M. V., Audus, K. L., and Borchardt, R. T., 1989, The application of bovine brain microvessel endothelial cell monolayers grown onto polycarbonate membranes *in vitro* to estimate the potential permeability of solutes through the blood-brain barrier, Pharm. Res., 6:624-627.
Suzuki, O., Oya, M., and Katsumata, Y., 1980, Characterization of N-methylphenylethylamine and N-methylphenylethanolamine as substrates for type A and type B monoamine oxidase, Biochem. Pharmacol., 29:2663-2667.
Suzuki, O., Katsumata Y., and Oya, M., 1982, Substrate specificity of type A and type B monoamine oxidase, in: "Monoamine oxidase, basic and clinical frontiers", K. Kanijo, E. Usdin and T. Nagatsu, eds., pp. 74-86 Excerpta Medica, New York.
Trammel, A. M., and Borchardt, R. T., 1987, Choline transport in cultured brain microvessel endothelial cells, Pharm. Res., 4:541 (Abstr.#PD428).
Wiseman-Distler, M. H., Sourkes, T. L., and Carabin, S., 1965, Precursors of 3,4-dihydroxyphenylacetic acid and 4-hydroxy-3-methoxyphenylacetic acid in the rat, Clin. Chim. Acta, 12:335-339.
Young, W. F. Jr., Okazaki, H., Laws, E. R. Jr., and Weinshilboum, R. M., 1984, Human brain phenol sulfotransferase: biochemical properties and regional localization, J. Neurochem., 43:706-715.

MEMBRANE RECYCLING, ADSORPTIVE AND RECEPTOR-MEDIATED ENDOCYTOSIS BY PRIMARY BOVINE CEREBRAL MICROVESSEL ENDOTHELIAL CELL MONOLAYERS IN VITRO

Thomas J. Raub and Carolyn R. Newton[a]

Drug Delivery Systems Research, The Upjohn Company and Department of Biology[a], Kalamazoo College, Kalamazoo, Michigan, USA

INTRODUCTION

Within recent years, we have realized that the blood-brain barrier (BBB) is an interactive, highly selective interface between the brain interstitial space and blood-borne molecules. It is the unique functional and morphological characteristics of the endothelial cells which comprise the majority of the BBB surface area that define this specialized structure. However, increasing evidence shows that communication between the endothelia and surrounding cell types is responsible for expression of the BBB phenotype[1].

The restricted passage of small hydrophilic and large molecules through the BBB is attributed morphologically to the epithelial-like occluding junctions of the continuous endothelium which limit diffusion through the paracellular route and to a paucity of pinocytic vesicles which are thought to shuttle molecules transcellularly. With the use of isolated brain microvessels, a number of macromolecular transport systems have been studied[2]. However, because these capillary fragments lack an anatomical polarity, fail to exclude trypan blue[3], and are ATP deficient[4,5], detailed studies on transendothelial transport pathways are not possible. Evidence for such pathways has been obtained by radiotracer administration in vivo followed either by autoradiography or tissue fractionation[6-8].

Development of cell culture models of the BBB consisting of primary cultures of brain (cerebral) microvessel endothelial cell (BMEC) monolayers[9,10] has enabled us to examine transport processes with regard to molecular cell physiology. Such studies are aimed at identifying and understanding those characteristics that make the BBB uniquely specialized. We have focused our attention on endocytosis and especially on the kinetics and pathways of membrane recycling and sorting. Current efforts involve searching for plasma membrane proteins that are unique to the BMEC including identification of additional receptors. In this paper we have reviewed what we know about membrane internalization by BMEC in culture.

With the idea that the magnitude of fluid accumulation reflects the extent of membrane internalization[11], fluid-phase endocytosis was examined to quantitate membrane dynamics. We concluded from our morphologic and kinetic results that BMEC in culture exhibited pinocytic activity that is consistent with the qualitative assessments of pinocytosis by BMEC _in vivo_. In the classic experiment by Reese and Karnovsky[12], accumulation of the fluid-phase tracer, Horseradish peroxidase (HRP), by BMEC following vascular perfusion was minimal and showed that pinocytic activity in these cells was low relative to endothelia in other capillary beds. _In vitro_, cultured BMEC also accumulated only low quantities of HRP as defined by the small number of HRP-containing endocytic compartments (Fig. 1).

Recently, fluid-phase endocytosis was quantified by measuring accumulation of the fluorescent molecule Lucifer yellow CH (LY)[13] by BMEC monolayers grown on a matrix-coated plastic surface[14]. Unlike most other cell types, BMEC accumulation of LY reached a steady-state after 15 min following a rapid rate of accumulation with a half-life of 5-6 min. In contrast, fibroblasts, macrophages and arterial endothelia continued to accumulate fluid with first-order rate kinetics for hours[13,15-17]. Since endocytosis had not ceased, steady-state accumulation of LY most likely indicates the rapid filling of most available compartments. Indeed, when pulse-chase experiments were carried out to measure efflux of the accumulated LY after steady-state had been reached, almost all of the intracellular fluid was lost from the BMEC with a half-life of approximately 0.5 min[14]. This result is consistent with the rapid recycling of fluid through a compartment that remains close to or in continuity with the plasma membrane.

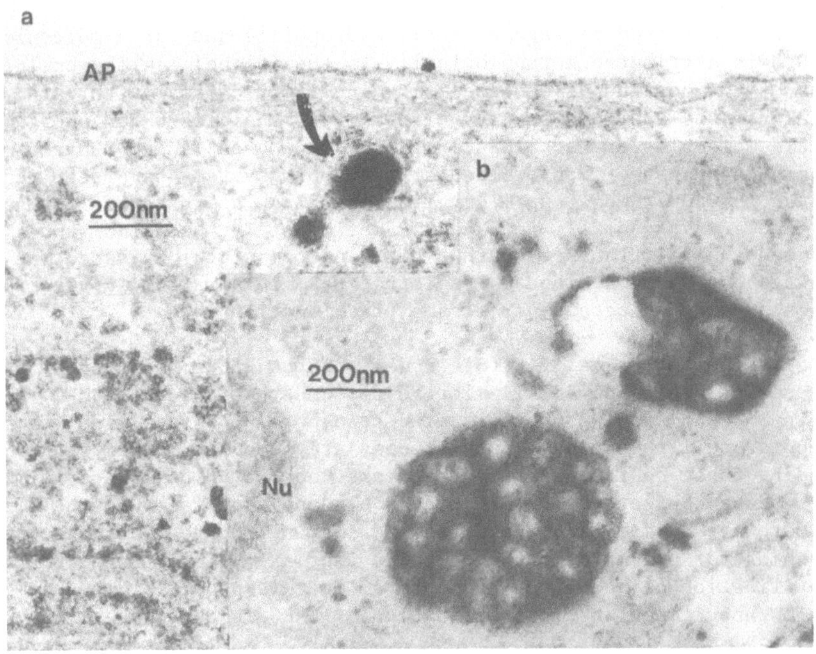

Figure 1. Distribution of HRP within cultured BMEC following accumulation by fluid-phase endocytosis at 37°C for (a) 10 min and (b) 60 min. Nu, nucleus.

Fluorescence microscopy following LY accumulation revealed numerous small cytoplasmic vesicles[14]. Unlike other cell types[13,16], BMEC in vitro apparently have limited involvement of a second, larger intracellular compartment as shown by kinetic analysis. This interpretation is consistent with morphological results showing that few large vesicles contain either fluid-phase marker, LY or HRP. Consequently, estimates of the turnover of the total endocytic compartment during LY accumulation by BMEC in vitro are 8- and 20-fold less than that calculated for fibroblasts and macrophages, respectively[14].

Although functional polarity is difficult to assess in plastic-grown monolayers, BMEC under these conditions appeared to be polarized based on their ultrastructure. The integrity of these cell monolayers was obvious following incubation of subconfluent BMEC monolayers with HRP which diffused into the intracellular space but not through the tight junction (Fig. 2). We recently found that BMEC when grown on matrix-coated semi-permeable filters exhibited an asymmetry with regard to fluid recycling. BMEC form confluent monolayers on Transwell™ polycarbonate membranes (0.4-μ pore size) that generate transendothelial electrical resistance (TEER) values of up to 160 ohm·cm^2 and that restrict the flux of sucrose (Raub, T., unpublished data). When these monolayers were allowed to accumulate [^{3}H]dextran (ave. MW of 70 kDa) from the apical or lumenal side, were rinsed and fresh medium replaced in both the apical and basolateral or ablumenal chambers, greater than 90% of the cell-associated dextran was returned to the apical compartment. This indicates that the majority of membrane recycling involved in accumulating blood-borne solutes is polarized[18].

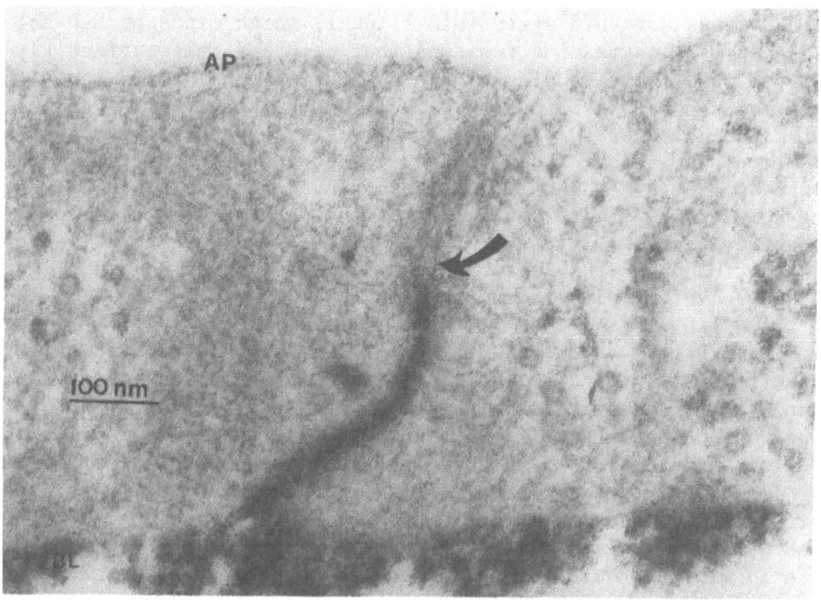

Figure 2. Functionally tight monolayers that are grown on a matrix-coated plastic surface are demonstrated by exclusion of HRP from the tight junction (arrow). HRP has gained access to the basolateral (BL) side of subconfluent BMEC monolayers. AP, apical side.

Like several epithelial cell lines in culture, a functional polarity appears to be expressed by BMEC *in vitro*. Such a phenotype is anticipated when one considers the function of the BMEC *in vivo* as a selective barrier. By maintaining polarity of the endocytic process, the two membrane domains will not be mixed and unnecessary transport of solutes will be prevented. Asymmetry with regard to pinocytic activity also has been indicated *in vivo* following ventricular injection of HRP where very little fluid is accumulated by BMEC from the ablumenal side[19]. We have not measured fluid accumulation by cultured BMEC from the basolateral side. However, the absence of HRP-filled vesicles at the basolateral surface following access by HRP suggests that fluid-phase endocytosis at this membrane domain is negligible (Fig. 2).

ADSORPTIVE ENDOCYTOSIS

We know from studies with other cell types that a large amount of plasma membrane is internalized during fluid-phase endocytosis[20]. In most cases, the fluid is either rapidly recycled or delivered to the lysosome whereas the membrane constituents are sorted from the fluid and either returned to the cell surface, delivered to the lysosome, or routed to other prelysosomal compartments within the endocytic pathway[21]. Since we found that BMEC in culture behave differently with respect to fluid-phase endocytosis[14], we were interested in examining the kinetics and pathway for membrane internalization. Although fluid accumulation by BMEC is small, membrane retrieval may not be so limited since it is likely that during endocytosis incoming membrane is rapidly and efficiently sorted from solutes that are recycled through a short-circuit pathway. In fact, such a mechanism is paramount for the maintainance of barrier integrity during transcytotic movement of specific ligand-receptor complexes.

To follow membrane traffic during endocytosis, we employed the plant lectin Ricinus communis agglutinin I (RCAI) which binds to ß-D-galactose and ß-D-N-acetyl-galactosamine residues that comprise cell surface glycoproteins and glycolipids. Our studies[22,23] and those of others[24-26] found that the plasma membrane of BMEC in a variety of animal species is rich in these sugar moieties. Therefore, RCAI can serve as an adsorptive tracer by binding to a mixed population of cell surface glycoconjugates which are then internalized as lectin-membrane complexes during endocytosis. The advantage of using a lectin is the reversibility of the interaction in the presence of the hapten sugar or, in this case, D-galactose. By quantitative removal of bound lectin, we were able to measure the kinetics of membrane internalization and recycling without compromising cell viability.

Binding sites on cultured BMEC for RCAI are distributed uniformly over the entire cell surface including membrane invaginations or caveolae (Fig. 3a) and the infrequent clathrin-coated pit. After allowing BMEC to bind HRP-conjugated RCAI at 4°C to inhibit endocytosis, the cells were warmed at 37°C to follow, by electron microscopy, the subsequent pathway of internalization. Immediately after warming the cells, the bound RCAI was localized within caveolae and tubules of 70-80 nm diameter (Fig. 3b). Membrane associated HRP-RCAI entered numerous endosomes of between 200-400 nm diameter within 10 min that were not acid phosphatase positive (Fig. 3c). HRP as a fluid-phase marker was found in vesicles, endosomes and, mostly, multivesicular bodies (Fig. 1b). It was not until 60 min that RCAI was found occasionally within acid phosphatase positive vesicles or lysosomes of >500 nm diameter. These results are consistent with our kinetic data obtained by measuring internalization of [^{125}I]RCAI. Degradation as defined by the appearance of acid-soluble radioactivity was not evident until 30 min and only

7% of the bound lectin was degraded after 4 h. We take this as evidence for a nondegradative pathway for most of the internalized RCAI binding sites.

Recently, Broadwell et al.[27] showed that BMEC in vivo internalized the lectin wheat germ agglutinin (WGA) into a number of different intracellular compartments including those associated with the trans-most Golgi cisternae. We too observed HRP-RCAI within small diameter tubules and vesicles within this region defined as the trans-Golgi reticulum[28,29], but did not see RCAI within the Golgi stack per se (Fig. 3d). In vivo, HRP-WGA also was found to be transcytosed after 6 h due to its appearance on the ablumenal side of the BBB[27]. In our cell culture model, HRP-RCAI did not appear on the basolateral side within 4 h after binding and internalization from the apical surface. The absence of RCAI is not due to the absence of the appropriate binding sites since these have been shown to be present on the basolateral membrane in vivo[25]. While this suggests that RCAI binding sites are not transcytosed, it shows that BMEC in culture apparently are polarized with respect to membrane traffic.

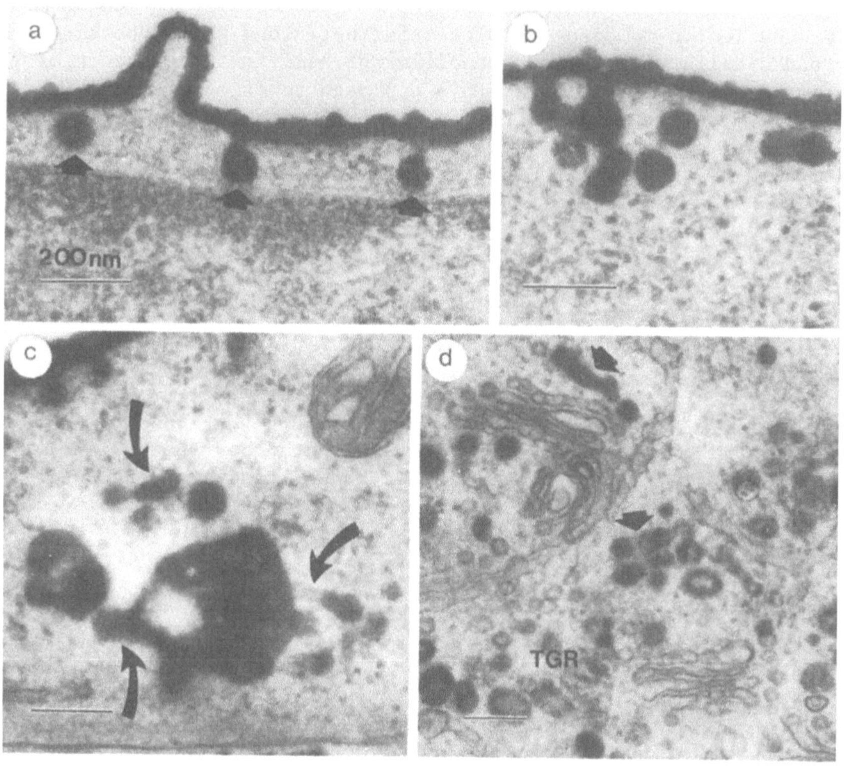

Figure 3. Endocytosis of HRP-RCAI that had bound to the apical surface of confluent BMEC. (a) After binding at 4°C, RCAI is distributed evenly over the cell surface including within caveolea (arrows). (b) After 5 min at 37°C, RCAI is present within vesicles that are near to and continuous with the cell surface. (c) By 10 min, RCAI appears within endosomes and numerous tubules (arrows). (d) RCAI becomes localized within vesicles and tubules that are located within the trans-Golgi reticulum (TGR).

Our kinetic results for RCAI recycling[22,23] are similar to those obtained using LY[14]. The majority of the RCAI binding sites are recycled rapidly and constitutively and efflux, like that of [^{3}H]dextran (see above and Table 2), is thought to be polarized. Approximately 30% of the RCAI that had bound after equilibrium at 4°C was internalized with a half-life of 6 min. Accumulation was temperature sensitive and energy dependent and was 95% inhibited in the presence of galactose. To show that recycling occurs, we measured efflux of accumulated RCAI into medium that either did or did not contain galactose. By design, the galactose removes the RCAI that has returned to the cell surface with >90% efficiency. In the absence of galactose, returned RCAI should be reinternalized since the measured "off" rate or dissociation constant for RCAI was longer than the endocytic rate. After 1 h, 60% of the intracellular RCAI was returned to the cell surface in the presence of galactose, 15% in the absence of galactose and 5% in the presence of galactose, but at 4°C. Therefore, at least 40% of the internalized membrane was recycled continuously.

We asked whether the recycling population of RCAI binding sites is in exchange with other intracellular compartments, especially since a number of different compartments were identified by electron microscopy. BMEC were allowed to accumulate RCAI for varying periods of time to "chase", if possible, the lectin-membrane complexes further along the endocytic pathway. The loaded cells were warmed and efflux of RCAI in the presence of galactose measured. In fibroblasts, this approach revealed a second nonrecycling or slowly recycling compartment[30]. With BMEC, a change in efflux of RCAI was not observed even after 2 h, suggesting that a second compartment is either absent, very small, or further removed temporally from the pathway.

Nonspecific adsorptive probes such as cationic bovine serum albumin (cBSA)[31,32], cationic immunoglobulin (cIgG)[8] and glycosylated BSA[32] have been used to implicate a transcytotic pathway in BMEC _in vitro_ and _in vivo_. As with RCAI, these macromolecules are internalized by the cell as a complex with plasma membrane components and, consequently, intracellular concentrations are increased significantly. Since a wide variety of different cell surface constituents are labeled by these tracers, the probability of observing transcytosis of a select population is increased. Alternatively, spill over into the transcytotic pathway might occur if the endocytic compartments are saturated. What is not known is how the endocytic/transcytotic pathway might be influenced by polycations or macromolecules, such as lectins or antibodies, that are capable of cross-linking.

RECEPTOR-MEDIATED ENDOCYTOSIS/TRANSCYTOSIS: TRANSFERRIN RECEPTOR

Only a few serum proteins have been found to bind to receptors on the BMEC surface (Table 1) and more work is required to identify other examples. Several reviews have described these various receptor systems at length with emphasis on insulin, the insulin-like growth factors and transferrin[2,6,33]. In the following section we will highlight the more recent receptor characterizations and focus on receptor-mediated endocytosis by BMEC through studies on the transferrin receptor (TfR) in cell culture.

The low density lipoprotein receptor (LDLR) is present on endothelia _in situ_, although it is most likely not responsible for transcytosis of LDL[34]. Only recently has the LDLR been shown to be expressed by BMEC _in vivo_ and _in vitro_[33,35]; however, subsequent degradation and/or transport of LDL are unknown. Expression of the scavenger receptor by BMEC is less certain. In contrast to the results of Pitas et al.[36], Fishman et al.[7] showed that BMEC _in situ_ accumulated Ac-LDL. The difference between these studies most likely is related to the method of administration. Although the site of IV injection of Ac-LDL was not stated for the first study[36], absence of uptake

Table 1. List of serum proteins that have been examined as ligands for specific receptors on brain capillary endothelia.

Ligand	Species	In Vivo	In Vitro		Refs.
			MVs	Cells	
Insulin	b,h,r,rb	Y	Y	Y	2,6,55,56
IGF-I	b,h	nd	Y	Y	2,6,56,57
IGF-II	b,h	nd	Y	Y	2,6,57
Tf	b,h,m,r	Y	Y	Y	2,6,7,18,44,47
LDL	b	Y	N?	Y	33,35
Ac-LDL	b,h,m,r	Y/N	nd	Y/N	7,36-39
NGF	b	nd	Y	Y	unpublished
Albumin	b,r	N	N	N	31,32,40

Abbrev.: Ac, acetylated; b, bovine; h, human; IGF, insulin-like growth factor; LDL, low density lipoprotein; m, murine; MVs, microvessels; nd, not determined; NGF, nerve growth factor; r, rat; rb, rabbit; Tf, transferrin.

by BMEC could be due to rapid and efficient removal in the liver. In contrast, the study by Fishman et al.[7] clearly showed that BMEC accumulated Ac-LDL following direct perfusion of the brain. Results with cultured BMEC also are conflicting[37-39]. The presence of the scavenger receptor is not universal to all endothelia and differences between in vitro and in vivo studies may be quantitative rather than absolute[36].

We have included in Table 1 preliminary results which show that isolated microvessels and cultured BMEC express an apparently small number of nerve growth factor receptors (NGFR). This tentative finding is based on specific binding of [^{125}I]NGF to BMEC monolayers (Buxser, S., and Raub, T., unpublished results) and on the detection of NGFR mRNA in isolated microvessels and cultured BMEC by using Northern analysis (Sampson, K., Abraham, I., and Raub, T., unpublished results). At this time, a biological response to NGF binding has not been found and a function for the NGFR is unknown.

Albumin apparently lacks a receptor on BMEC in vivo[40], in isolated microvessels[31] and in culture[32]. This is in contrast to studies with endothelia from peripheral capillaries where a high-affinity receptor for albumin has been identified[41,42]. It is proposed that this receptor is responsible for transcytosis of albumin across the continuous endothelium of mouse lung, heart and diaphragm[43].

In our laboratory, cellular mechanisms of receptor-mediated events are being defined by studying the expression and recycling of TfR in BMEC in culture. Transferrin receptors were demonstrated first within the BBB of the rat and human by immunohistochemistry and it was proposed that iron is supplied to the CNS by receptor-mediated transcytosis of the serum Tf-iron complex[44]. This concept has been examined further in spite of the disparity between Tf-TfR and iron distributions within the brain[45]. A similar regulatory role for iron availability has been proposed for the endothelium that comprises the blood-bone marrow barrier[46].

Pardridge et al.[47] were the first to demonstrate binding to and subsequent endocytosis of Tf by a high-affinity receptor on isolated microvessels, but transcytosis of the internalized Tf could not be demonstrated. By using an _in vivo_ approach, Fishman et al.[7] concluded that vascularly perfused [^{125}I]Tf was transported across the BBB by the BMEC; however, these results were obtained by separating rat brain into microvessel and "other" fractions following accumulation of radiotracer. Such approaches may lead to false conclusions since fractionation was conducted under conditions that would fail to prevent Tf efflux which would contaminate the "nonvascular" fraction. Although these studies raise the possibility for Tf transcytosis by BMEC, evidence is insufficient to assess if this pathway is adequate to meet the iron requirement of the brain and to discount the role of epithelial cells of the choroid plexus in Tf and/or iron transport especially since these cells express Tf[48].

To study receptor-mediated endocytosis by BMEC in detail, we used the Tf-TfR system as a model since it is so well characterized in other cell types, albeit mostly those that are transformed cell lines. Results with our _in vitro_ BMEC model suggest that Tf recycling is constitutive and polarized towards the apical surface although a small portion of the internalized Tf appears to be transcytosed[18]. We present some of these results here with the caution that cultured BMEC may not behave identically to their counterparts _in situ_.

Total or intracellular and surface TfR were measured by binding of bovine diferric [^{125}I]Tf in the presence of 0.5% saponin. Saponin permeabilizes cell membranes to allow access of Tf to the intracellular TfR pool. Such an approach was necessary because of the small number of TfR at the surface of BMEC. Binding of bovine Tf to BMEC monolayers was saturable and nonlinear curve fitting analysis of the binding isotherm gave a K_d of 4.8 ± 0.8 nM. A similar result has been reported for the human TfR in isolated microvessels[47]. We estimated there to be a maximum of approximately 100,000 total TfR per cell. By measuring Tf binding at 4°C in the absence of saponin, we found that only 10-15% of the TfR reside at the cell surface at any one moment. This distribution, while unusual, is similar to that found for TfR in HeLa cells where 20% are at the cell surface[49].

The number of TfR per BMEC monolayer was dependent upon age in culture. Maximal expression by the monolayer was found at confluence, although the number of TfR per cell was maximal at lowest cell density. By two days post-confluence, the number of TfR already had decreased by 40% from the total number at confluence. On a per mg basis, the number of TfR in cultured primary BMEC at confluence was 2.3-fold greater than that measured in passaged bovine adrenal endothelia and 3.5-fold greater than levels in passaged bovine kidney epithelia. Subconfluent passaged BMEC also had elevated levels of TfR. These results are consistent with the suggestion that TfR expression in BBB-derived endothelia is greater than it is in endothelia from most other capillary beds[44]. More importantly, this suggests that BMEC in culture retain characteristics that are reminescent of the _in vivo_ phenotype. Other examples of biochemical similarities have been defined in support of this _in vitro_ cell culture model[10].

Binding and uptake of Tf at 37°C was saturable within 60 min with a K_{uptake} of 6 nM which is identical to the K_d for binding. The half-time of TfR internalization is much slower than the several minute values reported for other cell types[50]. Pulse-chase experiments could not be done because of an insufficient number of cells per surface area and a paucity of receptors at the cell surface; therefore, the kinetics for a single round of endocytosis and recycling could not be measured. Specific accumulation was defined by 100% inhibition at 4°C or in the presence of 100-fold excess of Tf and by sensitivity to metabolic poisons. As expected, iron from [^{59}Fe]-loaded Tf accumulated with first order rate kinetics for as long as 3 h. Iron accumulation was inhibited with the weak base primaquine which raises endosomal pH and prevents dissociation of the iron[29]. Concomitantly, primaquine also decreased the accumulation of Tf by 30%, possibly by inhibiting receptor recycling as described below.

Since phorbol esters have been shown to induce surface expression of several receptors[51,52] and only 10-15% of the TfR are expressed at the surface of BMEC, we examined the effect of phorbol myristate acetate (PMA) on TfR recycling kinetics. After a 10 min preincubation in 100 mM PMA, the initial rate of Tf and [^{59}Fe] accumulation was increased 1.8-fold over untreated controls. To determine if this effect was due to a change in TfR recycling kinetics or to recruitment of TfR from an intracellular pool, BMEC that were at steady-state with regard to Tf accumulation were exposed to PMA. We found no change in subsequent accumulation suggesting that PMA acts by enhancing the rate of recycling (either uptake, efflux or both) and that it has no effect on the number of participating receptors. McGraw et al.[52] showed that PMA had a similar effect on the human TfR expressed in fibroblasts and further attributed this to an increase in the rate of exocytosis.

Rapid and extensive exocytosis of intact Tf has been demonstrated using capillaries isolated from rat epididymal fat[53], but similar studies have not been done with BMEC. In the study on Tf uptake by isolated cerebral microvessels[47], efflux experiments were not possible due to a defect in energy metabolism, i.e. efflux is energy dependent; however, it was unclear why receptor-mediated endocytosis occurred under the same conditions, since it too is energy requiring. In our cell culture system, the half-time of efflux of accumulated Tf was 38 min. This value was increased to 58 min in the presence of primaquine indicating that the associated decrease in Tf accumulation in the presence of primaquine was due to inhibition of TfR recycling. A similar effect was observed in HepG2 cells[29]. Degradation of Tf was minimal during accumulation since greater than 95% of the Tf was intact protein. It is obvious from these results that the slow transit time of TfR is responsible for the unusually large intracellular population of TfR in BMEC.

Can we assess the polarity of the Tf-TfR system in BMEC _in vitro_ and determine if transcytosis of Tf occurs? Although studies to date provide inferential support for transcytosis of Tf from blood to brain, direct evidence is lacking and difficult to obtain. Results with our BMEC culture model in Transwell™ membranes (see above) provide evidence for transendothelial efflux of intact Tf (Table 2). Because the BMEC monolayers were leaky enough to allow the slow diffusion of [^{125}I]Tf between chambers, we a pulse-chase design for our experiments and kept the chase time used short to minimize leakage following efflux. We wanted to determine if the TfR recycling pathway was polarized and if a transcytotic pathway existed.

Confluent BMEC monolayers were allowed to accumulate Tf from the apical side and after several quick washes the direction of subsequent efflux of Tf into either the apical or basolateral chambers was measured. The results are summarized in Table 2. While most of the Tf is recycled to the apical surface, as was true for constitutive fluid-phase endocytosis, a significantly greater fraction of intact Tf appeared in the basolateral chamber relative to the controls. For instance, accumulation in the presence of excess Tf should occur primarily by fluid-phase endocytosis and, therefore, efflux of Tf should be similar to efflux of a dextran of equivalent size. Indeed, this was the case since reappearance of Tf at the apical side was increased. Also, as expected, efflux of Tf was inhibited at 4°C. These results suggested that at least 8% of the Tf that had been endocytosed is

Table 2. Asymmetric efflux of transferrin and dextran (70 kDa) that had been accumulated by confluent BMEC monolayers grown on Transwell™ (0.4-μ pore size) membranes. Cells were allowed to internalize tracer with and without excess Tf from the apical chamber for 30 min at 37°C, washed and the reappearance of radioactivity in each chamber (AP, apical; cell, cell associated; BL, basolateral) measured after 30 min at either 37°C or at 4°C. Values represent the percent of radioactivity accumulated originally.

	TRANSFERRIN			DEXTRAN	
	37°C	100X excess Tf	4°C	37°C	4°C
AP	48	70	14	89	73
CELL	37	23	85	5	26
BL	15	7	1	6	1

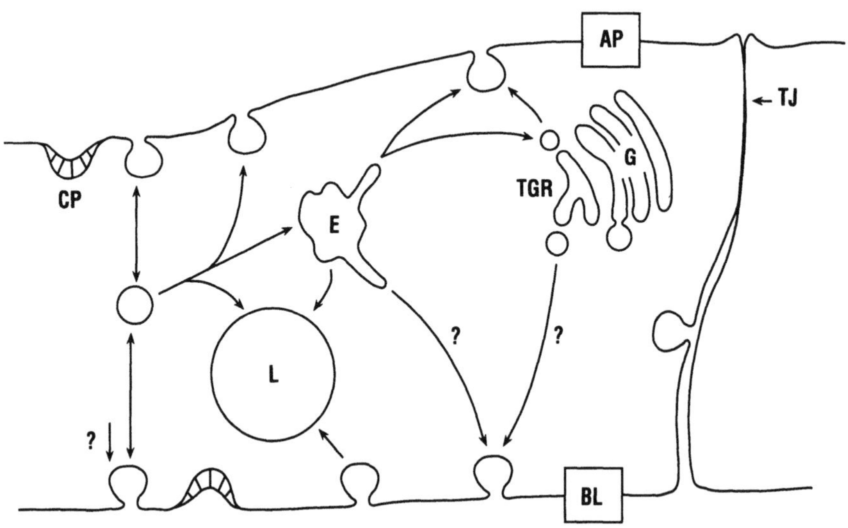

Figure 4. A general summary of the known, the possible and the proposed pathways for endocytosis and transcytosis within cultured primary BMEC monolayers is provided based on studies of fluid-phase, adsorptive and receptor-mediated endocytosis both <u>in vitro</u> and <u>in vivo</u>. AP, apical; BL, basolateral; CP, coated pit; E, endosome; G, Golgi; L, lysosome; TGR, trans-Golgi reticulum; TJ, tight junction.

transcytosed. Even if cross leakage of Tf occurred, the measured rates of diffusion were small for the duration (30 min) of these experiments.

Our results using cultured BMEC show that the majority of the TfR are recycled between the apical cell surface and yet to be identified endosomes. It is not known if TfR are internalized within clathrin-coated pits. At least in vitro, there is evidence for transcytosis of intact Tf from apical to basolateral side; however, we have not been able to demonstrate if this Tf contains iron.

SUMMARY AND FUTURE PROSPECTS

From a compilation of proposed vesicular pathways in endothelia in general and from our understanding of fluid-phase, adsorptive and receptor-mediated endocytic events, a conceptualized schematic for membrane traffic within BMEC is provided (Fig. 4). We actually know little about the pathways encountered by internalized molecules because of a lack of morphological and biochemical data. Aside from the numerous studies using HRP and the few demonstrating adsorptive endocytosis in vivo, results are scarce. To our knowledge, ultrastructural as well as uptake/cellular fractionation studies on receptor-mediated pathways in BMEC have not been performed and are needed to clarify these proposed phenomena.

A short-circuit recycling pathway is suggested from our results with fluid-phase endocytosis although the actual involvement of vesicles is unclear[14]. In most other cell types, receptor-mediated endocytosis via clathrin-coated pits appears to be responsible for the bulk of the accumumulated fluid[54]. In contrast, the number of coated pits in endothelia is small and their role in transport has not been examined. As in other cell types, the kinetics of membrane internalization in BMEC in culture is rapid and at equilibrium with intracellular compartments. Furthermore, this endocytic pathway, unlike that encountered by the bulk of the accumulated solutes, involves a variety of morphologically distinct endocytic compartments. Obviously, this pathway is vital to the survival of BMEC as a means to acquire needed nutrients.

The transcytotic pathway, in general, remains a mystery most likely due to the small number of compartments involved. Movement of solutes across the BBB endothelium would require direct vesicular shuttling to avoid the constitutive short-circuit pathway. If this occurs in undamaged BMEC, then it most likely is very infrequent. In contrast, transcytosis of membrane, including ligand-receptor complexes, must occur following passage through a sorting compartment such as the endosome[21]. An alternative route that has been proposed within BMEC[27] is passage of endocytosed apical membrane through the Golgi apparatus or the trans-Golgi reticulum enroute to the basolateral membrane. Aside from a specific sorting mechanism, such an event may be facilitated by the mixing of the endocytic pathway and a constitutive secretory pathway. These possibilities will be clarified as more investigators employ in vitro culture models for studies on membrane trafficking.

ACKNOWLEDGMENTS

The authors thank Dr. R. G. Ulrich for use of the Electron Microscopy facility, Ms. G. Sawada for printing the micrographs, Ms. S. Kuentzel for editorial assistance and Ms. B. Mussulman for secretarial assistance. Dr. Newton completed the transferrin receptor study while on sabbatical leave at The Upjohn Company.

REFERENCES

1. R. C. Janzer and M. C. Raff, Astrocytes induce blood-brain barrier properties in endothelial cells, Nature (Lond.) 325:253-257 (1987).
2. W. M. Pardridge, Recent advances in blood-brain barrier transport, Ann. Rev. Pharmacol. Toxicol. 28:25-39 (1988).
3. H. J. L. Frank and W. M. Pardridge, A direct in vitro demonstration of insulin binding to isolated brain microvessels, Diabetes 30:757-761 (1980).
4. A. L. McCall, J. Valente, R. Cordero, N. B. Ruderman, and K. Tornheim, Metabolic characterization of isolated cerebral microvessels: ATP and ADP concentrations, Microvasc. Res. 35:325-333 (1988).
5. I. Sussman, M. P. Carson, A. L. McCall., V. Schultz, N. B. Ruderman, and K. Tornheim, Energy state of bovine cerebral microvessels: comparison of isolation methods, Microvasc. Res. 35:167-178 (1988).
6. W. M. Pardridge, Receptor-mediated peptide transport through the blood-brain barrier, Endocrine Rev. 7:314-330 (1986).
7. J. B. Fishman, J. B, Rubin, J. V. Handrahan, J. R. Connor, and R. E. Fine, Receptor-mediated transcytosis of transferrin across the blood-brain barrier, J. Neurosci. Res. 18:299-304 (1987).
8. D. Triguero, J. B. Buciak, J. Yang, and W. M. Pardridge, Blood-brain barrier transport of cationized immunoglobulin G: enhanced delivery compared to native protein, Proc. Natl. Acad. Sci. (USA) 86:4761-4765 (1989).
9. P. D. Bowman, S. R. Ennis, K. E. Rarey, A. L. Betz, and G. W. Goldstein, Brain microvessel endothelial cells in tissue culture: a model for study of blood-brain barrier permeability, Ann. Neurol. 14:396-402 (1983).
10. K. L. Audus and R. T. Borchardt, Bovine brain microvessel endothelial cell monolayers as a model system for the blood-brain barrier, Ann. N. Y. Acad. Sci. 507:9-18 (1987).
11. D. N. McKinley and H. S. Wiley, Reassessment of fluid-phase endocytosis and diacytosis in monolayer cultures of human fibroblasts, J. Cell. Physiol. 136:389-397 (1988).
12. T. S. Reese and M. J. Karnovsky, Fine structural localization of a blood-brain barrier to exogenous peroxidase, J. Cell Biol. 34:207-217 (1967).
13. J. A. Swanson, B. D. Yirinec, and S. C. Silverstein, Phorbol esters and horseradish peroxidase stimulate pinocytosis and redirect the flow of pinocytosed fluid in macrophages, J. Cell Biol. 100:851-859 (1985).
14. F. L. Guillot, K. L. Audus and T. J. Raub, Fluid-phase endocytosis by primary cultures of bovine brain microvessel endothelial cell monolayers, Microvasc. Res., in press (1989).
15. P. F. Davies, S. C. Selden, and S. M. Schwartz, Enhanced rates of fluid pinocytosis during exponential growth and monolayer regeneration by cultured arterial endothelial cells. J. Cell. Physiol. 102:119-127 (1980).
16. J. M. Besterman, J. A. Aihart, R. C. Woodworth, and R. B. Low, Exocytosis of pinocytosed fluid in cultured cells: kinetic evidence for rapid turnover and compartmentation, J. Cell Biol. 91:716-727 (1981).
17. S. K. Williams and R. C. Wagner, Regulation of micropinocytosis in capillary endothelium by multivalent cations, Microvasc. Res. 21:175-182 (1981).
18. C. R. Newton and T. J. Raub, Characterization of the transferrin receptor in primary cultures of bovine brain capillary endothelial cells, J. Cell Biol. 107:770a.
19. R. D. Broadwell, B. J. Balin, M. Salcman, and R. S. Kaplan, Brain-blood barrier? Yes and no, Proc. Natl. Acad. Sci. USA 80:7352-7356 (1983).
20. R. M. Steinman, I. S. Mellman, W. A. Muller, and Z. A. Cohn, Endocytosis and the recycling of plasma membrane, J. Cell Biol. 96:1-27 (1983).

21. I. Mellman, Molecular sorting during endocytosis, Kidney Intl. 32:S-184-S-195 (1987).
22. T. J. Raub and K. L. Audus, Adsorptive endocytosis by cultured bovine brain endothelial cells, J. Cell Biol. 105:312a.
23. T. J. Raub and K. L. Audus, Adsorptive endocytosis and recycling of ricin agglutinin by cultured monolayers of primary bovine brain microvessel endothelial cells, submitted to J. Cell Sci.
24. D. Z. Gerhart, M. S. Zionis, and L. R. Drewes, Light and electron microscopic localization of D-galactosyl residues in capillary endothelial cells of the canine cerebral cortex, J. Histochem. Cytochem. 34:641-648 (1986).
25. A. W. Vorbrodt, D. H. Bodrogowska, A. S. Lossinsky, and H. M. Wisniewski, Ultrastructural localization of lectin receptors on the luminal and abluminal aspects of brain micro-blood vessels, J. Histochem. Cytochem. 34:251-261 (1986).
26. P. L. Debbage, H.-J. Gabius, K. Bise, and F. Marguth, Cellular glycoconjugates and their potential endogenous receptors in the cerebral microvasculature of man: a glycohistochemical study, Eur.J. Cell Biol. 46:425-434 (1988).
27. R. D. Broadwell, B. J. Balin, and M. Salcman, Transcytotic pathway for blood-borne protein through the blood-brain barrier, Proc. Natl. Acad. Sci. USA 85:632-636 (1988).
28. G. Griffiths and K. Simons, The trans Golgi network: sorting at the exit site of the Golgi complex, Science 234:438-443 (1986).
29. W. Stoorvogel, H. J. Geuze, J. M. Griffith, and G. J. Strous, The pathways of endocytosed transferrin and secretory protein are connected in the trans-Golgi reticulum, J. Cell Biol. 106:1821-1829 (1988).
30. T. J. Raub, J. B. Denny, and R. M. Roberts, Cell surface glycoproteins of CHO cells. I. Internalization and rapid recycling, Exp. Cell Res. 165:73-91 (1986).
31. A. K. Kumagai, J. B. Eisenberg, and W. M. Pardridge, Absorptive-mediated endocytosis of cationized albumin and a ß-endorphin-cationized albumin chimeric peptide by isolated brain capillaries, J. Biol. Chem. 262:15214-15219 (1987).
32. K. R. Smith and R. T. Borchardt, Permeability and mechanism of albumin, cationized albumin, and glycosylated albumin transcellular transport across monolayers of cultured bovine brain capillary endothelial cells, Pharm. Res. 6:466-473 (1989).
33. M. W. Brightman, 1989, The anatomic basis of the blood-brain barrier, in: "Implications of the Blood-Brain Barrier and its Manipulation, Vol. 1," E. A. Neuwelt, ed., Plenum Medical Book Co., New York, pp. 53-83.
34. P. Wang-Iverson, P. M. DeRosa, and W. V. Brown, 1988, Plasma lipoprotein interaction with endothelial cells, in: "Endothelial Cells, Vol. 1," U. S. Ryan, ed., CRC Press, Boca Raton, Florida, pp. 179-187.
35. S. Meresse, C. Delbart, J.-C. Fruchart, and R. Cecchelli, Low-density lipoprotein receptors on endothelium of brain capillaries, J. Neurochem. 53:340-345 (1989).
36. R. E. Pitas, J. Boyles, R. W. Mahley, and D. M. Bissell, Uptake of chemically modified low density lipoproteins _in vivo_ is mediated by specific endothelial cells, J. Cell Biol. 100:103-117 (1985).
37. J. Gaffney, D. West, F. Arnold, A. Sattar, and S. Kumar, Differences in the uptake of modified low density lipoproteins by tissue cultured endothelial cells, J. Cell Sci. 79:317-325 (1985).
38. M. P. Carson and C. C. Haudenschild, Microvascular endothelium and pericytes: high yield, low passage cultures, _In Vitro_ Cell. Develop. Biol. 22:344-354 (1986).
39. H. V. Vinters, S. Reave, P. Costello, J. P. Girvin, and S. A. Moore, Isolation and culture of cells derived from human cerebral microvessels, Cell Tiss. Res. 249:657-667 (1987).

40. M. Simionescu, N. Ghinea, A. Fixman, M. Lasser, L. Kukes, N. Simionescu, and G. E. Palade, The cerebral microvasculature of the rat: structure and luminal surface properties during early development, J. Submicrosc. Cytol. Pathol. 20:243-261 (1988).
41. J. E. Schnitzer, W. W. Carley, and G. E. Palade, Albumin interacts specifically with a 60-kDa microvascular endothelium glycoprotein, Proc. Natl. Acad. Sci. USA 85:6773-6777 (1988).
42. N. Ghinea, M. Eskenasy, M. Simionescu, and N. Simionescu, Endothelial albumin binding proteins are membrane-associated components exposed on the cell surface, J. Biol. Chem. 264:4755-4758 (1989).
43. L. Ghitescu, A. Fixman, M. Simionescu, and N. Simionescu, Specific binding sites for albumin restricted to plasmalemmal vesicles of continuous capillary endothelium: receptor-mediated transcytosis, J. Cell Biol. 102:1304-1311 (1986).
44. W. A. Jeffries, M. R. Brandon, S. V. Hunt, A. F. Williams, K. C. Gatter, and D. Y. Mason, Transferrin receptor on endothelium of brain capillaries, Nature (Lond.) 312:162-163 (1984).
45. J. M. Hill, M. R. Ruff, R. J. Weber, and C. B. Pert, Transferrin receptors in rat brain: neuropeptide-like pattern and relationship to iron distribution, Proc. Natl. Acad. Sci. USA 82:4553-4557 (1985).
46. R. Soda and M. Tavassoli, Transendothelial transport (transcytosis) of iron-transferrin complex in the bone marrow, J. Ultrastruc. Res. 88:18-29 (1984).
47. W. M. Pardridge, J. Eisenberg, and J. Yang, Human blood-brain barrier transferrin receptor, Metab. 36:892-895 (1987).
48. B. Bloch, T. Popovici, S. Chouham, M. J. Levin, D. Tuil, and A. Kahn, Transferrin gene expression in choroid plexus of the adult rat brain, Brain Res. Bull. 18:573-576 (1987).
49. J. E. Lamb, F. Ray, J. H. Ward, J. P. Kushner, and J. Kaplan, Internalization and subcellular localization of transferrin and transferrin receptors in HeLa cells, J. Biol. Chem. 258:8751-8758 (1983).
50. J. A. Hanover and R. B. Dickson, 1985, Transferrin: receptor-mediated endocytosis and iron delivery, in: "Endocytosis," I. Pastan and M. C. Willingham, eds., Plenum Press, New York, pp. 131-161.
51. S. S. Buys, L. H. Gren, and J. Kaplan, Phorbol esters and calcium ionophores inhibit internalization and accelerate recycling of receptors in macrophages, J. Biol. Chem. 262:12970-12976 (1987).
52. T. E. McGraw, K. W. Dunn, and F. R. Maxfield, Phorbol ester treatment increases the exocytic rate of the transferrin receptor recycling pathway independent of serine-24 phosphorylation, J. Cell Biol. 106:1061-1066 (1988).
53. R. C. Wagner, C. S. Robinson, P. J. Cross, and J. J. Devenny, Endocytosis and exocytosis of transferrin by isolated capillary endothelium, Microvasc. Res. 25:387-396 (1983).
54. M. Marsh and A. Helenius, Adsorptive endocytosis of Semliki Forest virus, J. Molec. Biol. 142:439-454 (1980).
55. W. M. Pardridge, J. Eisenberg, and J. Yang, Human blood-brain barrier insulin receptor, J. Neurochem. 44:1771-1778 (1985).
56. B. T. Keller, K. R. Smith, and R. T. Borchardt, Transport barriers to absorption of peptides, Pharm. Weekblad Sci. 10:38-39 (1988).
57. R. G. Rosenfeld, H. Pham, B. T. Keller, R. T. Borchardt, and W. M. Pardridge, Demonstration and structural comparison of receptors for insulin-like growth factor-I and -II (IGF-I and -II) in brain and blood-brain barrier, Biochim. Biophys. Res. Commun. 149:159-166 (1987).

LIPOPROTEINS AND RECONSTITUTED BLOOD-BRAIN BARRIER

Stéphane Méresse°, Marie-Pierre Dehouck°, Pierre Delorme*, Jean-Charles Fruchart° and Roméo Cecchelli°

°SERLIA et INSERM, U325 Institut Pasteur, 59019 Lille Cédex, FRANCE
*Laboratoire de Neurobiologie Fonctionnelle, Université de Lille I FRANCE

INTRODUCTION

Tight junctions between cerebral endothelial cells and the low rate of pinocytosis and vesicular transport are believed to constitute the structural basis of the blood-brain barrier (BBB) , that prevents passive diffusion of proteins from the blood into the brain (Bradbury, 1985). Carrier mediated transport systems which facilitate the uptake of hexoses, amino-acids, purines coumpounds and mono carboxylic acids have also been revealed in the cerebral endothelium (for review, see Betz and Goldstein, 1986), but, until now, little information has come to light regarding the cerebral uptake of lipids. Recently, the presence of apolipoproteins, lipoproteins (Roheim et al., 1979 ; Elshourbagy et al., 1985) and their receptors (Hofmann et al., 1987 ; Pitas et al., 1987) was detected in the brain. The fact that enzymes involved in lipoprotein metabolism are present in the brain microvasculature (Brecher and Kuan, 1979) suggests that a system designed to maintain cholesterol homeostasis must exist in the central nervous system. Thus, cholesterol could be derived from de novo synthesis within the brain or from the plasma via the blood-brain barrier. Whether plasma low density lipoprotein (LDL) which is the major carrier of cholesterol is involved in this exchange is not known. Furthermore, the fact that the entire fraction of drug bound to lipoproteins is available for entry into the brain (Urien et al.,1987) strongly suggests that the interactions of lipoproteins with endothelial cells could play an important role in drug transport across the BBB.

We have recently demonstrated (Méresse et al.,1989a) the occurrence of a low-density lipoprotein receptor on the endothelium of brain capillaries. In order to examine closely the interactions of lipoproteins with brain capillary endothelial cells, we have developed an "in vitro" system of BBB by growing endothelial cells on one side of a polycarbonate porous system and astrocytes on the other. The relatively ease with which such reconstituted blood-brain barrier can be produced in large numbers facilitates the "in vitro" study of drug transport across the blood-brain barrier.

LDL RECEPTOR ON ENDOTHELIUM OF BRAIN CAPILLARIES

When radioactive LDL in the absence or in the presence of an excess of unlabelled LDL, was injected immediately post-mortem into bovine brain circulation, the total amount of radioactivity that remained associated with the capillaries in the presence of excess unlabelled LDL was only 15% of the radioactivity recovered in the capillaries when ^{125}I-LDL was injected alone (0,54 vs 3,6 ng of bound LDL/mg of

Pharmaceutical Applications of Cell and Tissue Culture to Drug Transport
Edited by G. Wilson *et al.*, Plenum Press, New York, 1991

capillary protein). These results suggest that ^{125}I-LDL binds to a specific receptor on the endothelium of brain capillaries.

In order to confirm this possibility, we have examined "in vitro" the binding of ^{125}I-LDL on isolated brain capillaries. Irrespective of the incubation medium, incubation time or concentration of ^{125}I-LDL, no difference was observed between the binding in the presence or in the absence of excess unlabelled LDL, suggesting the absence of a high affinity binding process for LDL. But these results could likely be explained by the high non specific binding of LDL (approximately 20 fold higher than non specific binding on purified endothelial membranes), probably due to the large amount of connective tissue attached to the basement membrane. So, it was of interest to test the efficiency of the binding of both purified endothelial and basement membranes of brain capillaries.

The endothelial and basement membranes of freshly isolated brain capillaries were separated according to Lidinsky and Drewes (1983) and their own LDL binding capacity was recorded. The ^{125}I-LDL bound with a saturation kinetic, only in the presence of endothelial membranes. The saturability of this binding emphasized the concept that the interaction of LDL with a specific receptor has been detected. When transformed to Scatchard plots, the total binding data show evidence for a single binding site with an apparent dissociation constant (Kd) of 16 nM. The estimate of the affinity is similar to the Kd reported in cultured bovine aortic cells (Vlodasky et al.,1978) or human fibroblasts (Mahley and Innerarity, 1983). Incubations of basement membranes revealed no difference between the specific and the non-specific binding of ^{125}I-LDL. These results agree with those obtained "in situ" and strongly suggest a high affinity site for LDL on bovine cerebral capillary endothelial cells. Furthermore, trypsin treatment of the endothelial membranes abolished the LDL binding (data not shown) demonstrating that a membrane protein-LDL interaction was involved.

The specificity of the LDL binding was determined in several ways using a dot blot assay (Maggi and Catapano, 1987). The binding sites on endothelial cells membranes of brain capillaries exhibited a high specificity for LDL, as unlabelled LDL was found to achieve competitive inhibition while HDL_3 was a poor competitor. The specificity of the binding was confirmed by the lack of competition with methyl-LDL which is known not to abolish the specific binding (Weisgraber et al., 1978). Furthermore, the binding is Ca++ dependent.

This receptor exhibits the same characteristic properties as the LDL receptor on human fibroblasts (Goldstein and Brown, 1977) and steroid-secreting cells (Kovanen et al., 1979): 1)it binds human LDL, 2) HDL3 is a poor competitor, 3) it fails to bind LDL whose lysine residues have been methylated, 4) it requires a divalent cation, 5) it is extremely sensitive to destruction by proteolytic enzymes.

The molecular weight determined for bovine capillary receptor by ligand blotting is 132,000 in unreduced SDS-polyacrylamide gel (Figure 2). It is very similar to the 130-132,000 of the LDL receptor detected in bovine adrenal cortex by the same method (Soutar et al., 1986 ; Kroon et al., 1984).

A separate "scavenger" pathway of LDL uptake in macrophages and endothelial cells which differs from the LDL receptor in terms of specificity and biochemical properties has been described (Goldstein et al., 1979; Stein and Stein , 1980). This scavenger pathway recognizes only some chemically modified forms of LDL (derivatized by acetylation, acetoacetylation or treatment with malondialdehyde) and biologically modified LDL (incubated with cultured human endothelial cells). The requirement of Ca^{++} argues against the possibility that we are dealing with a scavenger receptor (Goldstein et al., 1979 ; Dresel et al., 1984). In addition, the molecular weight determined for the receptor on brain capillary endothelial cells (132,000) is very far from the molecular weight of the scavenger receptor (220-250,000) determined by Dresel et al., (1984). Moreover the presence of a scavenger receptor on brain capillary endothelial cells is very uncertain. It seems that these cells failed to take up acetylated LDL in vivo (Pitas et al., 1985) whereas in vitro the results are very conflicting (Gaffney et al., 1985 ; Carson and Haudenschild, 1986).

Pitas et al. (1987) have proposed a model for cholesterol transport and homeostasis within the central nervous system. Apolipoprotein E is secreted by astrocytes within the brain, transports and redistributes cholesterol via brain interstitial fluid to cells that require cholesterol and that express apo B,E (LDL) receptors (Hofmann et al., 1987). This cholesterol could be derived from de novo synthesis within the brain and from plasma (i.e. cholesterol that crosses the blood-brain barrier). Our results integrated in this model and showed, for the first time, the occurrence of a LDL receptor on brain capillaries.

Additional studies are required to determine if after binding, lipoproteins actually are degraded by the cerebral microvasculature and whether the subsequent metabolism of LDL is restricted to the vasculature or can occur in the underlying nervous tissue.

But freshly isolated microvessels or membrane preparations are not a suitable model to study the fate of lipoproteins in the endothelial cells. Cell cultures of brain capillary endothelial cells would be advantageous for studying these interactions. In vitro, normal differentiated cells often failed to proliferate or ceased to express differentiated cell properties beyond one or a very limited number of passages (Diglio et al., 1983). Some earlier reports claim longevity, persistence of characteristic properties and lack of contamination by non-endothelial cells. However, most work done up to now has been done with primary culture (Bowman et al., 1983 ; Hughes et Lantos, 1986). Although important information can be obtained in primary culture, the ability to form subcultures and to obtain a large number of cells while maintaining the differentiated characteristics would obviously facilitate biochemical studies as well as investigations of the blood-brain barrier.

We have recently reported (Méresse et al., 1989b) a method for culture of a large number of bovine brain capillary endothelial (BBCE) cells. These cells, absolutely free of pericyte contamination, can be repeatedly passaged (up to 50 generations) without the loss of endothelial cell markers. Furthermore, primary cultures and subcultures retain some of the blood-brain barrier features : low rate of pinocytosis, tight junctions and monoamine oxidase activity.

CELL CULTURE CONDITIONS

After isolation by mechanical homogenization from one hemisphere of bovine brain, microvessels were seeded onto dishes coated with an extracellular matrix (ECM) secreted by bovine corneal endothelial cells. Five days after seeding the first migrating cells emerged from attached capillary fragments (Figure 1A) . Two cell types were seen. The first was identified as pericytes, which grew as individual cells and were characterized by their irregular outlines and numerous processes (Figure 1D). These unwanted pericytes rapidly invaded the culture dishes when the floating capillary tufts were not discarded with the change of culture medium (2 h after microvessel seeding). The endothelial cells extruding from the capillary fragments (Figure 1B) grew in colonies of closely apposed and elongated cells (Figure 1C) as described by Bowman et al. (1984). These cells were found only on the ECM-coated dishes. In fact, no endothelial cells were encountered when the isolated capillaries were seeded on plastic dishes. As soon as the first bovine brain capillary endothelial cells emerged from the capillaries, cell proliferation was enhanced by addition of bFGF (1 ng/ml) to the culture medium. In these conditions, approximately 20% of the seeded fragments can initiate pericyte-free endothelial cell colonies. In each dish, three or four of the largest colonies were mechanically harvested and then seeded on gelatin-coated dishes in the presence of optimal culture medium (DMEM). bFGF (1 ng/ml) was added every other day, until BBCE cells became confluent, 7 to 8 days after seeding. They formed a monolayer of small, tighly packed, non overlapping and contact-inhibited cells (Figure 1E). Endothelial cells from one 35 mm-dish were harvested at confluence and seeded onto 60 mm gelatin-coated dishes. After 6 to 8 days, confluent cells were subcultured at the split ratio of 1:20. Cells at the third passage (about 50 dishes) were stored in liquid nitrogen. Cells frozen in passage 3 showed good recovery and growth after storage in liquid nitrogen. When endothelial cells and blood-brain barrier characteristics were studied, no differences between frozen and unfrozen cells were observed.

Under the optimal culture conditions, cells were passaged repeatedly at low cell density up to the passage 10. Each passage allowed a 20 fold increase of the cultured surface. The life span of the cultures was about 50 cumulative population doublings. Beyond the passage 10, cells became large and did not reach confluence.

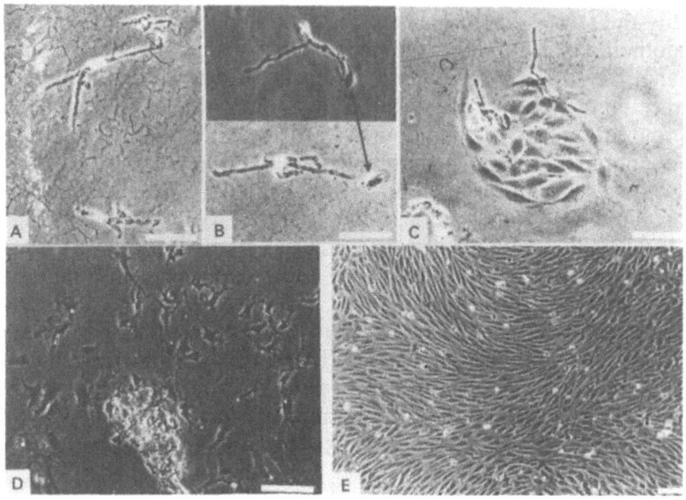

Fig. 1. Bovine brain endothelial cells in tissue culture. (A) Phase-contrast photomicrograph of isolated bovine brain capillaries. One hour after seeding and washing with fresh culture medium, the isolated capillaries adhere on extracellular matrix. (B) Five days after plating, emerging cells are seen at the end of the isolated capillaries (arrow). (C) Six days after plating, the migrating cells form a clump of endothelial cells. A remnant of the original capillary is present (arrow). (D) Five days after plating, contaminated cells, presumably pericytes, grow as individual cells which are characterized by irregular outlines and numerous processes. (E) Confluent monolayer of bovine brain capillary endothelial cells in their 3rd passage after cloning. The cells are grown on gelatin-coated dishes in the presence of DMEM supplemented with 15% CS and bFGF (1ng/ml). Bar, 100 μm.

This technique allowed us to circumvent the culture limitations of primary cultures and provide quickly a large quantity of monolayers. Furthermore, disruption of brain tissue by mechanical dispersion and filtration techniques, without enzymatic treatment, enabled us to isolate a microvascular network. Microvessels are typically defined as vessels with a diameter less than 100 μm and consequently consisting predominantly of capillaries with some arterioles and venules. Most methods of isolation of brain capillary endothelial cells involve collagenase digestion of capillary tufts (Joo, 1985). So, we cannot exclude the possibility that, after this treatment, migrating and proliferating cells might derive from the endothelial wall of arterioles or venules that contaminated the capillary preparation. There may be fundamental differences between endothelial cells derived from capillaries and those derived from large vessels (Wolff, 1977). These differences should preclude the use of large vessel endothelial cell cultures as a model system for the study of microvessel functions and pathology. Absence of the use of enzymes in our isolation procedure excludes this possibility while only the BBCE cell islands emerging from identifiable capillaries were cloned.

CHARACTERIZATION OF THE CELLS

Factor VIII-related antigen is the most widely used marker for identifying endothelial cells in culture (Jaffe et al., 1973). In our system, factor VIII related antigen could be detected within BBCE cells at early and later passages (10th), in contrast to what was observed by other authors (DeBault et al., 1981; Robinson et al., 1986). Although several non-endothelial cell lines, including bovine smooth muscle cells and pericytes, failed to stain with factor VIII antisera used for this experiments, the ability of human anti-factor VIII-related antigen to bind non specifically has been documented (Wilson, 1985).

Thus, demonstration of factor VIII-related antigen by this technique is useful, but other criteria must be used to identify these cells. That is why we have demonstrated i) the non-thrombogenic properties of the monolayer, ii) the production of PGI_2 by BBCE cells in response to bradykinin. iii) the presence of angiotensin converting enzyme (Bowman et al., 1983). Furthermore, we have shown that all these criteria are present on sparse, subconfluent and confluent BBCE up to the 10th passage (50 generations).

The structural basis of the blood-brain barrier is that the cells are sealed together by tight junctions. Spatz et al. (1980), Dropulic and Masters (1987) revealed that endothelial cells in culture were connected by many intercellular "gap" junctions. Bowman et al. (1983) showed that cells in primary culture are sealed by tight junctional complexes. Freeze fracture examination of confluent cultures at different passage numbers revealed, for the first time to our knowledge, intercellular junctional complexes consisting exclusively of tight junctions . This pattern is similar to that seen in the tight junctions of brain capillaries "in vivo" (Shivers et al., 1984).

In contrast to the observations of Tao et al. (1986) and Arthur et al. (1987) astrocytes in co-culture or astrocyte-conditioned medium and a "competent" extracellular matrix are not necessary to induce the formation of tight junctions. In a recent report Shivers et al. (1988) showed that astrocyte-conditioned medium and an endothelial-derived extracellular matrix can induce development of brain endothelium-like tight junctions in cultured bovine aortic endothelial cells. Nevertheless, in our model the presence of tight junctions in BBCE cells is related to a phenotype stabilization as demonstrated by the absence of such tight junctions in bovine aortic endothelial cells subjected to the same culture conditions as BBCE cells . In addition, the BBCE cells in culture have few pinocytotic vesicles, a further similarity to their in vivo counterpart.

The occurrence of monoamine oxidase (MAO) is one of the biochemical typical properties of the cerebral endothelium. Previous studies from Svendgaard et al. (1975) have shown that the barrier properties of cerebral capillaries towards amines and amine precursors are determined not by their origin but by the revascularised tissue. Glia foot induction of blood-brain barrier properties was suggested by these authors. But induction by glial end feet cannot be the only mechanism by which an endothelial barrier is initiated and maintained. Blood vessels in the iris (Raviola, 1977) and in peripheral nerves (Olson and Reese, 1971) form a barrier, and glial end feet are not present in these areas. In our system, we demonstrated that the cells expressed high levels of MAO activity only when they were confluent. ACE activity remained constant when it was detected on sparce, subconfluent and confluent cells. These results suggest that intimate contacts between cells are necessary to restore the level of MAO activity found in isolated brain capillaries and confirm once more that glial cells as in the case of formation of the tight junctions are not necessary for the expression of MAO.

An other enzymatic marker of the BBB was studied : gamma-glutamyl transpeptidase. Studies have shown that subcultivation of brain endothelial cells is associated with the loss of blood-brain barrier markers (Bowman et al., 1984). Cerebral vascular endothelium is the only endothelium that contains a histochemically or biochemically detectable concentration of gamma-glutamyl transpeptidase. Although the role of this enzyme in transport and metabolism of amino-acids is generally unexplained (Orlowski et al., 1974 ; Cancilla and DeBault, 1983), gamma-glutamyl transpeptidase is considered to be a marker of blood-brain barrier. The results concerning the occurence of gamma-glutamyl transpeptidase in cultured brain endothelial cells are very conflicting. DeBault (1981), DeBault and Cancilla, (1980 a, b) showed that the enzyme is often lost from the endothelial cells as soon as they migrate or proliferate from the isolated microvessels. However the synthesis of this enzyme can be induced in these endothelial cells by co-culturing them with astrocytes (C6 glioma cells). In contrast, the cultures described by Robinson et al. (1986) and Dropulic and Masters (1987) expressed high levels of gamma-glutamyl transpeptidase. These results suggesting that glial cells are not necessary for gamma-glutamyl transpeptidase expression were obtained with established cell lines. But, in these cases the gamma-glutamyl transpeptidase could be considered to be a marker of neoplastic cells instead of a marker of the blood-brain barrier (Caspers and Diglio, 1984). The specific activity of gamma-glutamyl transpeptidase in isolated brain capillaires (243 +/- 28 nmoles x mg prot^{-1}x min^{-1}) was more than forty fold measured over the starting homogenate (5.1 +/- 1.6 nmoles x mg prot^{-1} x min^{-1}), indicating the purity of the isolation. Low levels of gamma-glutamyl transpeptidase activity (2.9 nmoles x mg prot^{-1} x min^{-1}) were observed in primary culture and in passaged cells . Our results agree with those of DeBault and Cancilla (1980 a,b), which showed that cerebral endothelial cells lose their gamma-glutamyl transpeptidase in primary culture.

In order to reinduce the gamma-glutamyl transpeptidase in BBCE cells and to reconstruct some of the complexity of cellular environment that exits "in vivo" we have developed an "in vitro" model of BBB by growing endothelial cells on one side of a polycarbonate porous filter and astrocytes on the other.

Figure 2 illustrates the structure of confluent bovine brain endothelial cells cultured on an insert coated with collagen. Culture plate inserts (Millicell-CM 0.4µm, 30 mm diameter from Millipore) were coated on both sides with rat tail collagen prepared by a modification of the method of Bornstein (1958). For studies using phase contrast microscopy and histochemistry, prototypes Becton Dickinson were used, since the polycarbonate filter from the Millicell-CM did not allow adequate transmission of polarized light. BBCE cells form a monolayer of small, tightly packed, non-overlapping and contact-inhibited cells. Also noted in figure 2 is the close apposition of the cells and the absence of open spaces between them. As did the endothelial cells in our previous cultures on plastic coated with gelatin (Méresse et al.,1989b) , BBCE cells retain both endothelial and some of the BBB features (tight junctions, monoamine oxidase activity and few pinocytotic vesicles).

For the co-culture described by Dehouck et al. (1989), primary cultures of astrocytes, made from newborn cerebral cortex as described by Booher and Sensenbrenner (1972), were plated at the concentration of 2.5x10^5 cells/ml on the bottom side using filter up side down in DMEM supplemented with 10% fetal calf serum, 2mM glutamin, 50µg/ml gentamycin. After 8 days, filters were properly positioned and the medium changed twice a week. Three weeks after seeding, cultures of astrocytes became stabilized. Then BBCE cells, frozen at passage 3, were recultured on a 60 mm gelatin-coated dish. Confluent cells were trypsinized and plated on the upperside of the filters at the concentration of 4.10^5 cells/filter. The medium used for the co-culture was DMEM supplemented with 15% calf serum, 2 mM glutamin, 50 µg/ml gentamycin and 1 ng/ml bFGF added every other day. In these conditions, BBCE cells form a confluent monolayer in 8 days.

Figure 3 illustrates the ultrastructure of this co-culture. In transverse sections, endothelial cells form a monolayer while astrocytes form layers of overlapping cellular sheets. Astrocytes are GFAP^{+} and NF^{-} (results not shown). In these conditions, the presence of gamma-glutamyl transpeptidase activity is seen in BBCE cells as red granular staining. The cells demonstrate variability in the intensity of the staining but nearly all cells reveal some activity. Control cultures that contain ABAE cells and astrocytes show no activity.

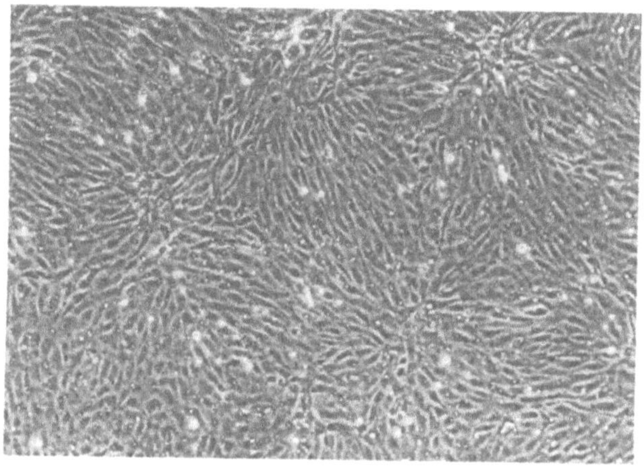

Fig.2. Phase contrast micrograph of confluent BBCE cells grown on the upper face of a collagen -coated filter of a prototype from Becton-Dickinson. (x340).

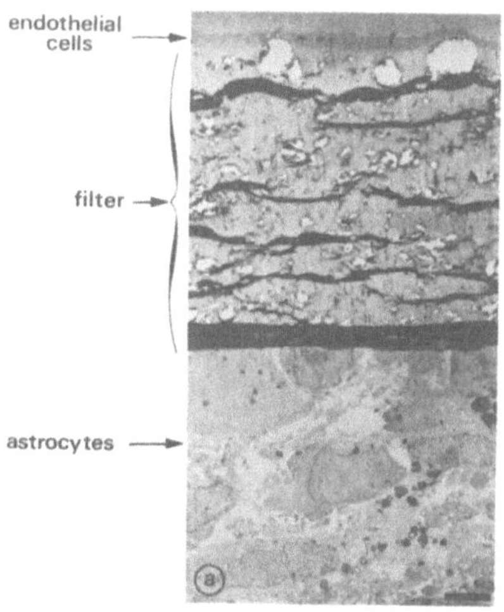

Fig.3. Electron micrograph showing ultrastructure of (a) co-culture of BBCE cells and astrocytes Bar : 5 µm.

Satisfied that monolayers of BBCE cells in co-culture possess many of the properties that are known to have important barrier functions "in vivo", we investigated the integrity of our monolayers by their ability to resist the transendothelial passage of electric current. After one week, the resistance of these monolayers in four experiments averages 661.5 ± 48 Ω.cm2 (n=20). This resistance is approximately 3 times less than that of frog brain capillaries (1,295 Ω.cm2 -Crone and Oleson, 1981) but 300 times greater than that of frog mesenteric capillaries (1.85 Ω.cm2) reported by Crone and Christiensen (1981). When BBCE cells are cultured alone, the resistance is 416 ± 57.6 Ω.cm2 . These results are in agreement with the studies of Tao-Cheng et al. (1986) in which primary capillary endothelial cells were co-cultured with an astrocyte-enriched preparation and showed significant enhancement of tight junction length, width and number as compared to various controls. Furthermore, when ABAE cells are co-cultured with astrocytes, the resistance of the monolayers averages 257 ± 25.6 Ω.cm2 (n=9). The average resistance found in the litterature for the monolayers of cultured microvascular endothelial cells was only 4.1 ± 2.3 Ω.cm2 by Furie et al. (1987), 4.91 ± 0.26 Ω.cm2 by Hennig et al. (1985) and 5 to 10 Ω.cm2 by Hashida et al. (1985).

The findings concerning the movement of extracellular markers ([^{3}H] inuline) across a co-culture of ABAE or BBCE cells grown on the filter shown that only the co-culture with BBCE cells provides a barrier to the diffusion of inuline.

The permeability of the reconstituted BBB to the [^{3}H]Leucine was investigated. In these experiments [^{14}C] sucrose was used as a control. In contrast to the barrier behaviour against sucrose, the leucine was seen to accumulate in the lower chamber and could be detected as early as 2 min after presentation of labelled compounds to the luminal face side of the reconstituted BBB.These results are in complete agreement with those of Audus et Borchardt (1986a, 1986b).

Preliminary studies in our laboratory have shown that BBCE cells in culture express a LDL receptor with the same apparent molecular weight as in vivo (Figure 4). Furthermore, the cells bind and internalize but degradation of LDL occur at very low rate. Similar to other ligand [Pardridge, 1986 (insulin, transferrin)] LDL receptor complexs seem also handled differently in BBCE cells than in other cells of the body. The role of this receptor in subserving the nutritional needs of the central nervous system can now be elucidated using our co-culture.

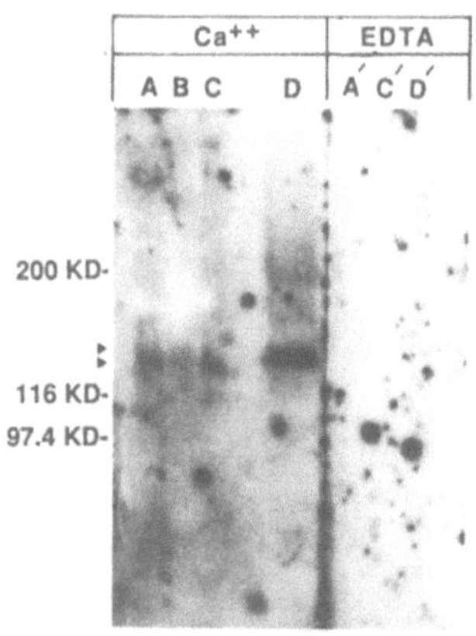

Fig. 4. Detection of bovine brain capillary endothelial cells, BBCE cells and human fibroblast LDL receptor by ligand blotting

Solubilized endothelial membrane proteins of brain capillaries (lane A and B), solubilized BBCE cells at passage 5 (lane C) and partially purified LDL receptors from human fibroblasts (lane D), were subjected to SDS-polyacrylamide gel electrophoresis and electrophoretically transferred to nitrocellulose paper. The blot was incubated with ^{125}I-LDL, washed and autoradiographied. The requirement for Ca^{++} (lane A', C', D') was studied by incubating the same preparations (A, C, D) in the same buffer in which $CaCl_2$ was omitted and EDTA was added.

In summary, we describe an "in vitro" blood-barrier model system that retains the morphological and enzymatical characteristics of the BBB "in vivo". This model not only facilitates the "in vitro" study of the brain capillary function by providing more easily representative monolayer of BBCE cells but also provides insights useful for studying transcellular transport in either the "blood to brain" or "brain to blood" directions with the same preparation.

Aknowledgements : We are very grateful to Dr F. Joo for hepful discussions and encouragements. We Thank Claudine Guichard for photographic work and Jacqueline Bernard for the preparation of electron microscopic sections. The secretarial assistance of Caroline Branlant is greatly appeciated.

REFERENCES

Arthur F.E., Shivers R.R. and Bowman P.D. (1987). Astrocyte-mediated induction of tight junctions in brain capillary endothelium : an efficient in vitro model. Develop. Brain Res. 36, 155-159.

Audus K.L. and Borchardt (1986a). Characterization of an "in vitro" blood-brain barrier model system for studying drug transfer and metabolism. Pharmaceutical Res. 3, 81-87.

Audus K.L. and Borchardt (1986b). Characteristics of the large neutral amino-acid transport system of bovine brain microvessels endothelial cell monolayers. J. Neurochem.47, 484-488.

Betz A.L. and Goldstein G.W. (1986).Specialized properties and solute transport in brain capillaries.Ann. Rev. Physiol.. 48, 241-250.

Booher J. and Sensenbrenner M. (1972). Growth and cultivation of dissociated neurones and glial cells from embryonic chick, rat and human brain in flask cultures. Neurobiology, 2, 97-105.

Bornstein M.B. (1958). Reconstituted rat tail collagen used substrate for time tissue cultures on coverslips in maximow slides and roller tubes. Lab. Invest. 7, 134-139.

Bowman P.D., Ennis S.R., Ravey K.E., Betz A.L. and Goldstein G.W. (1983). Brain microvessel endothelial cells in tissue culture : a model for study of blood-brain barrier permeability. Ann. Neurol. 14, 396-402.

Bowman P.D., Betz A.L. and Goldstein G.W. (1984). Culture of endothelial cells from neural capillaries. In Biology of endothelial cells, Jaffe E.A. (ed). Martinus Nijhoff Publishers, 27-33.

Bradbury M.W.B. (1985).The blood-brain barrier : Transport across the cerebral endothelium. Circulation Res. 57, 213-222.

Brecher P. and Kuan H.T. (1979). Lipoprotein lipase and acid lipase activity in rabbit brain microvessels.J. Lipid. Res. 20, 464-471.

Cancilla P.A. and DeBault L.E. (1983). Neutral amino-acid transport properties of cerebral endothelial cells ""in vitro"". J. Neuropath. Exp. Neurol. 42, 191-199.

Carson M.P. and Haudenschild C.C. (1986).Microvascular endothelium and pericytes: high yield, low passages cultures. In Vitro cell. Dev. Biol. 22, 344-354.

Caspers M.L., Diglio C.A. (1984). Expression of gamma-glutamyl transpeptidase in a transformed rat cerebral endothelial cell line. Biochim. Biophys. Acta, 803, 1-6.

Crone C. and Christiensen (1981). Electrical resistance of a capillary endothelium. J. Gen. Physiol. 77, 349-371.

Crone C. and Oleson S.P. (1981). The electric resistance of brain capillary endothelium. J. Physiol. 316, 53P-54P.

DeBault L.E. (1981). gamma-glutamyl transpeptidase induction mediated by glial foot process to endothelium contact in co-culture. Brain Res. 220, 432-435.

DeBault L.E., Henriquez E., Hart M.N. and Cancilla P.L. (1981). Cerebral microvessels and derived cells in tissue culture. In vitro, 17, 480-494.

DeBault L.E. and Cancilla P.A. (1980a). Induction of gamma-glutamyl transpeptidase into isolated endothelial cells. Adv. Exp. Med. Biol. 131, 79-88.

DeBault L.E. and Cancilla P.A. (1980b). gamma-glutamyl transpeptidase in isolated brain endothelial cells: induction by glial cells in vitro. Sciences, 207, 653,655.

Dehouck M.P., Méresse S., Delorme P., Delbart C., Fruchart J.C. and Cecchelli R. (1989). Co-culture of astrocytes and brain capillary endothelial cells : A potential model for "in vitro" studies of blood-brain barrier (Abstr.) J. Neurochem. 52 (Suppl.), S120.

Diglio C.A., Wolfe D.E., Meyers P. (1983). Transformation of rat cerebral endothelial cells by Rous Sarcoma Virus. J. Cell Biol. 97, 15-21.

Daniel T.O., Schneider W.J., Goldstein J.L. and Brown M.S. (1983).Visualization of lipoprotein receptor by ligand blotting. J. Biol. Chem. 258, 4606-4611.

Dresel H.A., Otto I., Weigel H., Schetter G. and Via D.P. (1984). A simple and rapid anti-ligand enzyme immunoassay for visualization of low density lipoprotein membrane receptors. Biochim. Biophys. Acta, 795, 452-457.

Dropulic B. and Masters C.L. (1987). Culture of mouse brain capillary cell lines that express factor VIII, gamma-glutamyl transpeptidase and form junctional complexes in vitro. In vitro cell. & Dev. Biol. 23, 775-781.

Elshourbagy N.A., Liao W.S., Mahley R.W. and Taylor J.M. (1985). Apolipoprotein E m RNA is abundant in the brain and adrenals, as well as in liver, and is present in other peripheral tissues of rats and marmosets. Proc. Natl. Acad. Sci. USA 82, 203-207.

Furie M.P., Naprstek B.L. and Silverstein S.C. (1987). Migration of neutrophil across monolayer of cultured microvascular endothelial cells. An in vitro model of leucocyte extravasation. J. Cell. Sci. 88, 161-175.

Gaffney J.,West D., Arnould F., Sattar A. and Kumar S. (1985). Differences in the uptake of modified low density lipoproteins by tissue cultured endothelial cells. J. Cell Sci., 79, 317-325.

Goldstein J.L., Ho Y.K., Basu S.K. and Brown M.S. (1979). Binding site on macrophages that mediates uptake and degradation of acetylated low density lipoprotein, producing massive cholesterol deposition. Proc. Natl. Acad. Sci. USA 76, 333-337.

Hashida R., Anamizu C., Kimura J., Ohkuma S., Yoshida Y. and Takano T. (1986). Transcellular transport of lipoprotein through arterial endothelial cells in monolayer culture. Cell Struc. and Func. 11, 31-42.

Hennig B., Shasby D.M. and Spector A.A. (1985). Exposure to fatty acid increases human LDL transfer across cultured endothelial monolayers. Cir. Res. 57, 776-780.

Hofmann S.L., Russel D.W., Goldstein J.L. and Brown M.S. (1987). mRNA for low density lipoprotein receptor in brain and spinal cord of immature and mature rabbits. Proc. Natl. Acad. Sci. USA 84, 6312-6316.

Hughes C.C.W., Lantos P.L. (1986). Brain capillary endothelial cells in vitro lack surface IgG Fc receptors. Neuroscience Letters 68, 100-106.

Jaffe E.A., Hoyer L.W. and Nachman R.L. (1973). Synthesis of antihemophilic factor antigen by cultured human endothelial cells. J. Clin. Invest. 52, 2757-2764.

Joo F. (1985). The blood-brain barrier in vitro : Ten years of research on microvessels isolated from the brain. Neurochem. Int. 7, 1-25.

Kroon P.A, Thompson G.M. and Chao Y.S. (1984). A comparison of the LDL receptor from bovine adrenal cortex, rabbit and rat liver and adrenal gland by lipoprotein blotting. Biochem. J. 223, 329-335.

Lidinsky W.A. and Drewes L.R. (1983). Characterization of the blood-brain barrier :Protein composition of the capillary endothelial cell membrane. J. Neurochem. 41, 1341-1348.

Maggi F.M. and Catapano A.L. (1987). A dot-blot assay for the low density lipoprotein receptor. J. Lipid Res., 28, 108-112.

Mahley R.W. and Innerarity T.L. (1983). Lipoprotein receptors and cholesterol homeostasis. Biochim. Biophys. Acta 737, 197-222.

Méresse S., Delbart C., Fruchart J.C. and Cecchelli, R. (1989a). Low density lipoprotein receptor on endothelium of brain capillaries. J. Neurochem. 53, 340-345.

Méresse S., Dehouck M.P., Delorme P., Bensaïd M., Tauber J.P., Delbart C., Fruchart J.C. and Cecchelli R. (1989b). Bovine brain endothelial cells express tight-junctions and monoamine oxidase activity in long term culture. J. Neurochem. (in press).

Olsson Y. and Reese T.S. (1971). Permeability of vasa nerverum and perineurium in mouse sciatic nerve by fluorescence and electron microscopy. J. Neuropathol. Exp. Neurol. 30, 105-110.

Orlowski M., Sessa G. and Green J.P. (1974). gammm-glutamyl transpeptidase in brain capillaries : Possible site of a blood-brain barrier for amino-acids. Science, 184, 66-68.

Partridge, V.M. (1986) Receptor-mediated peptide transport through the blood brain barrier. Endocrine Rev. 7,314-329.

Pitas R.E., Boyles J.K., Lee S.H., Hui D. and Weisgraber K.H. (1987). Lipoproteins and their receptors in the central nervous system. Characterization of the lipoproteins in cerebrospinal fluid and identification of apolipoprotein B, E (LDL) receptors in the brain. J. Biol. Chem. 262, 14352-14360.

Pitas R.E., Boyles J.K., Lee S.H., Hui D. and Bissel D.M.(1985). Uptake of chemically modified low density lipoproteins in vivo is mediated by specific endothelial cells. J. Cell. Biol. 100, 103-117.

Raviola G. (1977). The structural basis of brain-ocular barriers. Exp. Eye Res. Suppl. 25, 27-63.

Robinson R.A., Ten Eyck C.J. and Hart M.N. (1986). Establishement and preliminary growth characteristics of a transformed mouse cerebral microvessel endothelial cell line. Lab. Inves. 54, 579-587.

Roheim P.S., Carey M., Forte T. and Vega G.L. (1979). Apolipoproteins in human cerebrospinal fluid. Proc. Natl. Acad. Sci. USA 76, 4646-4649.

Shivers R.R., Arthur F.E. and Bowman P.D. (1988). Induction of gap junctions and brain endothelium-like tight junctions in cultured bovine endothelial cells: local control of cell specialization. J. Submicrosc. Cytol. Pathol. 20, 1-14.

Soutar A.K., Hardes-Spengel K., Wade D.P. and Knight B.L. (1986). Detection and quantitaion of LDL receptors in human liver by ligand blotting, immunoblotting and radioimmunoassay. J. Biol. Chem. 261, 17123-17127.

Spatz M., Bembry R.F., Dodson H., Hervonen H. and Murray M.R. (1980). Endothelial cell cultures derived from isolated cerebral microvessels. Brain Research 191, 577-582.

Stein O. and Stein Y. (1980). Bovine aortic endothelial cells display macrophage-like properties towards acetylated 125I-labelled low density lipoprotein. Bochim. Biophys. Acta 620, 631-635.

Svendgaard N.A., Bjorklund A., Hardebo J.E. and Stenevi U. (1975). Axonal degeneration associated with a defective blood-brain barrier in cerebral implants. Nature 225, 334-336.

Urien S., Pinquier J.L., Paquette B., Chaumet-Riffaud P., Kiechel J.R. and Tillement J.P. (1987). Effect of the binding of Isradipine and darodipine to different plasma proteins on their transfer through the rat blood-brain barrier. Drug binding to lipoproteins does not limit the transfer of drug. J. Pharm. Exp. Theraup. 242, 349-353.

Tao C., Jung H., Nagy Z. and Brightman M.W. (1986). Tight junctions of cerebral endothelium in vitro are greatly enhanced in the compagny of astrocytes. Anat. Rec. 214, 131A.

Vlodavsky I., Fielding P.E., Fielding C.J. and Gospodarowicz D. (1978). Role of contact inhibition in the regulation of receptor-mediated uptake of low density lipoprotein in cultured vascular endothelial cells. Proc. Natl. Acad. Sci. USA 75, 356-360.

Weisgraber K.H., Innerarity T.L. and Mahley R.W. (1978). Role of lysine residues of plasma lipoproteins in high affinity binding to cell surface receptors on human fibroblasts.J. Biol. Chem. 253, 9053-9062.

Wilson A.J. (1985). Factor VIII related-antigen staining by immunoperoxidase technique in smaller laboratories: a potentiel problem. Am. J. Clin. Pathol. 81, 117-120.

Wolff J.R. (1977). Ultrastructure of the terminal vascular bed as related to function. Kaley G., Altura B.M. eds Microcirculation,Baltimore University Park Press, 1, 95-130.

REGULATION OF THE MYOCARDIAL BLOOD SUPPLY BY ADENOSINE AND ADENINE NUCLEOTIDES: CONSEQUENCES OF NEWLY DETECTED BARRIER FUNCTIONS OF THE CORONARY ENDOTHELIUM

S. Nees and A. Dendorfer

Dept. of Physiology
University of Munich
Pettenkoferstr. 12
D-8000 Munich 2, F.R.G.

INTRODUCTION

In the classical physiological view coronary flow is regulated mainly by certain metabolic products that are liberated by the cardiomyocytes, in amounts that increase with increasing cardiac work-load. Like chemical transmitters, these products are thought to develop their dilatory action directly at smooth muscle cells of the coronary resistance vessels. Adenosine, which is mainly derived by stepwise dephosphorylation of the central energy carrier of the myocardial metabolism, 5'-ATP, is still considered to be the most important mediator substance of this metabolic regulation of coronary flow (1) (fig 1).

Attractive though this physiological concept is, it is difficult to confirm by unequivocal experimental facts. We still lack a non-traumatic, direct access to the interstitial spaces of the heart, so that the exact composition of the interstitial fluid in the myocardium remains unknown. It was formerly assumed, however, that the coronary endothelium was generally permeable to small solutes. This suggested a possible solution to the problem, namely the idea that after intracoronary application such substances rapidly attained a concentration in the interstitium which largely equalled that in the intravascular space. For this reason, attempts were made early on to apply adenine nucleotides and adenosine intracoronarily and, on the basis of the biochemical and physiological reactions thus induced, to draw conclusions about the mode of action of these substances, in the interstitial space. In good agreement with the hypothesis described above, intracoronary applications of adenosine in concentrations above 0.1 µM do indeed induce immediate increases in coronary flow. Proof, then, for a direct action of adenosine at the vascular smooth muscle? Hardly, for these considerations and many other explanatory models completely ignore a tissue that is ubiquitous in the heart, separates the intravascular from the interstitial space and, though extremely thin, functions as a multi-facetted biochemical catalyst at the boundary of the two fluid compartments (2, 3). The tissue in question is the coronary endothelium.

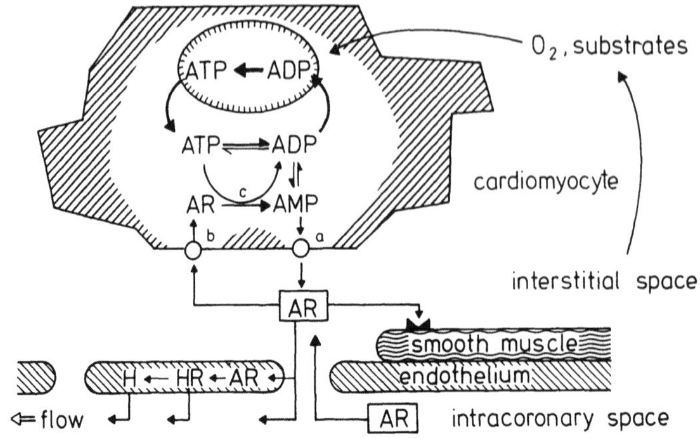

Fig. 1. Schematic presentation of the adenosine hypothesis of coronary flow regulation in its classical form. In the event of coronary flow limitation (=insufficient supply of O_2 and substrates), AMP is increasingly generated, dephosphorylated to adenosine (AR) by 5'-nucleotidase (a), the nucleoside being released into the interstitial space to exert its coronary dilating and flow increasing effect directly at the smooth muscle cells of the resistance vessels. This "endogeneous" AR may be partially taken up by the cardiomyocytes via a sarcolemmal nucleoside transporter (b), whence it can be rephosphorylated by the action of AR-kinase (c). To another portion it is catabolized by the endothelium. In all other respects, the endothelium is regarded as a passive, porous tissue. Hence, both the action and the metabolic fate of infused, i.e. "exogeneous" AR are taken to be directly comparable to those of endogeneously formed AR.

Below, a number of properties of this remarkable tissue are described, which enable the coronary endothelium not only to determine the purine-composition of the interstitial fluid of the myocardium and of the plasma of the coronary blood, but also to play an active role in the mediation of the vasodilatory effects of adenosine and its nucleotide derivatives to the smooth muscle cells of the coronary resistance vessels.

In an appendix, methodological information is also given on how coronary endothelial cells of microvascular origin can be isolated and cultivated for a variety of different investigational purposes.

EVIDENCE FOR AN UNEXPECTEDLY HIGH TIGHTNESS OF CONTINUOUS ENDOTHELIA

The entire vascular endothelium of the heart ventricles is of the so-called contiuous type. Accordingly, it is characterized by a continuous, single layer of cells with very narrow intercellular clefts, and a, likewise, continuous basal lamina. Also the large coronary arteries and all the other arteries of the vascular tree are lined by a continuous endothelial sheet. Under physiological conditions, such endothelia are apparently much tighter than was formerly assumed on the basis of theoretical and ultrastructural pore models for the transendothelial transport of substances. The cells of continuous endothelia are linked by well ordered gap and tight junctions and the intervening clefts are filled with an apparently rather complex but, as yet, biochemically undefined "cellular cement". It is therefore becoming increasingly doubtful that these narrow and intimately organized intercellular spaces can serve as routes of a fast and non-selective transport for all small hydrophilic molecules.

Neither the existence of this "small pore system", nor that of a rapid "large pore transcytosis system", implicated by the detection of what seemed to be vesicles in the endothelial cells, can be taken at face value. In their large majority these electron microscopically observable structures apparently do not serve transport purposes. As the pioneering and systematic studies by C. Crone, J. Frøkjaer-Jensen and M. Bundgaard from the Panum Institute, University of Copenhagen (i.e. (4)) have revealed, they are not really vesicles at all, but rather cross sections of branched invaginations of the luminal and abluminal plasmalemma of the endothelial cells. These invaginations neither form hydraulically conductive channels, nor do they carry their contents across the cells via fission or fusion events. Particular sources of error, and difficulty with the fixation and contrasting of endothelial ultrastructures that have recently come to light, underscore all these discrepancies.

In particular, more and more pharmacological and physiological studies are providing evidence that continuous endothelia must not simply be considered porous filter layers that permit rapid diffusive or convective passage of all small molecular substances. For example, compounds such as 5'-ATP and 5'-ADP, infused into large coronary arteries, induce vascular relaxation, provided the smooth muscle is shielded by an intact endothelium. If, however, the endothelial sheet is damaged, the same nucleotides induce a contraction of the now directly accessible vascular myocytes, and a resulting constriction of the vessel. Actually, a very intensively pursued modern area of pharmacological research, the investigation of endothelium-mediated vascular reactions (5), depends on, and in every instance reflects, a pronounced, selective impermeability, especially of the arterial endothelium, to many solutes.

In sophisticated studies on isolated perfused arterioles it has recently convincingly been demonstrated that also the endothelium of these tiniest arteries represents a substantial diffusional barrier to various vasoactive substances as well as their pharmacological antagonists. This was the more so, the more hydrophilic the compounds involved (6).

Recently, we found that even in myocardial capillaries the endothelial lining can represent a considerable barrier to the diffusion of hydrophilic, small molecules.
Isolated, but otherwise intact, fragments of capillaries from guinea pig heart rapidly hydrolyze 2'- and 3'-AMP. This is due to the action of a rather unspecific ecto-phosphomonoesterase,

which in the guinea pig heart we could demonstrate exclusively in the basal membrane of capillaries and on capillary pericytes (V_{max}= 196 nmol 2'-AMP/min/g ventricular tissue and K_M= 30 µM at pH 7.4). If, on the other hand, 2'- or 3'-AMP is selectively applied into the blood-perfused coronary system of intact hearts, it is impossible to detect phosphate ester hydrolysis. Consequently, even the capillary endothelium effectively shields the intravascular space from the interstitial space and the basal membrane-enzyme concentrated there, and must hence be virtually devoid of transcellular pores or open clefts.

ACTIVE CONTROL OF SOLUTE TRANSPORT ACROSS THE VESSEL WALL BY THE CORONARY ENDOTHELIUM

The observations just described prompt us to reconsider the mechanism of transendothelial exchange of hydrophilic solutes in the myocardium. If pores directly connecting the intracoronary and interstitial space of the heart do not exist, the essential exchange of substances must occur through the cells, via transport systems and pathways that are not revealed by electron microscopy. Components of the luminal and abluminal plasmalemma of the endothelial cells are, in fact, of major importance for accomplishing this transport, and may already determine whether certain substances are transported or not, simply on the grounds of their specificity. The complex endowment of the endothelial cells with all kinds of enzymes, establishes an additional, and very specific, level of transport control. Many substances could thus be completely metabolized during transport.

The results of our studies on the intracoronary metabolism of adenine nucleotides and adenosine provide strong supporting

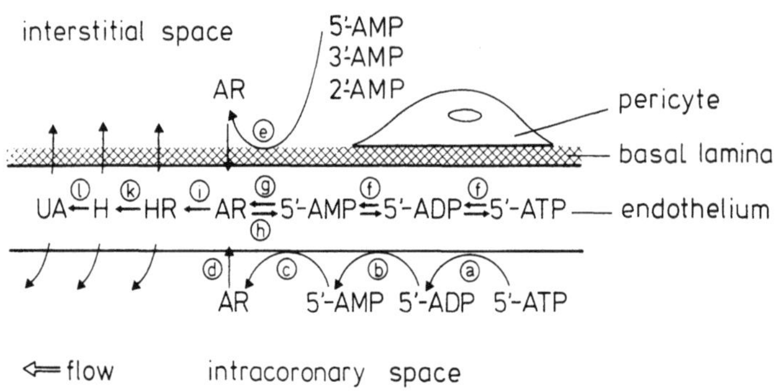

Fig. 2. Illustration of the metabolic barrier function of the coronary endothelium with respect to purine compounds. AR=adenosine, HR=inosine, H=hypoxanthine, UA=uric acid. a=ecto-ATPase, b=ecto-ADPase, c=ecto-5'-nucleotidase, d=purine nucleoside transporter, e=alkaline phosphatase, f=adenylate kinase (myokinase), g=soluble 5'-nucleotidase, h=adenosine kinase, i=adenosine deaminase, k=purine nucleoside phosphorylase, l=xanthine dehydrogenase.

evidence for these views (2, 7). Like vasoinactive 2'- and 3'-AMP, the coronary dilating adenine nucleotides 5'-ATP, 5'-ADP and 5'-AMP are negatively charged hydrophilic phosphate esters, for which the coronary system lacks direct transendothelial pores or open clefts. Thus, the coronary endothelium represents a virtually impenetrable physical barrier against the diffusional or convective passage of these nucleotides from the intracoronary to the interstitial space. In contrast to 2'- and 3'-AMP, however, during only a single passage through the coronary system, all the 5'-adenine nucleotides mentioned above are

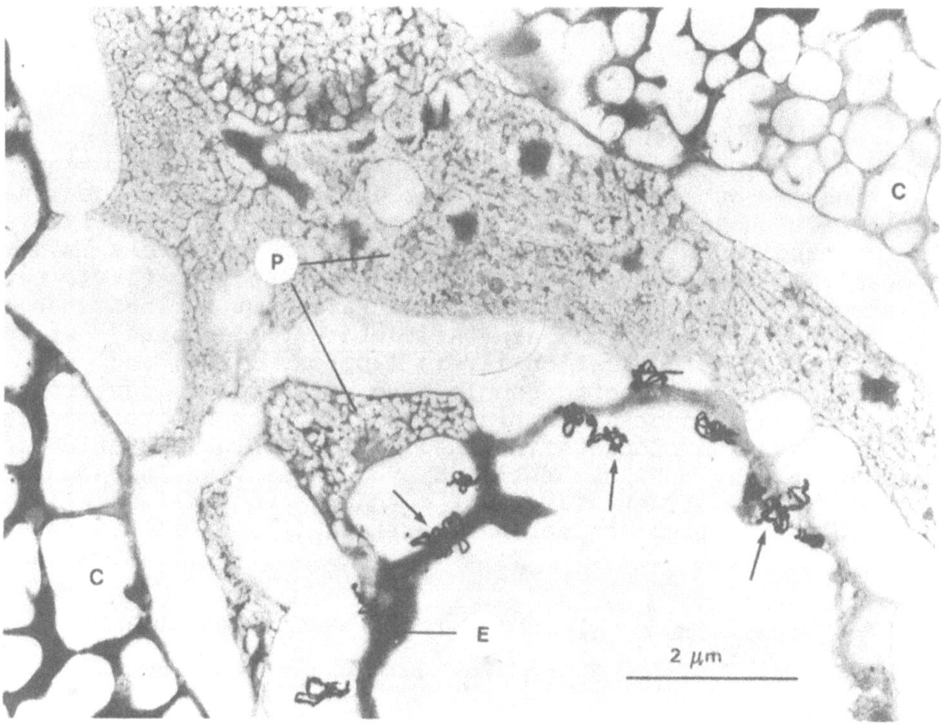

Fig. 3. Selective incorporation of radioactive adenosine into the coronary endothelium of an isolated perfused guinea pig heart as evidenced by autoradiography. The picture is an electron micrograph of myocardial tissue perfused for 15 min with $5x10^{-7}$M [2,5',8-^{3}H]-adenosine. Silver grains reflecting the distribution of radioactive purines can be seen on the extremely thin endothelial cell (E) of a capillary (3 of them are indicated by arrows); there are no grains on the closely associated pericyte (P) and the surrounding cardiomyocytes (C). In order to prevent excessive diffusion of low molecular weight metabolites, the tissue was rapidly frozen at -90°C and freeze-dried for 3 days at -100°C. The author is indebted to Prof. Dr. V. Herzog, Zellbiologisches Institut Universität München, for his kind help. For further experimental details see (2).

considerably dephosphorylated to adenosine via a cascade of specific and highly active ecto-5'-nucleotidases at the luminal surface of the endothelium (fig. 2). Consequently, the endothelium also acts as a metabolic barrier to 5'-adenine nucleotides in the coronary perfusate.

But the story does not end here. The vasoactive nucleoside adenosine, formed at the luminal endothelial surface by the action of the ecto-nucleotidase cascade, can now be taken up into the endothelial cells via a relatively non-selective plasmalemmal transporter (fig. 2). In the endothelial cell, adenosine is preferentially rephosphorylated, thus remaining trapped within the cell in form of nucleotides incapable of crossing the cell membrane. Autoradiographic studies using ^{3}H-adenosine demonstrate in a particularly direct manner how, in this way, coronary dilating adenosine is shielded from contact with other myocardial cells, such as smooth muscle cells, or even the very closely adjacent pericytes (fig. 3).

We have recently been able to show that not merely more than 80% of the adenosine kinase activity present in ventricular tissue, but also virtually the entire adenosine deaminase, nucleoside phosphorylase and xanthine dehydrogenase activity of the myocardium is contained in the coronary microvessels, very probably in the endothelial cells (unpublished results, report in preparation). This amazing concentration of purine catabolizing enzymes serves to explain the fact that adenosine, at least up to certain concentration levels (approx. 1 µM), and even when the endothelial nucleotide pools have been filled, can cross the extremely thin endothelial tissue only in form of its vasoinactive degradation products inosine, hypoxanthine and uric acid. This fact has been demonstrated by quantitative studies performed with confluent layers of coronary endothelial cells established in culture on porous substrata (2), by identifying the

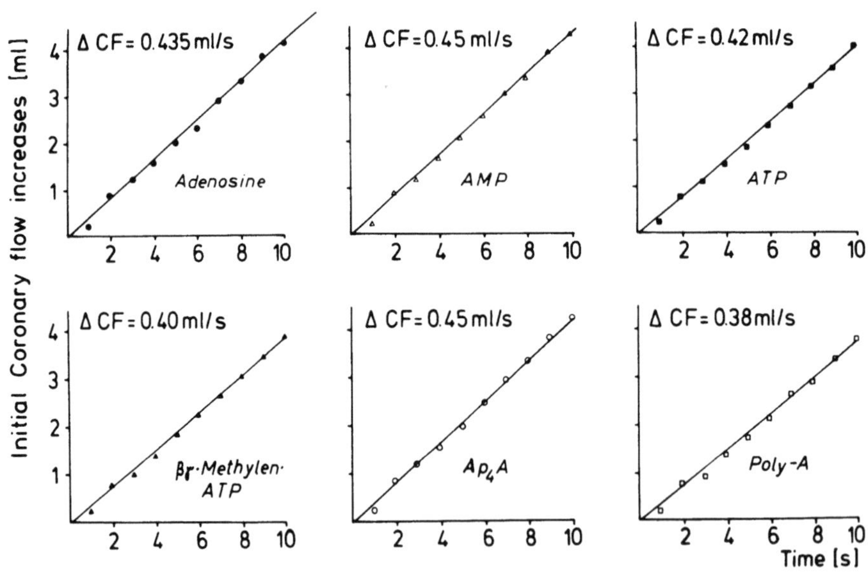

Fig. 4. Coronary flow increases 10 s after intracoronary application of $5x10^{-7}$M adenosine, AMP, ATP, β,γ-methylene-ATP, diadenosine-tetraphosphate (Ap_4A), or polyadenylic acid (Poly-A).

various purines released with the venous effluent of isolated perfused hearts (8), and by comparative studies on the purine metabolism in cultured aortic and coronary endothelial cells (9).

DILATORY EFFICACY OF ADENOSINE AND ADENINE NUCLEOTIDES

Now, the flow increases induced by the intracoronary application of equimolar adenosine, 5'-AMP, 5'-ADP and 5'-ATP, are all virtually identical in terms of the 1-2 seconds required for their onset, the initial velocity, concentration dependency and amplitude.

As fig. 4 shows, these results can be expanded to include such metabolically stable low molecular weight nucleotide derivatives as β,γ-methylene-5'-ATP and 5'-,5'-diadenosine-tetraphosphate, as well as high molecular weight polyadenylic acid, all of which are distinguished by a single vasoactive adenosine moiety per molecule (7). Also of interest in this connection are older studies involving high molecular weight oligosaccharide or protein conjugates of adenosine or AMP, which likewise proved to be fast-acting coronary dilators (see citations in reference (2)).

MEDIATION OF ADENOSINE- AND ADENINE NUCLEOTIDE-INDUCED DILATATION VIA THE ENDOTHELIUM

On the basis of all the data presently available, we draw three important conclusions:

1) The dilatory action of infused adenosine and its nucleotide derivatives cannot be elicited directly at the smooth muscle cells of the coronary resistance vessels; the relaxation signal must somehow be mediated by the endothelial barrier.
2) Since adenosine and low and high molecular weight adenine nucleotides all exhibit a similar dilatory efficacy, irrespective of whether they are metabolized or not, this effect must be induced via extracellular receptors localized on the luminal surface of the endothelial cells.
3) The postulated receptor type recognizes the adenosine moiety of the agonist, irrespective of whether it is the nucleoside itself, a low molecular weight 5'-nucleotide derivative, or a high molecular weight adduct of these compounds.

Besides an adenosine A_2-receptor (10) which, however, apparently is not involved in conveying the signal responsible for vascular relaxation (11), we were able to identify another, as yet only poorly characterized purinergic receptor on the surface of cultured coronary endothelial cells (12). Stimulation of this particular receptor with either adenosine or various adenine nucleotides leads to immediate activation of the endothelial phosphoinositol metabolism. Further studies are, however, required to confirm the putative significance of this structure for the regulation of coronary flow in the intact heart.

DO EDRFs HAVE A ROLE TO PLAY?

What are the underlying mechanisms that permit the coronary endothelium to mediate the relaxing effect of adenosine-containing compounds on the smooth muscle cells of the coronary resistance vessels?

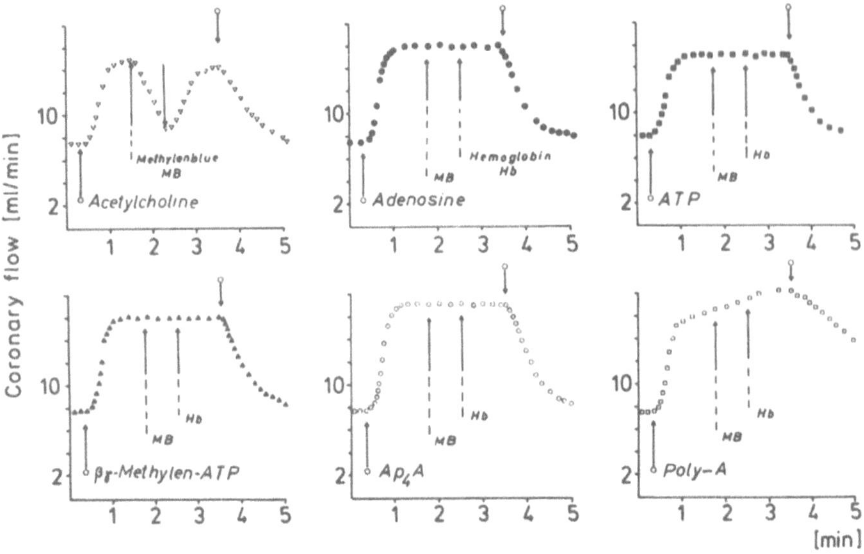

Fig. 5. Changes in coronary flow after intracoronary application of $5x10^{-7}$M acetylcholine, adenosine and various nucleotides. Influence of methylene blue ($5x10^{-5}$M) and hemoglobin (10^{-5}M). While the EDRF-dependent coronary flow increase induced by acetylcholine was strongly inhibited by these substances, the EDRF-independent effects of the purines remained unchanged.

With the aim of finding answers to such questions, we studied the influence of adenosine and various adenine nucleotides on the coronary system of isolated perfused guinea pig hearts, first in the absence and then in the presence of typical inhibitors of nitric oxide (NO), the best characterized of the Endothelium-Derived-Relaxing-Factors (EDRFs) (5). To our surprise, neither 10 µM methylene blue and hemoglobin, nor albumin (up to 0.5 mM) proved to have any influence on the effect of the adenosine compounds (fig. 5). Nor did pre-perfusion of the hearts for 30 minutes with 1 µM indomethacin, which inhibits the sythesis of other possible EDRFs such as prostacyclin PGI_2, result in any decrease in the effect of adenosine or its nucleotide derivatives in the coronary system (7).

Recently, we were able to further diminish the possibility that the endothelium-mediated coronary flow increase induced by adenosine might be mediated by a humoral endothelial factor. A specially designed bio-assay system was used, which comprised two isolated guinea pig hearts sequentially perfused at constant pressure. The first heart was exposed to 1 µM adenosine in the Krebs-Henseleit perfusate. Immediatly after emerging from the coronary sinus the venous effluent was supplemented with adenosine deaminase (1 U/ml) and passed through a teflon coil heated to 37 °C and immersed in carbogen gas (95% O_2, 5% CO_2) at a pressure of 1 atm. This sufficed both to fully rid the perfusate of the remaining adenosine and to reoxygenate the solution within only 10 seconds. Subsequent perfusion through the second heart (indicator heart) did not change coronary resistance in comparison with fresh perfusion medium. If the addition of adenosine deaminase was interrupted, vascular resistance in the indicator heart fell after approximately 11 seconds to a level corresponding to an intracoronary infusion of about 0.5 µM adenosine. In the heart preparation used, coronary dilatation by infused adenosine becomes detectable at roughly 0.1 µM, i.e. at 1/10th of the concentration applied to the first heart. Thus if a humoral factor is essential for transmitting endothelium-mediated dilatation by adenosine, and if a substantial portion of this factor is washed out in the coronary effluent, then it must have a half-life of less than 3 seconds for its level to fall below 1/10th and thus remain undetected in the indicator heart. All the EDRFs discussed to date, however, are much more stable under comparable conditions.

CONSEQUENCES AND PERSPECTIVES

The results described only briefly here indicate, in the first instance, that the coronary endothelium is of very central significance in the entire span of purine metabolism in the myocardium. Owing to its strategic location at the interface of the intravascular and interstitial compartments (100 and 20 µl/g ventricular tissue, respectively (13)) and a large cross sectional area (about 500 cm^2/g ventricular tissue (14)) the coronary endothelium is in extremely intimate contact with both the blood and the interstitial fluid (the respective quotient of endothelial surface to fluid volume is 5,000 and 25,000 cm^{-1}!). Together with its very efficient endowment with purine metabolizing enzymes, the coronary endothelium is able to determine the purine content not only of the various myocardial fluid spaces, but also of the different myocardial tissues. This being the case, it is difficult to accept the view presented at the beginning of this article, that adenosine liberated by cardiomyocytes is able to diffuse over long distances to exercise a direct effect on the smooth muscle cells of the coronary resistance vessels.

In this context it is very interesting that physiologically relevant adenine nucleotides, and also adenosine can induce an endothelium-dependent rise in coronary flow. Our experiments with these purines indicate an especially close communication between the endothelium and the myocytes of the microvascular coronary system. Since humoral EDRFs hardly seem to be involved in mediating the purinergically induced rises in coronary flow, the communication would appear far more likely to take place via direct cellular contacts within the vascular walls.

In good agreement with such an interpretation, we have been able to demonstrate the ability of microvascular coronary endothelial cells to transmit signals electrotonically, and therefore almost instantaneously, over large distances (15). This property stems from a variable resting membrane potential (33 ± 6 mV), a very high input resistance (R_i = 1.7 ± 0.4 GΩ), and an impressive electrical length constant (λ = 250-550 µm). With the mediation of myoendothelial gap junctions so common in the coronary system the endothelium and the smooth muscle cells of the resistance vessels might be electrically coupled (fig. 6).

Adenosine and/or adenine nucleotides at the luminal surface of the arterial endothelium could very well induce intraendothelial electric relaxation signals that are then instantaneously conveyed to the arteriolar smooth muscle downstream of the stimulated endothelium. In the sense of temporal and/or spatial entrainment of similar signals, it even seems possible that the capillary endothelium might be substantially involved in such a mechanism of flow induction, in this instance upstream of the stimulated endothelium. This bidirectional operating signal system would be largely independent of the fluid flow in either the intracoronary or the interstitial spaces. Provided that appropriate receptors are also present on the abluminal side of the capillaries, the endothelium of these microvessels would also respond very rapidly to purine mediators released from nerve endings, interstitial cells, or cardiomyocytes. This would be a constellation ideally suited for the realisation of metabolic control of coronary flow. Moreover, the rapid metabolism of adenosine by the coronary endothelium that we have been able to show, would prove to be of functional benefit for such a mechanism. Very much like the synaptic space, where transmitters are usually removed by up-take and metabolism, the interstitial space of the myocardium would always be cleared of the vasoactive nucleoside, readying the coronary system for a new, distinct adenosine signal.

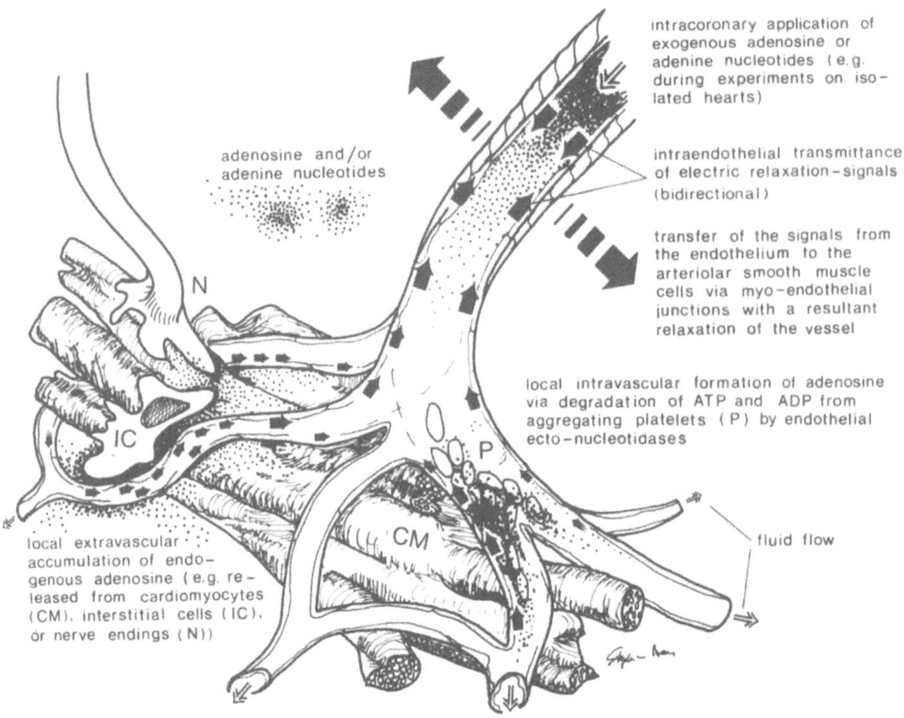

Fig. 6. How purinergically induced coronary flow increases mediated by the coronary endothelium may come: a working hypothesis. Exogeneous adenosine and/or adenine nucleotides can be infused into the coronary system or can be released from aggregating platelets (P) in the lumen of the coronary system. Endogeneous adenosine and/or adenine nucleotides can be released from nerve endings (N), interstitial cells (IC) or cardiomyocytes (CM).

As fig. 7 schematically illustrates, it is possible to largely disintegrate isolated perfused guinea pig or rat hearts within 20-30 min at 37 °C. For this purpose we make use of a protease solution made up of dispase I (Boehringer, Mannheim, FRG; 0.012 % w/v), collagenase II (Sigma, Deisenhofen, FRG; 400 U/ml), trypsin 1:250 (Serva, Heidelberg, FRG; 0.012 % w/v) and fatty acid free serum albumin (Sigma, Deisenhofen, FRG; added to a final protein concentration of 0.27 % w/v) in Ca^{2+}-free Krebs-Henseleit solution (KHS). In a subsequent phase of post-digestion, the network of the cardiomyocytes is completely dissociated into single cells. This process is best carried out in a funnel during gentle gassing from below with carbogen (95 % O_2, 5 % CO_2). Since, on account of their rigid basement membranes, the capillaries remain largely preserved under these conditions, an ideal point of departure is provided for the isolation of both pure cardiomyocytes and capillaries. Following removal of connective tissue, larger vessels and debris with the aid of a coarse-meshed nylon net (mesh size 200 µm), the myocardial cell and tissue suspension is now further separated with the aid of Percoll density gradient centrifugation. This is best achieved by using a novel centrifuge shown in cross-section in fig. 8. A major feature of this device ("Centrifuge ZS", E. Bühler, Rottenburger Str. 3, 7454 Bodelshausen, FRG) is the fact that it permits access to the separation glass via an axially

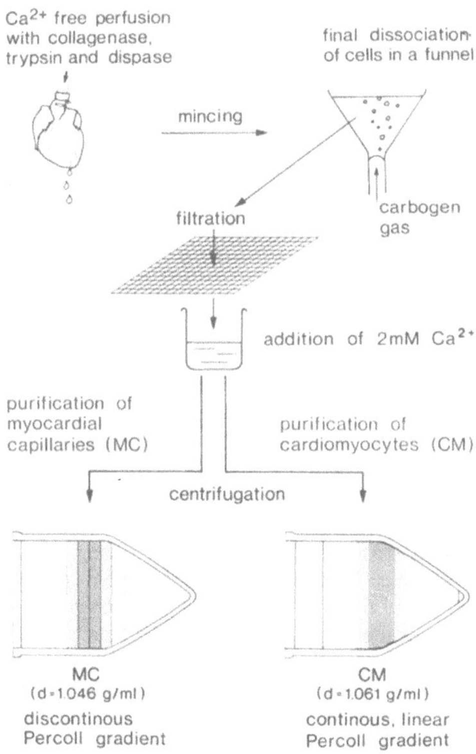

Fig. 7. Schematic representation of the isolation of extremely pure cardiomyocytes and myocardial microvessels from guinea pig or rat hearts.

arranged cannula, even while centrifugation is being carried out. In this manner, highly reproducible density gradients can be established, and, subsequently, the separated cell fractions can also be removed with no mixing effects. A separate cannula is available for applying the sample. Since the entire rotor can be autoclaved in superheated steam at a temperature of 120 °C, cell fractionation on sterile Percoll gradients can also be utilized at the same time for the complete removal of contaminating microorganisms (fungi, bacteria, viruses). These features ensure that subsequent cultivation of the isolated cells, too, presents no problems, even when the isolation steps just mentioned are carried out in the non-sterile environment of a usual laboratory. Some details of the extremely efficient separation of cardiomyocytes and myocardial capillaries on Percoll gradients are shown in fig. 9.

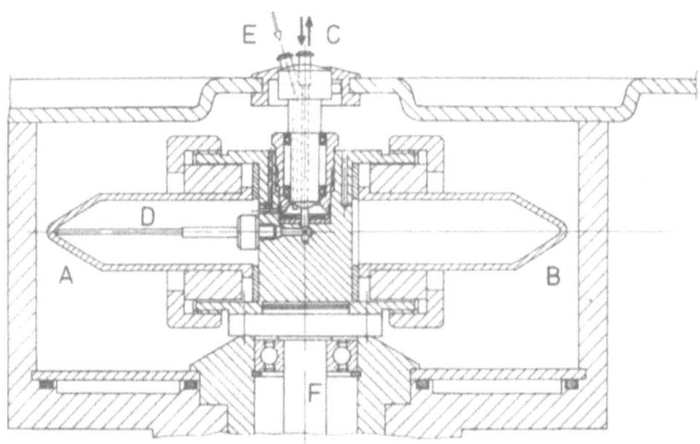

Fig. 8. Cross-sectioned view through the rotor of the ZS-centrifuge manufactured by E. Bühler. A=separating glass, B=balancing glass, C=axially arranged cannula, D=cannula in the separating glass, connected with cannula C, E=cannula for sample application, F=rotor axis.

Within the scope of the topic of this article, the identification and culture of coronary endothelial cells is of particular interest. For this purpose, the isolated microvessels, repeatedly washed with KHS, are first stained for alkaline phosphatase using conventional histochemical procedures. This enzyme proves to be a specific marker for the capillaries in the guinea pig heart (unpublished results) and was often used in the past to specifically stain the capillaries in the heart and skeletal muscles of the rat (16). In this way typical microvessel preparations can be shown to comprise more than 98 % capillaries. After suspending in complex culture media (i.e. Medium 199 supplemented with 20 % FCS) and seeding into petri dishes, rapid adhesion of the capillaries to the bottom and proliferation of the vascular cells takes place. One week later, on the basis of the presence of cellular receptors for acetylated LDL (17), the cultures can be shown to comprise more than 98 % endothelial cells. A phase contrast representation of such a culture is shown in fig. 10.

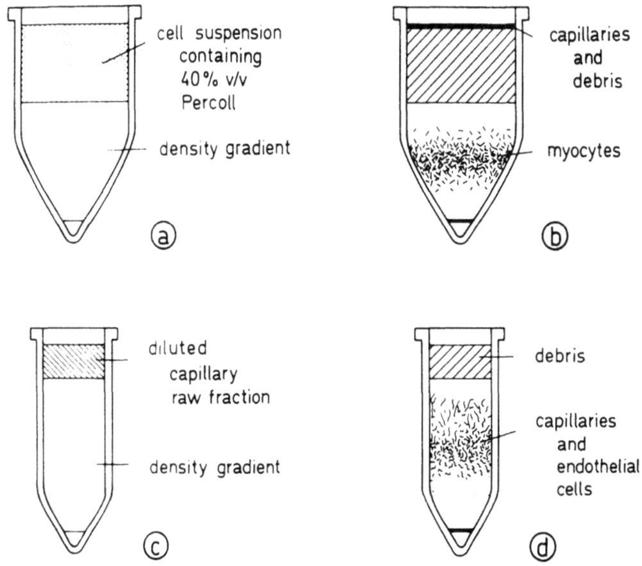

Fig. 9. Fractionation of myocardial cells and tissues by Percoll density-gradient centrifugation. a) application of the crude cell- and tissue-suspension on a Percoll gradient (density range: 1.05-1.07 g/ml), b) pattern of bands after centrifugation (10 min at 1000 rpm, 20 min at 2500 rpm). In addition to a band of pure, rod-shaped cardiomyocytes, a crude fraction of myocardial capillaries is also obtained. On account of its low density it forms the uppermost layer. c) application of the crude fraction of myocardial capillaries on a second Percoll gradient (density range: 1.026-1.055 g/ml), d) arrangement of the microvessels after centrifugation (same protocoll). Larger capillary coils are characterized by a lower density than shorter ones.

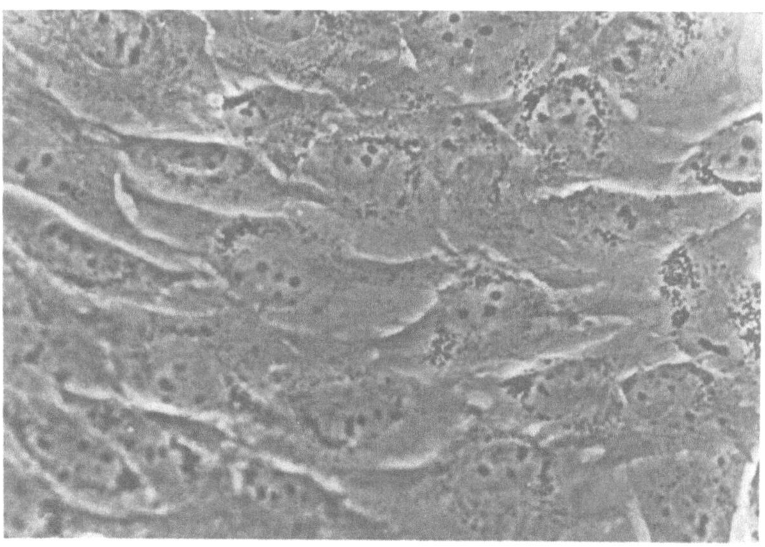

Fig. 10. Confluent layer of cultured microvascular coronary endothelial cells (phase-contrast micrograph).

Petri dish cultures are extremely suitable for numerous biochemical investigations. For the bulk production of microvascular endothelial cells from the coronary system, in contrast, microcarrier culture employing Cytodex-1-beads (Pharmacia, Uppsala, Sweden) has proved highly successful. This technique, with the aid of which gramm-amounts of endothelial cells can be obtained, is easily employed by using a special culture system ("Cell culture system ZKA", E. Bühler).

Finally, the establishment of confluent sheets of capillary cells on porous filters (i.e. Millipore filter, pore size 0.45 μm) should be mentioned. These filters can be used in subsequent experiments as the actual barrier between two fluid-filled spaces in a dual perfusion chamber apparatus. Such tissue culture systems are very suitable for the study of the transport of a wide range of substances through the vascular endothelium, and also for investigating numerous other directed processes at or within the endothelium. Further suitable devices are presently being developed in our laboratory.

ACKNOWLEDGEMENTS

These investigations were possible only with the collaboration of numerous colleagues and the financial support of the Wilhelm Sander and the Friedrich Bauer Stiftung.

REFERENCES

1. R. M. Berne, The role of adenosine in the regulation of coronary blood flow, Circ. Res. 47: 807-813 (1980).
2. S. Nees, V. Herzog, B. F. Becker, M. Böck, C. des Rosiers, and E. Gerlach, The coronary endothelium: a highly active metabolic barrier for adenosine, Bas. Res. Cardiol. 80: 515-529 (1985).
3. S. Nees, The adenosine hypothesis of metabolic regulation of coronary flow in the light of newly recognized properties of the coronary endothelium, Z. Kardiol., in press (1989).
4. M. Bundgaard, P. Hagman, C. Crone, The three-dimensional organization of plasmalemmal vesicular profiles in the endothelium of rat heart capillaries, Microvasc. Res. 25: 358-368 (1983).
5. P. M. Vanhoutte, G. M. Rubanyi, V. M. Miller, and D. S. Houston, Modulation of vascular smooth muscle contraction by the endothelium, Ann. Rev. Physiol. 48: 307-320 (1986).
6. M. J. Lew, and B. R. Duling, Differential arteriolar responses to luminally and extraluminally applied agonists and antagonists: evidence for an endothelial cell diffusion barrier, FASEB-Journal 2: A944 (abstr. 3774) (1988). Paper in press, Am. J. Physiol. (1989).
7. S. Nees, Coronary flow increases induced by adenosine and adenine nucleotides are mediated by the coronary endothelium: a new concept of the metabolic regulation of coronary flow, Europ. J. Cardiol., in press (1989).
8. B. F. Becker, and E. Gerlach, Uric acid, the major catabolite of cardiac adenine nucleotides and adenosine, originates in the coronary endothelium, in: "Topics and Perspectives in Adenosine Research", E. Gerlach and

B. F. Becker, eds., Springer, Berlin, Heidelberg, New York, pp. 209-222 (1987).
9. C. des Rosiers, S. Nees, E. Gerlach, Purine metabolism in cultured aortic and coronary endothelial cells, Biochem. Cell Biol. 67: 8-15 (1989).
10. C. des Rosiers, and S. Nees, Functional evidence for the presence of adenosine A_2-receptors in cultured coronary endothelial cells, Naunyn-Schmiedeberg's Arch. Pharmacol. 336: 94-98 (1987).
11. W. H. Newman, B. F. Becker, M. Heier, S. Nees, and E. Gerlach, Endothelium-mediated coronary dilation by adenosine does not depend on endothelial adenylate cyclase activation: studies on isolated guinea pig hearts. Pflügers Arch. 413: 1-7 (1988).
12. S. Nees, C. des Rosiers, and M. Böck, Adenosine receptors at the coronary endothelium: functional implications, in: "Topics and Perspectives in Adenosine Research", E. Gerlach and B. F. Becker, eds., Springer, Berlin, Heidelberg, New York, pp. 454-469 (1987).
13. P. Anversa, V. Levicky, C. Beghi, S. L. McDonald, and Y. Kikkana, Morphometry of exercise-induced right ventricular hypertrophy in the rat, Circ. Res. 52: 57-64 (1983).
14. J. B. Bassingthwaighte, T. Yipintsoi, and R. B. Harvey, Microvasculature of the dog left ventricular myocardium, Microvasc. Res. 7: 229-249 (1974).
15. J. Daut, G. Mehrke, S. Nees, and W. H. Newman, Passive electrical properties and electrogenic sodium transport of cultured guinea-pig coronary endothelial cells, J. Physiol. 402: 237-254 (1988).
16. O. Hudlická, Growth of capillaries in skeletal and cardiac muscle. Circ. Res. 50: 451-461 (1982).
17. P. A. Netland, B. R. Zetter, D. P. Via, J. C. Voyta, In situ labelling of vascular endothelium with fluorescent acetylated low density lipoprotein, Histochem.J. 17: 1309-1320 (1985).

CULTURED HUMAN UMBILICAL VASCULAR ENDOTHELIAL CELLS AND THEIR UNDERLYING MATRIX AS MODELS TO STUDY CELL ADHESION

Eva Bastida
Cell Biology Laboratory, R&D
Laboratorios Menarini
Barcelona, Spain

Vascular endothelium is the blood compatible container that covers blood vessels. The endothelial cell layer lining of the circulatory system, although structurally simple, is functionally complex and its integrity is essential for normal vascular function[1]. Although tissue culture techniques for mammalian cells were established about two decades ago, the specific requirements of vascular endothelial cells hampered their routine culture until rather recently. Jaffe and co-workers were the first who succeeded in growing vascular endothelial cells in the laboratory[2]. The source of cells were veins from human umbilical cords. The cells were isolated by incubation with the enzyme collagenase and cultured on glass or plastic surfaces in the presence of specific culture media supplemented with growth factors. More recently, other types of vessels have been used to obtain vascular endothelial cells, including bovine and swine aorta and human adult saphenous vein. Moreover, microcapillary endothelium has also been isolated and cultured. This represents an interesting advance that allows the study of the differences in the structure and properties of endothelial cells from large and small vessels.

Structurally, vascular endothelial cells show many typical organelles, including mitochondria, microtubules, microfilaments, as well as a unique organelle called the rod-shaped tubular body or Weiber-Palade body, of primary importance in the hemostatic function of the endothelial cell. It has been demonstrated by ultrastructural and immunocytochemical techniques that cultured endothelial cells possess the same characteristics of those freshly isolated, indicating that the basic morphological and secretory properties are kept under culture conditions.

In culture, vascular endothelial cells generate an extracellular matrix whose components are the same as the basal membrane of the blood vessels. The major components of the extracellular matrix are collagenous proteins, mainly collagens types I, II and IV, and the non-collagenous fibronectin and laminin. Other proteins such as von Willebrand Factor, thrombospondin and nidogen are also present[4]. Under physiological conditions the extracellular matrix of the vascular endothelial cells is not exposed to circulating blood; however, under certain pathological conditions the components of the subendothelium come in close contact with circulating blood cells and soluble plasma factors. The specific interaction of blood platelets and certain subendothelial components is the initiating step of the hemostatic mechanism. As it is well known, the uncontrolled amplification of the hemostatic mechanism leads to the development of thrombosis. In the laboratory exposure of the extracellular matrix is achieved by a variety of physical and chemical procedures.

Recent studies suggest that the thrombogenic properties of the matrix depend upon the method

Pharmaceutical Applications of Cell and Tissue Culture to Drug Transport
Edited by G. Wilson *et al.*, Plenum Press, New York, 1991

of its preparation[5]. Endothelial cell culture has also permitted the study of synthesis, storage and release of different cell products which are important in regulating vascular reactivity. Vascular endothelial cells are very active in synthesizing substances of different chemical nature, including prostacyclin (PGI_2), endothelial-derived relaxing factor (EDRF), platelet-aggregating factor (PAF), proteoglycans, thrombomodulin, tissue-plasminogen activator (t-PA), endothelin, and certain products of the enzyme lipooxygenase pathway; namely, 15-hydroxyoctadecadienic acid (15-HETE) and 13-hydroxyoctadecadienoic acid (13-HODE)[6].

Important pathologies other than thrombosis are also involved in the process of cell-cell interaction. Thus, focal adhesion of leukocytes to the blood vessel lining is a key step in inflammation and the attachment of malignant cells to capillaries represents one of the crucial events in the metastatic cascade. Studies on the interactions between tumor cells and the endothelium indicate that, at least in part, they depend upon the properties of the endothelial cells[7]. The combination of endothelial cell culture with the use of perfusion chambers has allowed for the first time the study of cellular interactions under dynamic conditions. Specifically, the use of a parallel-plate perfusion chamber, designed to mimic the rheological conditions of flowing blood has represented an interesting approach to elucidate the relative contribution of the endothelial and subendothelial components in promoting tumor cell adhesion[8].

REFERENCES

1. J. J. Sixma, "Role of blood vessel, platelet, and coagulation interactions." In: Haemostasis and Thrombosis A. L. Bloom and D. P. Thomas (Eds.). Churchill Livingstone: Edinburgh 1982; 252-267.
2. E. A. Jaffe, R. L. Nachmann, C. G. Becker, "Culture of human endothelial cells derived from umbilical veins identified by morphometrical and immunological criteria." J. Clin. Invest. 1973; 52:2745-2752.
3. M. A. Gimbrone, Jr., "Culture of vascular endothelium." Prog. Hemost. Thromb. 1976, 3:1-28.
4. P. G. De Groot, J. H. Reinders, J. J. Sixma, "Perturbation of human endothelial cells by thrombin or PMA changes the reactivity of their extracellular matrix towards platelets." J. Cell Biol. 1987; 104:697-704.
5. J. Aznar-Salatti, E. Bastida, G. Escolar, P. De Groot, M. R. Buchanan, A. Ordinas, "The method of exposure of endothelial cell extracellular matrix influences its thrombogeneicity." Atherosclerosis 1989 (In Press).
6. M. R. Buchanan, E. Bastida, "The role of 13-HODE and HETEs in vessel wall/circulating cell interactions." Agents and Actions 1987; 22:1-4.
7. G. L. Nicolson, "Metastatic tumor cell attachment and invasion assay using vascular endothelial cell monolayers". J. Histochem Cytochem 1982; 30:214-220.
8. E. Bastida, L. Almirall, A. Ordinas, "Platelet and shear rate promote tumor cell adhesion to human endothelial extracellular matrix - Absence of a role for platelet cyclooxygenase." Thrombosis Haemost 1989; 61:485-489.

HUMAN NASAL EPITHELIAL CULTURES

Richard C. Boucher and Elaine H.C. Cheng

University of North Carolina
School of Medicine
Chapel Hill, NC

I. Introduction

The proximal airway epithelia act in a coordinated manner to defend the lung from the effects of inhaled foreign materials. The defense mechanisms provided by these airway epithelia are usually analyzed within the framework of the mucociliary clearance system. Nasal epithelia are similar to those of the proximal intrapulmonary airways as both surfaces are populated by relatively equal numbers of ciliated, secretory, and basal cells. The motive force required to move materials trapped in airway surface liquids (ASL) to the oropharynx is provided by the cilia of the ciliated cell.

The composition of the liquids that line airway surfaces is complex and is regulated by distinct components of the epithelium. Most active ion transport appears to be a function of ciliated cells. Active transport of ions, via the generation of osmotic gradients, is thought to be important for moving bulk volume (water) across the epithelium. In addition, active ion transport systems control the ionic composition of this liquid. ASL also contains macromolecules that are regulated by two general processes located within the epithelium. First, macromolecules that are relatively specific to the region, e.g., mucins, are secreted by secretory epithelial cells. Second, macromolecules contained in serum permeate cross the epithelium, typically through the tight junctions that adjoin epithelial cells. The permeability of the epithelial barrier to serum macromolecules is a property that defines the "barrier" characteristics of the epithelium. This more simple notion reflects the fact that the epithelium is a relatively impermeable structure that heavily retards the movement of macromolecules from blood to airway lumen.

The epithelia that line the proximal airways appears to be deranged in the number of human diseases. Inherited diseases such as dysmotile cilia syndromes and cystic fibrosis (CF) appear to be localized to this tissue. Probably the best studied disease of the airways epithelium is cystic fibrosis (1). It appears that the clinical

Pharmaceutical Applications of Cell and Tissue Culture to Drug Transport
Edited by G. Wilson *et al.*, Plenum Press, New York, 1991

phenotype of cystic fibrosis reflects the ion transport abnormalities that are consequent to the central genetic lesion. These transport defects perturb the volume and composition of the airway surface liquid, which may contribute to the chronic bacterial infection these patients suffer. Although not rigorously proven, it appears that other diseases such as asthma involve the airways epithelium. This supposition is based on studies that have shown that airways hyperreactivity is induced by inhalation of solutions that may perturb the composition of ASL, e.g., hypotonic solutions (2). Finally, it is clear that the epithelium of the proximal airways is attacked by numerous infectious agents. Specific pathogen attachment to the proximal airway epithelium has been described for mycoplasma pneumonia and influenza. Consequently, much of the clinical sequellae of these diseases reflects the direct interaction of the pathogen and host at the airway epithelial barrier.

The proximal airway epithelium has also been a target for drug delivery. Delivery of drugs to the nasal mucosa by direct installation and to the proximal airways by aerosols have been widely employed. Despite the widespread use of these routes for drug delivery, relatively little is known about the quantitative aspects of drug translocation across airways epithelium. Little work has been done on characterizing the "barrier" characteristics of the human airways epithelium with regard to size dependent solute translocation. Similarly, virtually no work has been done on human airways epithelium with regard to active transcellular drug transport. Given the wide variety of drugs given via this route, these lack of data are somewhat surprising.

A common problem that limits the investigation of the human proximal airways epithelium with regard to the perspectives described above is the limitation of adequate material for study. Because of the requirement that human airway epithelia be studied for functional properties rapidly after excision (less than 20 min.), the amount of material available for study is severely limited. It is because of these requirements, and the information stemming from studies of the culture of animal airway epithelia, that in collaboration with Dr. Reen Wu we sought to establish methods for culture of human nasal epithelium (3).

II. General features of cell culture of human nasal epithelium (HNE)

A. Tissue sources: The principal requirement for culture of human nasal epithelium is freshly excised surgical material. Generally, normal nasal epithelium is obtained from surgical resections of the inferior turbinate for the treatment of sleep apnea syndromes or plastic reconstructive surgery. Nasal polyps are also available as tissue sources and are extremely useful because they are associated with pertinent diseases, e.g., aspirin-induced asthma, cystic fibrosis, and dysmotile cilia syndromes. Surgical material is most useful if it is directly placed into modified Eagle's medium (MEM) at 4°C prior to cell disaggregation. Despite the use of antibiotics and antifungal agents, relatively little success has been accrued employing maxillary sinus tissue that is typically colonized/infected with bacterial organisms.

B. Cell disaggregation and isolation: The nasal specimens are washed in MEM solution and then placed in an MEM solution containing 0.1% protease 14 (Sigma). Specimens can be left in this solution at 4°C for 16-48 hours. For cell isolation, the MEM containing disaggregated cells is decanted from the specimen, the protease neutralized by 10% fetal bovine serum, and the cells washed twice in MEM-10% fetal bovine serum. On average, $5x10^6$ cells are obtained from nasal turbinate and nasal polyps specimens with a viability that exceeds 85%. The analysis of the disaggregated cell populations indicates that approximately 10% of the cells are ciliated with the remainder being basal and secretory cells. This distribution indicates that the ciliated cells are preferentially lost during the disaggregation and isolation procedure.

C. Media: For nasal cell cultures, a serum free culture medium has been developed (3). This medium is F12 based and is supplemented with the following hormones: insulin (5 μgs/ml); ECGS (7.5 μgs/ml); cholera toxin (10 ng/ml); hydrocortisone (10^{-7}M); T3 (10^{-10}M); EGF (1 ng/M); and transferrin (7.5 μgs/ml). The most rigorous studies relating the effects of the individual hormone/growth substances on cell proliferation, morphology, and transport function have been performed by Van Scott and co-workers in dog tracheal epithelial cells (4). However, some work has been performed to investigate these relationships in human cells. With regard to growth studies on tissue culture plastic substrates, Wu and co-workers identified that insulin and ECGS were the most potent mitogens (3). On plastic substrates, EGF did not appear to exhibit mitogenic activity for human nasal epithelia. However, discrepancies with regard to these observations do appear and reflect the nature of the substrate on which the cells are grown. For example, EGF appears to induce cell proliferation and multilayering of human nasal epithelia when they are grown on suspended collagen substrates (see below).

Specific studies have focused on the role of cholera toxin in the human nasal cell culture system. Cholera toxin exhibits little mitogenic activity when HNE are cultured on plastic supports (3). Because of interests in studying beta agonist actions on HNE, the interactions between cholera toxin and the beta adrenergic path have been examined. In an initial series of studies, the responses of cell cAMP levels of HNE cells grown with or without cholera toxin upon exposure to isoproterenol were explored. As shown in Fig. 1, cholera toxin appears to increase the basal level of cAMP in human nasal epithelial cells. Paradoxically, it appears that cells grown in cholera toxin are more sensitive to induction of cAMP by the beta agonist isoproterenol than HNE cultured without cholera toxin: IC_{50} is equal to $9x10^{-10}$ for HNE grown in cholera toxin; $IC_{50} = 3x10^{-8}$ for HNE grown without cholera toxin. Thus, cholera toxin appears to have effects on cell cultures that are complex and deserve careful consideration.

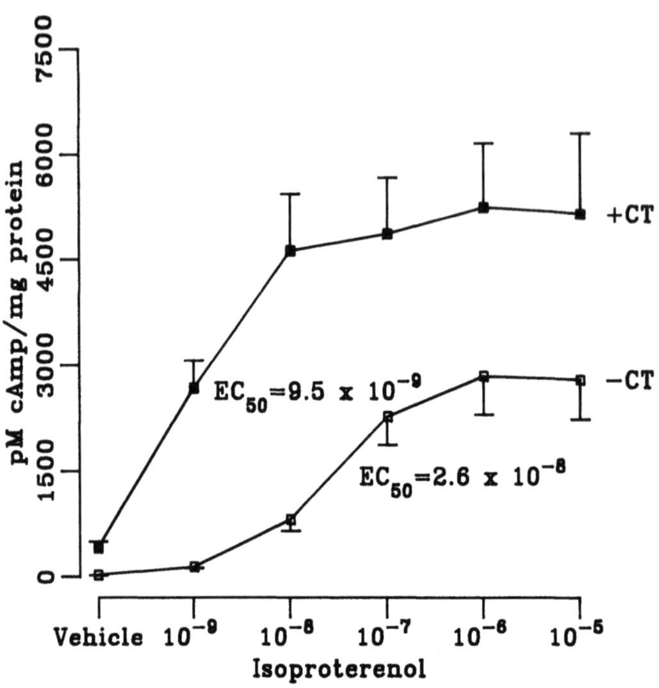

Fig. 1. Effect of cholera toxin in culture media on human nasal epithelial cell cAMP content before and during 10 minute exposure to isoproterenol. Mean ± SEM; N=7 or more per point.

Specific manipulations are currently being evaluated to increase the differentiation of HNE in culture. Retinoic acid has been employed by Wu and co-workers to increase the differentiation of a number of animal airway cells in culture, and it would appear in preliminary studies to exhibit similar effects on human nasal epithelium. The effects of TGF beta are also being evaluated with regard to inducing the differentiated phenotype in HNE. TGF beta may be one of the prime components of 3T3 fibroblast conditioned media that appears to inhibit cell proliferation and increase differentiation in HNE (5).

D. Substrates for cell culture

1. Tissue culture plastic: Human nasal epithelial cells readily attach to standard tissue culture grade plastic supports. The attachment efficiency is approximately 30% for human nasal epithelial cells on this substrate. All growth studies for HNE have been performed on this substrate and indicate that HNE cells exhibit a doubling time of approximately 24 hours in primary culture. HNE can be passaged approximately two-three times from one plastic substrate to another employing one to four splits. The disadvantages of culturing HNE in tissue culture plastic are that only one side of the cells is readily accessible for manipulation and the cells exhibit a very de-differentiated, flat morphology.

2. Suspended collagen matrices: An extremely versatile substrate is the suspended collagen matrix. This culture type was originally described by Steele and co-workers and involves cross-linked collagen (6). For HNE, bovine type I collagen that is cross-linked with ammonium hydroxide and glutaraldehyde is adequate (5). The collagen matrix can be affixed across an orifice drilled in a cup that has had legs affixed to its sides that creates a insert that fits into a standard tissue culture cup. The cells are plated on the collagen covering the orifice at high density ($3x10^5$ cells/0.15 cm^2). This matrix can also be made in larger sizes that can be affixed to polystyrene tubing to create surface areas of 5 cm^2. This substrate configuration is ideal for many studies of human nasal epithelia because of the enhanced differentiation of the epithelial cells on this substrate and the accessibility to both the apical and basolateral surfaces of these cells. For ion transport studies, cells are typically monitored for the development of a transepithelial potential difference by serial measurements directly in the culture dishes as a function of time in culture (5). Because the collagen is highly translucent, the development of morphologic confluence and differentiation can also be monitored.

3. Denuded tracheal graft: The approach of Terzaghi and Nettesheim, employing rat or rabbit tracheas denuded of the endogenous epithelium by freezing and thawing, has been utilized for human nasal epithelium (7). Cells (10^4-10^5) are inoculated into the denuded trachea, both ends of the tracheal cylinder are tied off, and the graft is placed in the nape of a nude mouse. The graft can then be harvested at 1-8 weeks for assessment of the cellular repopulation of the trachea. This graft technique is ideal for generating the most differentiated morphologic phenotype. It suffers from the disadvantage of being limited to single measurements.

III. Representative studies of the functions of human nasal epithelium

A. Ion transport studies. The ion transport activities of the cultured nasal epithelium have been compared with those of freshly excised intact tissues. It appears that, like the native tissue, electrogenic sodium absorption is the major basal ion transport system expressed in cultured HNE (5). Studies with intracellular microelectrodes have confirmed that the intracellular electrical profile is similar in the cultured cells to that reported in the native tissue (8).

The nasal epithelial cell cultures have also been shown to exhibit chloride secretion. The chloride secretory current is induced routinely by the addition of amiloride (9). However, a number of receptor gated mechanisms appear to also induce chloride secretion. Perhaps the best studied agonist has been bradykinin which raises intracellular calcium and induces chloride secretion over parallel time courses (10). The cell culture system reproduces the findings derived from freshly excised material.

B. Macromolecule secretion: In organ culture the freshly excised human nasal epithelium has been shown to secrete high molecular weight glycoconjugates that are considered to be airway mucins. In the organ culture specimen, the mucins are likely derived from both the goblet cells located within the superficial epithelium and the mucin secreting cells in the glandular region. Using the suspended collagen system, radiolabeled precursors of high molecular weight glycoconjugates, e.g., ^{3}H-glucosamine, ^{3}H serine, $^{35}SO_4^{=}$, have been added to culture media and the material secreted into the apical and basolateral media compartments harvested and analyzed (11). The cultures release into media a variety of high molecular weight macromolecules, including proteoglycans and mucin-like macromolecules. These latter glycoconjugates, "high molecular weight mucin-like glycoconjugates", are characterized by the following qualities: molecular mass more than 10^6 daltons; resistance to degradation by bovine testicular hyaluronidase; and a sedimentation coefficient of ˜1.51 on cesium chloride density gradients. Amino acid analysis of these compounds, however, did not show them to be typical mucins with low threonine and proline contents. Trypsinization of the cell surface, a maneuver that releases cell surface glycoconjugates, appears to release similar macromolecules. Thus, it appears that at present the nasal cell culture system releases into culture media high molecular weight, glycocalyx-associated glycoconjugates and does not secrete typical airway mucins. The biochemical data are consistent with histological studies that reveal an absence of typical goblet cells containing Ab-PAS positive granules. Consequently, the development of a culture system that contains differentiated goblet cells that maintain a mucin secretory capacity is a necessary goal for this cell culture system.

C. Biochemical characterization: An extensive characterization has been performed of the intracellular kinases involved in signal transduction in cultured HNE. Recent studies have shown that the predominant isoform of the cyclic AMP dependent kinase (PKA) is the R_{II} form (10). Subcellular fractionation has shown that the R_{II} form appears to be membrane associated. Because beta agonists can rapidly activate chloride channels in this tissue, this may be one mechanism for compartmentalization of the cAMP response.

In addition, protein kinase C (PKC) activities have been measured in human nasal epithelial cells in culture (10). The PKC activities appear to be relatively small (10%) compared with the activity of PKA. The PKC activity has been shown to translocate from the cytosol to membrane compartment in response to stimulation with PMA.

Biochemical studies have also been performed to describe mediators formed and released by epithelial cells into their environment. Much attention has focused on arachidonic acid (AA) metabolism. In freshly excised human nasal epithelium, arachidonic acid is predominantly metabolized by the 15-lipoxygenase system with relatively little or no cyclooxygenase activity detectable (12). Human nasal epithelial cells cultured on suspended collagen matrices at high densities appear to exhibit identical patterns of AA metabolism as the fresh material. However, cells grown on tissue culture plastic appear to express cyclooxygenase activity as the major metabolic route for arachidonic acid metabolism. Thus, whereas studies of mediator formation may be useful in cultured cells, the nature of the substrate may be extremely important with regard to the accuracy of the cell culture system to mimic freshly excised material for these investigations.

IV. Transformed cells

A. Development: A variety of transforming proto-oncogenes have been employed in attempts to extend the proliferative potential of human airway epithelia. With the exception of the report of Yoakum et al (13), initial attempts to transform HNE employing Harvey and Kirsten ras were unsuccessful. Preliminary reports have indicated that some success has been achieved using the adenovirus gene E1A. However, the most consistent success in transforming human airway epithelial cells has been reported using the SV40T gene. Success has been reported using the origin minus SV40 virus and using retroviral-SV40T constructs.

In collaboration with Dr. Anton Jetten at the NIEHS, we have recently generated transformed cells that were selected for expansion on the basis of preservation the active ion transport phenotype (14). In these studies, pZIP-neoSV(X)1/SV40T retroviral constructs were employed. Colonies were selected on the basis of neomycin resistance and proliferative potential. The colonies were subsequently screened for ion transport phenotype by transferring cells to the collagen matrix preparation and monitoring for development of transepithelial resistance (R_t) and ionic current flow. Colonies that exhibited the capacity to generate these properties were further studied with regard to their sensitivity to maneuvers that increase differentiation. Of the variables studied, it was found that vitamin A, raised calcium media (1.2 mM), and 3T3 cell conditioned media appeared to increase ion transport capacity. Subsequently these materials, supplementing Keratinocyte Growth Media, were used for ion transport studies.

B. Phenotypic characterization: The transformed cells were shown to express the SV40T antigen by immunoprecipitation of this antigen with monoclonal antibodies. Further, the transformed cells were shown to express the pattern of keratins that was typical of this cell in the pre-transformed state.

The development of the transepithelial potential (V_t) was monitored serially in the cultures and was found to reach peak magnitudes approximately six to eight days after seeding. These preparations were then characterized with regard to their transepithelial bioelectric parameters and their response to chloride secretogogues. These responses are summarized in Table I. In brief, it was found that both

Table I. Comparison of Basal Bioelectric Parameters and Qualitative Respones to Drugs of Primary Cultures and Transformed Cells from CF and Normal Subjects

Source	Vt (mV)	Rt ($Ohm \cdot cm^2$)	Ieq ($\mu A \cdot cm^{-2}$)	Amil	Iso	A23187
				(▲Ieq)		
Normal Primary	-10.1 ± 1.0	400 ±35	28.2 ±3.0	↓	↑	↑
Transferred Normal	-0.9 ±0.1	79 ±12	12.8 ±3.0	↓	↑	↑
CF Primary	-29.2 ± 4.4	435 ±42	76.2 ±12.9	↓	0	↑
Transferred CF	- 2.3 ± 0.3	125 ±12	19.2 ± 2.1	↓	0	↑

normal and CF transformed cells developed active ion transport as indicated by the development of a short circuit current (Ieq). The initial characterization of this current was performed with use of pharmacologic inhibitors. It was shown that the majority of the current for both the CF and normal cells reflected a sodium absorptive current because of the inhibition by amiloride. In airway cells, the amiloride resistant current is chloride secretion. Therefore, we characterized the ability of two classes of chloride secretogogues to modulate this current. In the normal cells, both agents that raise cell cAMP (isoproterenol) and agents that raise intracellular Ca^{++} (calcium ionophore A23187) increase short circuit current. In contrast, like freshly excised tissue and primary cultures, only A23187 induced a chloride secretory current in CF. Thus, the basal transepithelial bioelectric parameters and the regulation of the chloride secretory current appear to be phenotypically preserved both in the normal and CF transformed cells.

The cells were also characterized with regard to their intracellular bioelectric parameters. Studies with double barreled chloride and sodium selective electrodes revealed that the intracellular potentials and the distribution of the cellular resistances in both CF and normal cells were similar to those in primary cell cultures (9). Direct measurements of the basal apical membrane permeability employing the use of chloride ion substitutions in the luminal bathing solution showed that the normal cells exhibited a conductive chloride permeability in the resting state whereas CF cells did not. These findings also paralleled thcse described in fresh tissues.

Finally, these cells were characterized at the single channel level. For these studies, the patch clamp technique was employed and the focus was on the regulation of the secretory chloride channel. In brief, the secretory chloride channel was expressed in the transformed cells (14). The biophysical characterization of the resting properties of these channels, both in CF and normal cells, approximated that obtained in the primary cultures. The chloride channel exhibited an outward rectifying behavior with a single channel conductance at a zero pipette potential of approximately 35 pS. The kinetic analysis of the channels revealed them to be a relatively "fast flickering" type of channel. Importantly, the abnormal regulatory features of the chloride channel reported in cultured primary cells were preserved in the transformed cells. Exposure of excised inside out patches from normal cells to the purified catalytic subunit of the cyclic AMP dependent kinase initiated activation of chloride channels. In contrast, parallel experiments performed in inside out excised patches from the CF transformed cells indicated no activation of channels by the purified subunit of PKA. In summary, phenotypic features of both CF and normal cells are well preserved after the transformation event.

V. Future considerations

It appears that in general the use of cultured cells for the study of proximal airway epithelia is well established. Primary cell cultures are routinely successful and sufficiently versatile so that definitive studies with regard to specific proximal airway epithelial functions can be conducted. The major limitation to this class of studies is the scarcity of tissue material. The development of transformed epithelial cells with increased proliferative potential promises to alleviate this problem. However, these cell lines presently do not quantitatively reproduce the transport features of the primary cells or the intact epithelium. Further substantial phenotypic "drift" as a function of passage number has been a feature of the SV40T transformed cells. Perhaps other transforming proto-oncogenes, or proto-oncogenes under the control of temperature sensitive promoters, will produce in quantity the cell numbers required for the future.

References

1. Knowles, M.R., M.J. Stutts, J.R. Yankaskas, J.T. Gatzy and R.C. Boucher. Abnormal respiratory epithelial ion transport in cystic fibrosis. (Review) Clin. Chest Med. 7:285-297 (1986).
2. Finney, M.J.B., S.D. Anderson, J.L. Black. The effect of non-isotonic solutions on human isolated airway smooth muscle. Respir. Physiol. 69:277-286 (1987).
3. Wu, R., J. Yankaskas, E. Cheng, M.R. Knowles and R. Boucher. Growth and differentiation of human nasal epithelial cells in culture: serum-free, hormone-supplemented medium and proteoglycan synthesis. Am. Rev. Respir. Dis. 132:311-320 (1985).

4. Van Scott, Michael R., N.P. Lee, J.R. Yankaskas, and R.C. Boucher. Effect of hormones on growth and function of cultured canine tracheal epithelial cells. Am. J. Physiol. 255 (Cell Physiol. 24): C237-245 (1988).
5. Yankaskas, J.R., C.U. Cotton, M.R. Knowles, J.T. Gatzy, and R.C. Boucher. Culture of human nasal epithelial cells on collagen matrix supports. Am. Rev. Respir. Dis. 132;1281 -1287 (1985).
6. Burg, M., N. Green, S. Sohraby, R. Steele, J. Handler. Differentiated function in cultured epithelia derived from thick ascending limbs. Am. J. Physiol. 242:229-233 (1982).
7. Yankaskas, J.R., J.T. Gatzy and R.C. Boucher. Persistence of abnormal Cl permeability in cystic fibrosis nasal epithelial cells in heterologous culture. Lancet 1:954-956 (1985).
8. Boucher, R.C., C.U. Cotton, J.T. Gatzy, M.R. Knowles, and J.R. Yankaskas. Evidence for reduced Cl^- and increased Na^+ permeability in cystic fibrosis human primary cell cultures. J. Physiol. 405:77-103, 1988.
9. Willumsen, Niels J., C. William Davis, and Richard C. Boucher. Intracellular Cl^- activity and Cl^- pathways in cultured human airway epithelium. Am, J. Physiol. 256 (Cell Physiol. 25): C1033-C1044 (1989).
10. Boucher, Richard C., Elaine H.C. Cheng, Anthony M. Paradiso, M. Jackson Stutts, Michael R. Knowles, and H. Shelton Earp. The chloride secretory response of cystic fibrosis (CF) human airway epithelia: Preservation of calcium but not protein kinase C and A - Dependent mechanisms. (In Press J. Clin. Invest., 1989).
11. Cheng, Pi-Wan, Thomas F. Boat, Karen Cranfill, James R. Yankaskas, and Richard C. Boucher. Increased sulfation of glycoconjugates by cultured nasal epithelial cells from patients with cystic fibrosis. J. Clin. Invest. 84:68-72 (1989).
12. Henke, David, Roberta M. Danilowicz, John F. Curtis, Richard C. Boucher and Thomas E. Eling. Metabolism of arachidonic acid by human nasal and bronchial epithelial cells. Arch. of Biochem. and Biophysics 267:426-436 (1988).
13. Yoakum, George H., John F. Lechner, Edward W. Gabrielson, Brent E. Korba, Louise Malan-Shibley, James C. Wiley, Marion G. Valerio, Abulkalam M. Shamsuddin, Benjamin F. Trump, Curtis C. Harris. Transformation of human bronchial epithelial cells transfected by harvey ras oncogene. Science 227:1174-1179 (1985).
14. Jetten, Anton M., James R. Yankaskas, M. Jackson Stutts, Niels J. Willumsen, and Richard C. Boucher. Persistence of abnormal chloride conductance regulation in transformed cystic fibrosis epithelia. Science 244:1472-1475 (1989).

EXPERIMENTAL MODELS FOR BUCCAL AND SUBLINGUAL DRUG ABSORPTION INCLUDING EPITHELIAL CELL MULTILAYERS AND MONOLAYERS

Alan Rogerson

Reckitt & Colman Pharmaceutical Division
Dansom Lane Kingston-upon-Hull U.K. HU8 7DS

Like the skin the oral mucosa is constantly exposed to the environment thus its fundamental property is probably a barrier function. Pathogens and other potentially harmful materials, which may be borne in air, water or food, are excluded by the stratified epithelial cell layers, which seem to be more closely related to skin than to the columnar epithelia of the gastrointestinal tract (Squier and Johnson, 1975). Thus while the barrier properties are important in the resistance of disease they may also compromise potential drug absorption via the buccal or sublingual routes.

Many models for drug absorption have been proposed including partition coefficient, mathematical models and diffusion studies utilising excised tissue and cultured cell monolayers. With particular reference to the buccal route there appear to be few tried and tested models. Consequently some current models under investigation at Reckitt & Colman are reviewed here, and some attempt made to correlate these with each other, with cultured cell systems and, ultimately, with clinical data.

There are many advantages offered by drug delivery via the buccal or sublingual route :

a. Absorbed drug passes directly from the oral cavity into the systemic circulation and thus avoids hepatic first-pass metabolism. This may subsequently give rise to an increase in bioavailability, e.g. with buprenorphine (Bullingham et al., 1981) or reduce inter-patient variation, e.g. with triazolam (Scavone et al, 1986).

b. Labile drugs which would normally be inactivated by gastric acid and digestive enzymes may be delivered by the buccal route such that these processes are avoided. This has been successfully utilised for oxytocin (Anders et al, 1985).

c. During periods of nausea there is often accompanying gastric stasis. Any medication subsequently administered is thus likely to be very slowly and poorly absorbed. Successful treatment via the buccal or sublingual route, which bypasses the G.I. tract, would then be capable of producing relatively rapid and extensive drug absorption.

d. Side-effects and toxicity are potential hazards of any drug treatment. In the event of acute toxicity crises buccally or sublingually administered dosage forms may easily be removed.

Thus while these advantages may appear to make the buccal route very appealing there are also some inherent disadvantages that must be addressed :

a. Taken together the buccal and sublingual mucosae have a surface area of only some $200cm^2$ (Danhof and Breimer, 1978). Obviously this is a fairly insignificant proportion of the $200m^2$ which is generally recognised as an approximate surface area for the intestine.

b. The problem of delivery to such a small area may be circumvented by increasing the exposure time, and this approach has been fairly successful for transdermal medication systems. And while solid buccal and sublingual preparations are not debilitating while they are in position, it is often inconvenient for patients to retain tablets for long periods, especially if eating, drinking or speaking are compromised.

c. Similarly there is an upper limit to tablet size. Garen and Repta (1989) calculated that, for a "hypothetical" drug which displayed a molecular weight of 250, a solubility of 1mM and log P (octanol:water) of 2.5, absorption from a saturated solution would occur at the rate of 1.6 mg cm^{-2} day^{-1}. It is clearly not feasible to retain dosage forms in the oral cavity for days on end, thus drugs must be extremely potent to be feasible for administration by these routes.

d. The positioning of any formulation within the buccal cavity is of vital importance to the plasma profiles it subsequently generates. Saliva production, and dissolution of tablet formulations, would appear to be greater in the lower vestibule than the upper, and similarly posterior is greater than anterior (Weatherell et al., 1986). Thus the location of the tablet is a critical factor in choosing dosage forms for rapid onset of action or sustained release.

Thus there are many advantages and disadvantages associated with the buccal and sublingual route of delivery. The barrier properties of the mucosa are an important component of disease resistance. However despite the relative impermeabilities to many substances these routes appear to offer some distict advantages over oral, percutaneous and nasal routes.

It is not unreasonable to suggest that little is known about drug absorption across the buccal epithelium. In general polar drugs tend to be absorbed by means of leakage through intercellular lateral spaces (the paracellular route). More lipophilic drugs would be expected to partition into the cell membrane then undergo lateral diffusion within the membrane followed by partitioning out of the membrane on the serosal side. By this method these drugs may display depot effects as their rate of partitioning out of the tissue is likely to be inversely proportional to their partition coefficient. So while it is reasonable to assume that buccal absorption will follow these guidelines there appear to be little hard and fast data to substantiate this.

There are many absorption models and it is likely that most will provide results to indicate that drug diffusion is proportional to partition coefficient. In addition models incorporating tissue sections or cell monolayers are probably capable of providing predictive rank order values for a group of drugs or penetration enhancers. However what is more difficult to determine is the finite quantity of drug likely to be absorbed i.e. rank orders are not sufficient to predict the likely extent of a drug's in vivo uptake. Thus it is advisable to use a drug whose buccal or sublingual absorption has been well characterised as a marker for comparative purposes.

The following models have been undertaken, with varying degrees of success, as means of assessing the "gross" absorption of drug from the oral cavity. A limited amount of work has also recently been undertaken investigating madin darby canine kidney (MDCK) epithelial cell monolayers as potential models for buccal drug diffusion, and this will be addressed later.

A few animal models have been reported for assessment of buccal and sublingual absorption of drugs and they tend to suffer from a problem which is almost unavoidable i.e. dry mouth induced by the anaesthesia. This may be circumvented, to some extent, by delivering the dose as a small volume sublingual solution, but this then dictates that hydratable, solid dosage forms cannot be tested.

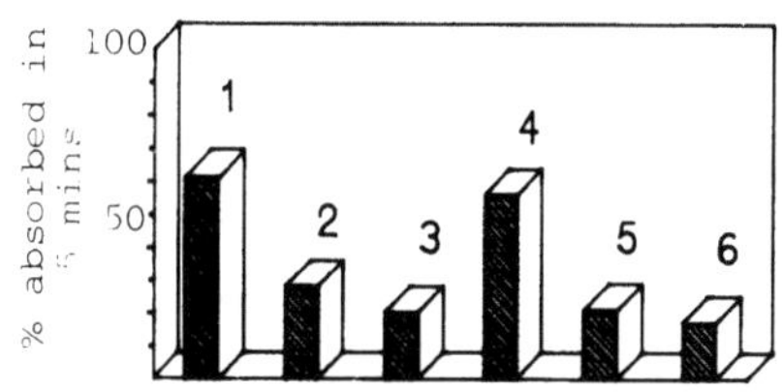

Figure 1. The uptake of six drugs, in a five minute period, following sublingual administration of 5μl aliquots to anaesthetised rats. Values are means of at least three determinations.

1. buprenorphine 2. haloperidol
3. naloxone 4. prochlorperazine
5. RX451372 6. morphine

An anaesthetised rat model has been routinely used in these laboratories for candidate sublingual drugs. Various analgesics, anti-emetics and antidepressants were tested and some results are shown in Figure 1. On this evidence all six drugs must be considered good candidates especially as the five minute exposure time could be substantially increased by judicious tablet formulation.

Most extensive work has been undertaken for buprenorphine, a strong analgesic which is extensively degraded by the gut wall and liver. The drug is highly lipophilic and, while its rate of partitioning into sublingual mucosa is rapid, its subsequent absorption and appearance in plasma is not. A significant quantity of the dose tends to depot in the head and neck region and thus any potential model for sublingual absorption of equally lipophilic drugs must be able to account for such a phenomenon. This rat model is particularly useful for drugs which may undergo extensive metabolism, however extraction of active drug and metabolites from plasma can often be difficult and time-consuming, and alternatives have been employed as primary screening procedures.

The stratum corneum, closely packed layers of terminally differentiated squamous epithelial cells which comprise the epidermis, possesses a specialised barrier function and thus controls the rate of absorption of many topically-applied drugs. However the dermis, like the buccal mucosa, consists of a series of essentially unkeratinised epithelial cells and has been utilised as a model for buccal drug absorption.

Original work in this area was undertaken by Galey et al (1976) who characterised the relative permeabilities of whole human skin, human dermis and dog buccal mucosa. Their results, which account for the relative tissue thicknesses, clearly indicate a degree of correlation between dermis and buccal mucosa (Figure 2). This technique has since been shown to be rapid and reproducible for screening drugs for their potential in buccal delivery (unpublished). Skin is excised from the back of hairless mice and dekeratinised by stripping twelve times with sticky tape (twelve is a sufficient number to completely remove the stratum corneum). Dekeratinised tissue is then loaded into diffusion cells and the subsequent rate of drug diffusion determined by assay of radiolabel or HPLC.

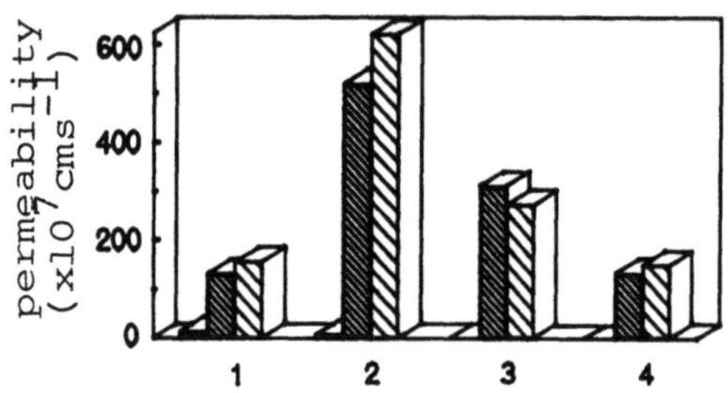

Figure 2. The relationship between the permeabilities of whole human skin, human dermis and dog buccal mucosa. The values for amphetamine and ouabain in whole human skin are 0.039 and 0.011 respectively (Galey et al, 1976).

1. oestradiol 2. water
3. amphetamine 4. ouabain

There are certain advantages associated with this technique, not merely its speed and reproducibility. Diffusion values correlate very well with <u>in vivo</u> data in human subjects (not shown) and the depot potential of lipophilic drugs may be assessed by determining the extent of partitioning of drug into the tissue.

An alternative to the anaesthetised rat model has been developed in these laboratories and utilises conscious dogs. Beagles are trained to accept the application of muzzles and stand in restraining harnesses for two hours. During this time gum-based, muco-adhesive tablets may be applied to the buccal mucosa, as the tongue is immobilised, and subsequent blood samples are then assayed to determine the rate and extent of drug absorption.

Results from <u>in vitro</u> diffusion cells, utilising dekeratinised hairless-mouse skin, suggested that buccal absorption of RX77368, a tripeptide TRH analogue, would not be extensive (less than 20% of the initial dose diffused in 4 hours) and this appeared to be largely confirmed in the dog model (Figure 3). Retrospectively it is probable that the low plasma levels associated with the buccal route may be due to the use of the upper buccal mucosa as the site of administration. It has since been suggested that positioning of the tablet in the lower posterior region of the mouth would have induced more rapid tablet dissolution and the likelihood of more extensive drug absorption. Indeed these results tend to suggest that the buccal route would not be viable if rapid absorption were a prerequisite, and consequently it may be more appropriate to consider the buccal route for more sustained-release profiles.

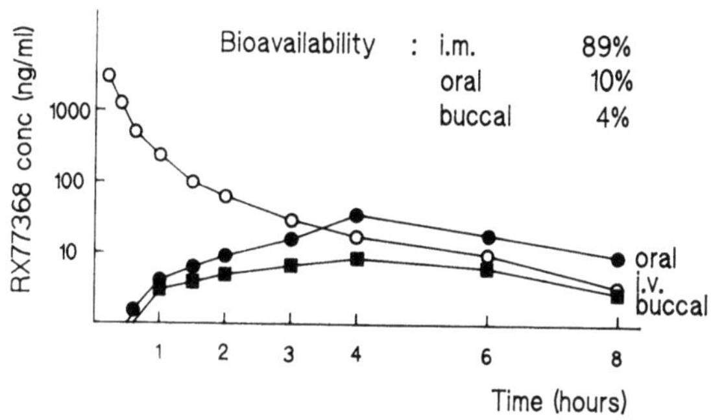

Figure 3. Plasma-time profiles for RX77368 following administration of $1mgkg^{-1}$ to a beagle dog by three routes. Bioavailability figures were generated by comparing the area under each curve with that of the i.v. dose (which was taken to be 100% bioavailable).

There are many advantages associated with this model, especially the fact that absorption is being determined through intact, in situ buccal mucosa. In addition it is relatively straightforward to simultaneously assess pharmacokinetic data. However training of dogs is extremely time-consuming and dictates that only small sample sizes may be utilised. Irrespective of the correlation between a model system and the in vivo absorption it mimics, these problems of saliva production could not have been predicted by either in vitro models (as they utilise only solutions of drug) or anaesthetised-animal models (as salivary output is reduced as a side-effect of anaesthesia).

No matter how closely the absorption model mimics drug absorption across human buccal or sublingual mucosa, human volunteers will be required at some stage during the development of a drug. Large scale, controlled clinical trials are extremely time-consuming and expensive, and consequently many groups have utilised a non-invasive model, comprising in situ absorption of drug from the oral cavity which does not require blood sampling.

This technique was originally described by Beckett and Triggs (1967) as a "buccal absorption test" and involved circulating a solution of drug around the oral cavity for a period of time then expelling the solution which was then assayed for drug content. The difference between the dose and the assay was attributed to absorption. The technique may be further refined by incorporating a non-absorbable marker such as carmine or phenol red and has been used to investigate drug interactions (McElnay and Mooney, 1983) and absorption kinetics (Tucker, 1988). The model provides information relating to the probability of a drug being absorbed, but cannot distinguish between different absorption sites within the oral cavity. This would be of extreme importance in the case of solid dosage forms, where the position within the cavity would profoundly affect subsequent plasma levels.

Various drug solutions, and potential penetration enhancers, have been tested in these laboratories using this technique. Solutions of not more than 20ml are taken into the mouth, retained for 5 minutes, then expelled. Subsequent assay reveals a loss of drug which may be attributed to absorption, after swallowing has been quantitated with carmine (Figure 4).

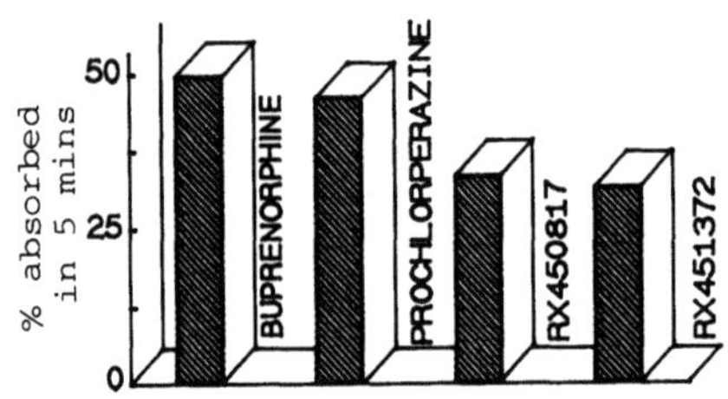

Figure 4. "Buccal absorption" of four drugs in human volunteers, following exposure to 20ml of buffered solution for 5 minutes. Swallowing was quantitated by pre-treatment with carmine. RX451372 and RX450817 are anti-emetic drugs.

All of the techniques discussed to date have their uses, although they all provide values for "gross" absorption which may not be representitive of absorption at a cellular level. Epithelial cell multilayers and monolayers are an interesting new development. However these techniques are likely to be embraced by the Pharmaceutical Industry only if they can be shown to provide reliable, reproducible, predictive information from which assessments of the therapeutic potential of drugs and drug-enhancer combinations may be made.

Tavakoli-Saberi and Audus (1989) recently proposed a model comprising hamster pouch buccal epithelial (HPBE) cell primary cultures supported by collagen-coated membrane filters. After three days' incubation the entire surface of the collagen matrix was covered with a continuous layer (usually five cells deep) of epithelial cells, with only the uppermost layer apparently displaying terminal differentiation. These cultured cells maintained morphological characteristics of stratified epithelia, and there were no significant differences in the specific activity of certain enzymes between cultured and excised cheek pouch epithelium. Permeability to water, fluoresceine and dextrans decreased rapidly to a minimum at day three, then increased more slowly to plateau at day seven, indicating the sloughing of the uppermost keratinised layer.

It is difficult to determine exactly what effect this keratinisation will exert on drug diffusion. Intercellular barriers, in the form of tight junctions, appear to limit the penetration of polar molecules, while lipophilic drugs are more susceptible to the retarding influence of keratin layers (Tavakoli-Saberi and Audus, 1989). And there have been reports suggesting that the permeability of mucosal surfaces is profoundly affected by keratinisation, indicating that keratinised mucosa possesses greater barrier properties (Reid et al, 1986). However despite the terminal differentiation displayed by the uppermost cell layer, this model of Tavakoli-Saberi and Audus (1989) appears to offer some potential as a model for human buccal absorption.

The importance of epithelial keratinisation was also addressed by Garen and Repta (1989). They investigated the relative permeabilities of full thickness hamster cheek pouch and cheek pouch epithelium, isolated by means of collagenase treatment, and they observed no significant difference in the diffusion rates of four substituted acetanilides (Figure 5). As would be anticipated the rates of diffusion across excised epithelium appear to be generally proportional to log P values, and these rates correlate very well with relative amounts of drug absorbed from solution, in 5 minutes, in human volunteers (Dearden and Tomlinson, 1971).

	permeability $(x10^{6}\ cms^{-1})$	
log P	full thickness pouch	isolated epithelium
1.39	5.48 + 1.46	4.37 + 0.93
1.87	6.56 + 1.21	6.36 + 0.82
2.29	7.63 + 1.32	7.29 + 0.49
2.46	8.58 + 1.32	8.38 + 1.40

Figure 5. Comparison of permeability values of full-thickness hamster cheek pouch and isolated epithelial cell layer as a function of log P for four substituted acetanilides.

Garen and Repta (1989) ultimately concluded that the keratinisation of the hamster cheek pouch, which is not apparent to such an extent in the buccal mucosa in man, may limit the utility of their model. Yet the results generated by Tavakoli-Saberi and Audus (1989) suggest that HPBE cultures may have some contribution to make to buccal absorption studies, and the advantages of utilising a system of this nature are immediately apparent. This would be of particular importance in the study of peptide absorption, where biochemical processes, which may degrade labile drugs during diffusion, would remain intact and could be closely monitored.

Epithelial cell monolayers, derived from madin derby canine kidney (MDCK) cells, have been utilised for a number of years in investigation of drug absorption and drug-membrane interactions. Confluent monolayers of this nature, formed with occluding junctions, provide convenient models which "within certain limitations, may be adapted for studying drug transport across renal, gastrointestinal, nasal, rectal and other epithelia of the body" (Cho et al., 1989). This report is not intended to be an exhaustive description of such cell systems. Rather it is intended to assess their potential as tools in the elucidation of drug absorption by the buccal or sublingual routes.

Cho et al., (1989) exhaustively investigated the formation of MDCK monolayers, monitoring the transepithelial electrical resistance (TEER), which provides an indication of the integrity of the monolayer, and their results suggested that maximum TEER was apparent some seven days after the cells were first incubated. Similar work has been undertaken by Al Jabar and Baillie (unpublished) at the University of Strathclyde, and they have developed confluent monolayers of electrically "tight" MDCK cells, with modelling of buccal drug absorption being one of their ultimate aims.

Stock MDCK cells are incubated until required then seeded onto collagen-coated polycarbonate filters (Millipore) with 3μm pore diameter. Subsequently cells settle, then adhere to the collagen base, and thereafter cell division generates a confluent, sealed layer with electrically tight junctions. The integrity of the monolayer may be monitored throughout its development by means of light microscopy or TEER (Figure 6). Resistance values rapidly increase up to day 6-8 as the tight junctions are formed. Thereafter the TEER plateaus at day 8-12, and diffusion studies would be undertaken at this time. The shape of the graph, and the finite TEER values, are very similar to those reported by Cho et al (1989).

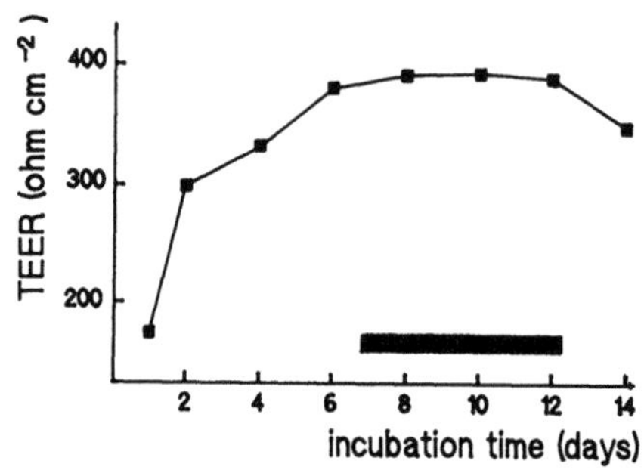

Figure 6. Transepithelial electrical resistance of MDCK epithelial cell monolayers as a function of incubation time.

Work on this buccal model is at a very preliminary stage. However it has been demonstrated that the monolayers formed are sealed and confluent, and initial studies utilising buprenorphine have been generated.

In the anaesthetised rat model buprenorphine is rapidly cleared from the oral cavity (42% absorbed in 5 minutes), but thereafter appearance of the drug in plasma is significantly less rapid, indicating depot formation. Thus rapid clearance from the oral cavity, which is not accompanied by early peak plasma levels, was predicted by <u>in vitro</u> models and is observed clinically.

Diffusion of buprenorphine across hairless mouse skin was essentially linear after an initial lag phase of some 45 minutes. Thereafter diffusion was monitored up to four hours and reached a maximum of 9.6% (Figure 7). At the same time depot values (i.e. the quantity of drug which partitions into the tissue from the mucosal solution but does not partition back out into the serosal solution) would be expected to be of the order of 25 - 30% (unpublished). This depot is responsible for the delayed onset of peak plasma levels of buprenorphine and is a phenomenon which may be anticipated in other very lipophilic drugs.

The diffusion of buprenorphine through MDCK epithelial cell monolayers was very similar (Figure 8). In this instance there was no lag phase, then diffusion was similarly linear over the entire time-course of the study. The absence of the initial diffusion lag is probably indicative of the absence of a depot effect (although the depot was not quantitated in these preliminary experiments). There is great clinical significance in the fact that buprenorphine displays a marked depot effect. Any potential model for buccal or sublingual absorption must thus be capable of determining the extent of the drug depot. In this respect it would appear that the MDCK epithelial cell monolayers may not possess the capacity to adequately model absorption and depot of highly lipophilic drugs simultaneously.

The terms "buccal" and "sublingual" are not interchangeble, as each region within the oral cavity is profoundly different in terms of its epithelial thickness and keratinisation (Squier and Hall, 1985) and saliva production (Weatherell et al., 1986). Clinical trials undertaken with buprenorphine were able to provide blood levels for both the sublingual and buccal routes of administration, and it is unlikely that any absorption model would be able to predict such differences. The

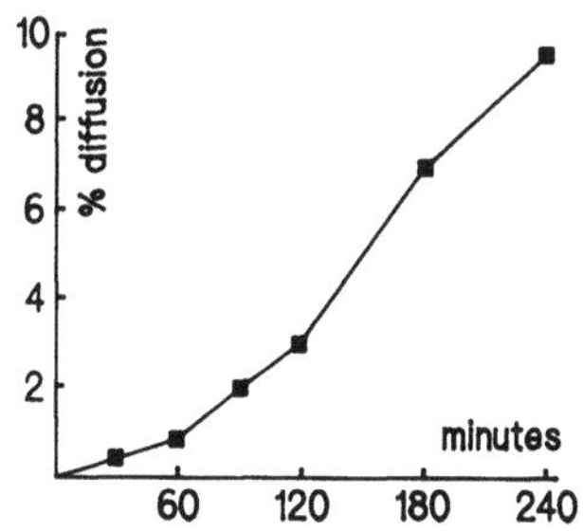

Figure 7. Diffusion of ^{3}H-buprenorphine through dekeratinised hairless mouse skin as a function of time. Initial drug concentration was 50μgml^{-1} at pH 5.4, surface area of tissue was 44.2 mm^2.

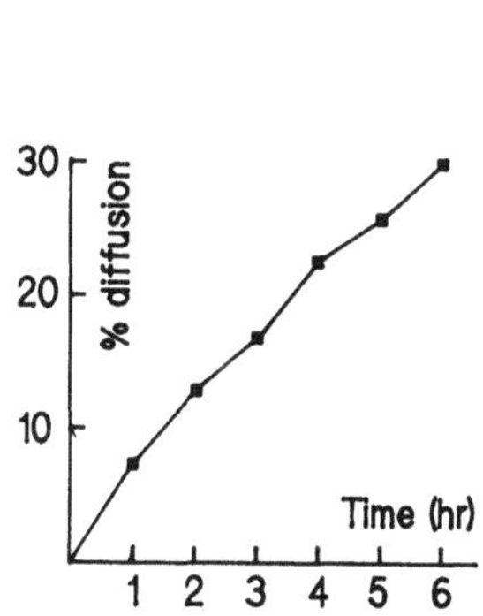

Figure 8. Diffusion of ^{3}H-buprenorphine across MDCK epithelial cell monolayers, after 8 days incubation, as a function of time. The surface area of the monolayer was 132 mm^2.

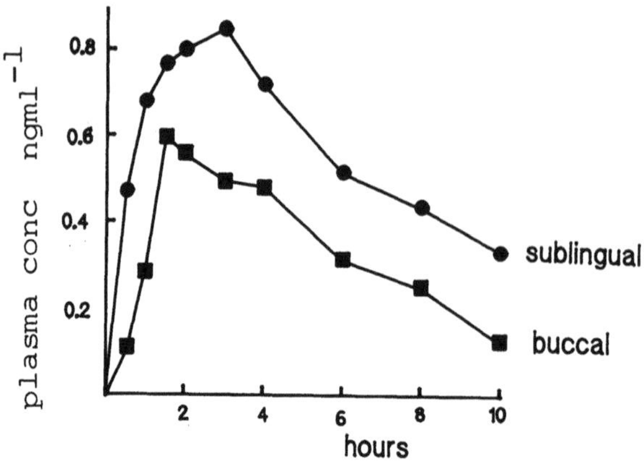

Figure 9. Plasma-time profiles for buprenorphine following administration of 0.4mg buccally and sublingually to a human volunteer.

		C_{max} ($ngml^{-1}$)	T_{max} (h)	K_{el} (h^{-1})	AUC_{0-6h} ($mgml^{-1}.h$)	clearance ($mlmin^{-1}$)
buccal	mean	0.43	2.00	0.228	1.603	3264
	+ SEM	0.09	0.35	0.042	0.370	1004
sub-lingual	mean	0.63	2.63	0.221	2.239	2671
	+ SEM	0.15	0.38	0.026	0.583	1001

Figure 10. Pharmacokinetic parameters for buprenorphine following administration of 0.4mg buccally and sublingually to human volunteers. (n = 12).

sublingual route was superior in terms of plasma concentration profiles, maximum plasma concentration and bioavailability (Figure 9, 10).

The higher plasma levels from the sublingual route, in conjunction with almost identical elimination rate constants, indicate that this route provides superior buprenorphine bioavailability. Clearly it would be impossible for any potential model to provide predictive information of this nature.

There are not many models for buccal and sublingual drug absorption, and there are obvious inherent disadvantages associated with each. Results from the MDCK epithelial cell monolayers are very preliminary, and the apparent absence of a drug depot (although this has yet to be confirmed) may be indicative of one of this system's limitations. In terms of drug diffusion it is probably safe to suggest that, for such lipophilic drugs as buprenorphine, any epithelial cell line would probably provide similar results.

Despite these problems there are certainly many uses for MDCK cell monolayers. The mechanisms of absorption of hydrophilic drugs may provide one potential use, and these monolayers certainly appear to be more easily managed than buccal cell monolayers derived from humans and animals. In addition there has recently been considerable interest in the use of penetration enhancers, some of which may be capable of promoting the absorption of poorly absorbed drugs. Candidate enhancers have included salicylates, surfactants and chelating agents. By their very nature molecules which are utilised as penetration promoters may be anticipated to alter the barrier function of the oral mucosa, and thus may prove to irreversibly alter its integrity. Thus monolayer systems may be particularly useful in elucidating some of the mechanisms of absorption enhancement at a cellular level.

There is now considerable interest in the administration of biologically-active peptides and proteins by non-parenteral routes. Epithelial cell monolayers may find particular application in the study of peptide and protein absorption, and its enhancement. Degradation of these labile species during absorption, due to the influence of epithelial aminopeptidases, may compromise uptake of sufficient quantity of native drug. Thus development of living, biochemically intact cell monolayers may be able to quantitate these potential problems, and help provide practical solutions.

Epithelial cell monolayers are interesting in their own right, from the point of view of studying mechanisms of drug absorption and mass transport of nutrients. The cell monolayer of MDCK are not truly representitive of the squamous epithelia of the buccal and sublingual mucosae, but this may ultimately be of little significance. A barrier to drug absorption, especially one which possesses the capacity to adequately assess the potential depot of lipophilic drugs, may be sufficient in a model. Ultimately what is required of an absorption model, of any description, is that it be able to determine what drugs, drug/enhancer combinations or drug dosage forms are likely to be viable for the buccal or sublingual routes.

There is ultimately no substitute for clinical data (Figure 9), but expensive clinical trials cannot be undertaken routinely. None of the models described here are likely to be able to predict the plasma levels in humans, or to determine the effects of subtle alterations in dosage form or site of administration. But by comparison of test drugs with drugs of known absorption characteristics, estimations of the rate and extent of absorption, and the effects of penetration enhancers, may be obtained. These values will be estimations, but they should be sufficiently accurate to provide any subsequent clinical investigations with a high probability of success.

I am deeply indebted to Drs Alan Baillie and Zara Al-Jabar at the University of Strathclyde in Glasgow, and to my colleagues Mr Michael Havler, Dr Graham Parr and Miss Tracey Seaton at Reckitt & Colman.

References

Anders, R., Sandow, J. and Merkle, H.P., 1985, Self-adhesive patches for buccal delivery of peptides, 12th Int. Symp. Contr. Rel. Bioact. Materials, Geneva.

Beckett, A.H. and Triggs, E.J., 1967, Buccal absorption of basic drugs and its application as an in vivo model of passive drug transfer through lipid membranes, J. Pharm. Pharmacol., 19:31S.

Bullingham, R.E.S., McQuay, H.J., Dwyer, D., Allen, M.C. and Moore, R.A., 1981, Sublingual buprenorphine used postoperatively : clinical observations and preliminary pharmacokinetic analysis, Brit. J. Clin. Pharmacol., 12:117.

Cho, M.J., Thompson, D.P., Cramer, C.T., Vidmar, T.J. and Scieszka, J.F., 1989, The madin darby canine kidney (MDCK) epithelial cell monolayer as a model cellular transport barrier, Pharm. Res., 6:71.

Danhof, M. and Breimer, D.D., 1978, Therapeutic drug monitoring in saliva, Clin. Pharmacokin., 3:39.

Dearden, J.C. and Tomlinson, E., 1971, Physico-chemical studies of analgesics. The protein-binding of some p-substituted acetanilides, J. Pharm. Pharmacol., 23:73S.

Galey, W.R., Londsdale, H.K. and Nacht, S., 1976, The in vitro permeability of skin and buccal mucosa to selected drugs and tritiated water, J. Invest. Dermatol., 67:713.

Garen, K.W. and Repta, A.J., 1989, Buccal drug absorption II : in vitro diffusion across the hamster cheek pouch, J. Pharm. Sci., 78:160.

McElnay, J.C. and Mooney, C., 1983, The use of buccal partitioning as a model of drug absorption interactions, J. Clin. Hosp. Pharm., 8:137.

Scavone, J.M., Greenblatt, D.J., Friedman, H. and Shader, R.I., 1986, Enhanced bioavailability of triazolam following sublingual versus oral administration, J. Clin. Pharmacol., 26:208.

Squier, C.A. and Hall, B.K., 1985, The permeability of skin and oral mucosa to water and horseradish peroxidase as related to the thickness of the permeability barrier, J. Invest. Dermatol., 84:176.

Squier, C.A. and Johnson, N.W., 1975, Permeability of oral mucosa, Brit. Med. Bull., 31:169.

Tavakoli-Saberi, M.R. and Audus, K.L., 1989, Cultured buccal epithelium : an in vitro model derived from the hamster pouch for studying drug transport and metabolism, Pharm. Res., 6:160.

Tucker, I.G., 1988, A method to study the kinetics of oral mucosal drug absorption from solution, J. Pharm. Pharmacol., 40:679.

Weatherell, J.A., Strong, M., Robinson, C. and Ralph, J.P., 1986, Fluoride distribution in the mouth after fluoride rinsing, Caries Res., 20:111.

EFFECT OF CULTURE ENVIRONMENT ON TERMINAL DIFFERENTIATION OF HUMAN EPIDERMAL KERATINOCYTES

Fiona M. Watt

Keratinocyte Laboratory, Imperial Cancer Research Fund
P.O. Box 123, Lincoln's Inn Fields, London WC2A 3PX

INTRODUCTION

The epidermis is the outer covering of the skin and consists of several different cell types: keratinocytes, melanocytes, Langerhans cells and Merkel cells. Keratinocytes are the most abundant of the epidermal cells and are organised into distinct layers, as illustrated in Fig. 1a.

The basal layer of the epidermis is attached to a basement membrane and contains almost all the keratinocytes that are capable of proliferating. Cells that move upwards out of the basal layer have lost the ability to divide and they undergo terminal differentiation as they move towards the tissue surface. In the spinous layers keratinocytes increase in size and protein content, synthesizing a range of new molecules characteristic of this stage in the terminal differentiation process. Cells next move into the granular layers, so named because the cells contain cytoplasmic keratohyalin granules. Finally, as the cells enter the cornified layers their nuclei and cytoplasmic organelles are destroyed and an insoluble protein envelope, the cornified envelope, is laid down underneath the plasma membrane. Cornified cells (squames) are continuously shed from the surface of the epidermis, but the tissue remains a constant thickness because the rate of cell loss is balanced by the rate of new cell production in the basal layer.

The aim of this chapter is to review the extent to which it is possible to reconstitute normal human epidermis in culture and what is known about factors controlling terminal differentiation. There are many applications for these cultures, ranging from basic research into growth control, through screens for drug toxicity and delivery, to clinical therapy, as in the grafting of burns victims.

When assessing cultured epidermis, much emphasis is placed on the degree of differentiation achieved. However, it is important to realise that the word 'differentiation' has three quite distinct meanings, depending on context. Firstly, from the perspective of an embryologist, all keratinocytes in adult epidermis are 'differentiated', because when cells in the basal layer divide they always give rise to more keratinocytes; they are not the progenitors of other cell types. Secondly, for the histologist, the degree of differentiation is assessed by morphological criteria, such that a 'fully differentiated' cultured epidermis must contain the four distinct layers illustrated in Fig. 1a. Finally, terminal differentiation at the level of individual cells refers to a whole series of events, starting with commitment in the basal layer and ending with a cornified squame. The earliest event that can be defined at this level is currently cell

Pharmaceutical Applications of Cell and Tissue Culture to Drug Transport
Edited by G. Wilson *et al.*, Plenum Press, New York, 1991

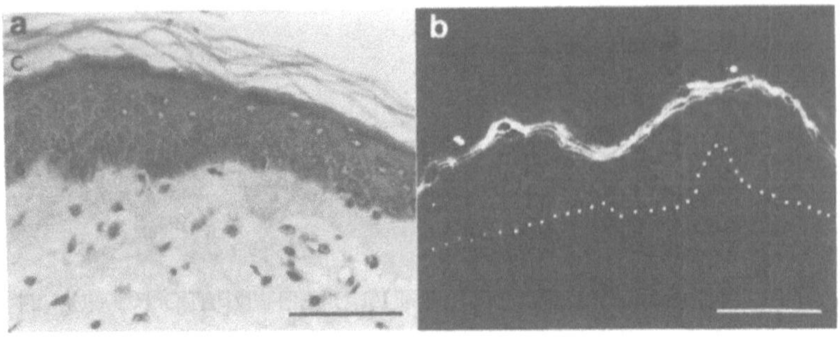

Figure 1 Sections of adult forearm epidermis. a) Haematoxylin and eosin (H and E) stained paraffin wax section. B : basal layer; S = spinous layers; G = granular layer; C = cornified layers. b) Immunofluorescence staining of unfixed frozen section with antiserum to involucrin. Dotted line indicates position of basement membrane. Scale bars = 100μm. Reproduced from Watt et al. (1987), with permission.

cycle withdrawal, which occurs in the basal layer (reviewed by Potten and Morris, 1988), but there is growing evidence that the dividing population of keratinocytes is heterogeneous and thus that the first event may be when an epidermal stem cell generates a daughter that has finite self-renewal capacity (reviewed by Hall and Watt, 1989).

In summary, one's goals in keratinocyte culture depend on which aspects of differentiation one is interested in.

Clonal growth of keratinocytes with a 3T3 feeder layer

Human epidermal keratinocytes can be grown at clonal densities in the presence of a 3T3 feeder layer (Rheinwald and Green, 1975) in a mixture of Ham's F12 and Dulbecco's MEM, supplemented with adenine, foetal calf serum, hydrocortisone, cholera toxin, insulin and epidermal growth factor (Rheinwald, 1989). Keratinocytes from newborn foreskin usually undergo 80-100 cell generations before senescence; those from children or adults senesce after about 40-70 cell generations (Rheinwald, 1989). Thus the number of cells that can be grown from a single biopsy is enormous.

At confluence (Fig. 2a) the epidermis formed by this culture method resembles the basic features of normal epidermis, in that the cells are stratified; mitosis is restricted to the basal layer; and cells undergo terminal differentiation as they move through the suprabasal layers, culminating in the assembly of cornified envelopes. However, distinct granular and cornified layers are absent and the morphology of cells in the basal layer is spread rather than cuboidal. Clearly, the degree of morphological differentiation is poor compared with epidermis in situ.

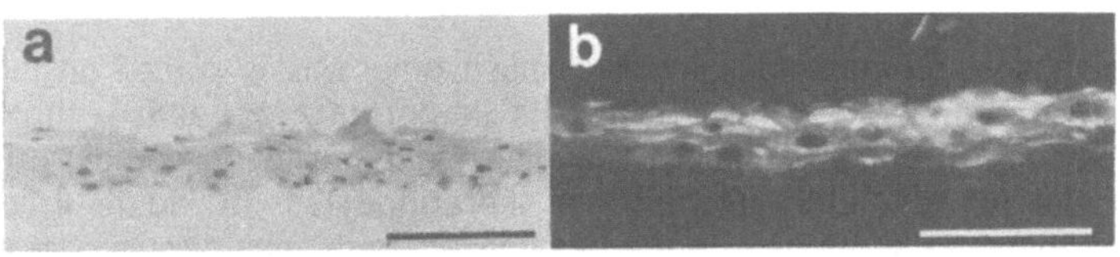

Figure 2 Sections of epidermal keratinocytes grown to confluence in the presence of a 3T3 feeder layer and harvested with dispase. a) H and E staining of unfixed frozen section. b) Immunofluorescence staining of paraffin wax section with antiserum to involucrin. Scale bars = 100μm.

Melanocytes survive and proliferate in these cultures, at least at low passage number (De Luca et al., 1988; Compton et al., 1989). Rare Langerhans cells have been observed in primary cultures, but these are lost on passaging (Compton et al., 1989). The fate of Merkel cells in culture is at present unclear.

Improvements in the histological appearance of the cultures

The poor histological appearance of keratinocytes grown with a 3T3 layer is not due to defects in the cells themselves, but is a response to deficiencies in the culture environment. The evidence for this comes from histological analysis of cultures grafted as autografts onto burns victims (Compton et al., 1989): within 6 days distinct granular and cornified layers form and within 1 month a basement membrane is assembled at the dermal-epidermal junction. One approach to improving the histological appearance of the cultures is therefore to mimic different aspects of the epidermal environment: specifically, to culture the cells on extracellular matrix; to reconstitute the dermis; to expose the cells at the air-liquid interface and feed them from below.

The effects on histology of some of these changes are shown in Fig. 3 (Mackenzie and Fusenig, 1983; Watt et al., 1987). When cells are grown on raised collagen gels at the air-liquid interface distinct cornified layers accumulate (Fig. 3a), although a granular layer is lacking and most of the cornified cells retain their nuclei. When viable dermis is inserted underneath the collagen layer there is further improvement in the histological appearance of the epidermis: cells in the basal layer have a cuboidal, instead of flattened, morphology; there is some evidence of a granular layer and there are more layers of cornified cells, some of which have lost their nuclei (Fig. 3c). Full histological differentiation is observed after transplantation of human keratinocytes onto nude mice (Fig. 3e,g).

Two other strategies for providing a cultured equivalent to dermis are to seed keratinocytes directly onto dead deepidermised dermis (which retains the lamina densa of the basement membrane) (Pruniéras et al., 1983) or to form gels of type I collagen containing dermal fibroblasts. In the second type of culture the keratinocytes are seeded some days later, after the fibroblasts have contracted the gel (Bell et al., 1981). In each case, when the keratinocytes are exposed at the air-liquid interface the histological appearance of the cultures is better than that of cells grown on feeders.

Molecular markers of terminal differentiation

So far, I have described the effect of different culture conditions on morphological differentiation of keratinocytes. For studies of terminal differentiation at the molecular level it is necessary to have markers that distinguish different cells at different stages in the process. Of the many available, keratins have been most widely studied (reviewed by Fuchs et al., 1988; and see chapter in this volume by Breitkreutz et al.), but I shall concentrate on the proteins that form the cornified envelope.

Involucrin is the major precursor of the cornified envelope (Rice and Green, 1979; Simon and Green, 1984). It is synthesized after cells have stopped dividing, but some time before the envelope is assembled: in the epidermis, this is a variable distance through the spinous layers (see Fig. 1b). Involucrin has been cloned and sequenced (Eckert and Green, 1986); it contains a 10 codon repeat which is rapidly evolving (Teumer and Green, 1989). Involucrin is cross-linked via glutamyl-lysine isopeptide bonds, through the action of a keratinocyte-specific transglutaminase, which, like involucrin, is first expressed in the spinous layers of the epidermis (Thacher and Rice, 1985). At least two other envelope precursors are expressed during terminal differentiation: these are membrane glycoproteins of 195 and 210 kD, which may serve to anchor the envelope to the plasma membrane (Simon and Green, 1984).

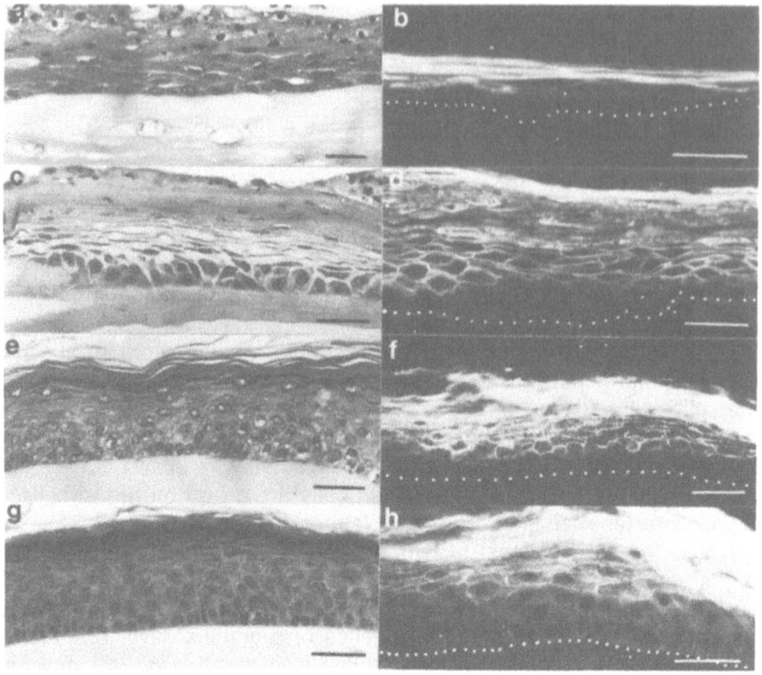

Figure 3 Sections of cultured human keratinocytes. a, c, e, g : H and E staining of paraffin wax sections. b, d, f, h : Immunofluorescence staining of unfixed frozen sections with antiserum to involucrin. Dotted lines indicate boundary between epidermis and collagen. a, b : cells cultured on collagen at the air-liquid interface; c, d : as a, b but with dermis placed under the collagen; e, f : culture 1 week after transplantation onto nude mouse; g, h : culture 3 weeks after transplantation. Scale bars = 50µm. Reproduced from Watt et al. (1987), with permission.

In cultures grown with a feeder layer as described by Rheinwald and Green (1975), involucrin, the keratinocyte-specific transglutaminase and the 195 and 210 kD glycoproteins are all expressed immediately above the basal layer (Fig. 2b; Banks-Schlegel and Green, 1981; Simon and Green, 1984; Watt, unpublished observations). Improved histological differentiation is not associated with an upward shift in the site of involucrin expression (Fig. 3b,d,f; Watt et al., 1987); this has only been noted with time after grafting keratinocytes onto nude mice (Fig. 3h; Banks-Schlegel and Green, 1980; Watt et al., 1987). Thus morphological differentiation and expression of molecular markers of terminal differentiation can be regulated independently (Asselineau et al., 1986).

Factors that regulate terminal differentiation

In this section, I shall discuss the evidence that three specific factors can regulate keratinocyte terminal differentiation. These are the concentration of calcium ions or retinoids in the culture medium and the interaction of keratinocytes and extracellular matrix proteins.

Calcium In 1980 Hennings et al. showed that when mouse keratinocytes are grown in medium containing a low concentration of calcium ions (0.05 - 0.1mM), the cells proliferate, but are forced to grow as a monolayer because desmosome formation and, hence, stratification is inhibited. When the cells are switched to medium containing a standard calcium ion concentration (1.2mM), desmosomes assemble, the cells stratify, DNA synthesis is inhibited and after 3 days cells start to detach from the dish. They interpreted these results as showing that terminal differentiation is induced by 1.2mM calcium (Hennings et al., 1980; Hennings and Holbrook, 1983).

Human keratinocytes can also be forced to grow as a monolayer in low calcium medium, but in these cultures the cells are able to initiate terminal differentiation, as measured by cell cycle withdrawal and initiation of involucrin synthesis (Watt and Green, 1982; Dover and Watt, 1987) and peanut lectin binding (Morrison et al., 1988). Low calcium monolayers of rat keratinocytes also contain terminally differentiating keratinocytes, identified by lectin binding and cell cycle withdrawal (Bernstein et al., 1987). When human keratinocytes are switched to medium containing the standard calcium ion concentration, they assemble desmosomes and stratify, with selective migration of the involucrin-positive cells to the suprabasal layers; there is, however, no inhibition of proliferation and such cultures with time resemble normal confluent cultures grown from clonal density in standard medium (Watt and Green, 1982; Watt, 1984; Watt et al., 1984; Magee et al., 1987; Watt, unpublished observations; Fig. 4).

Is there a fundamental difference between mouse and human keratinocytes in their response to changes in extracellular calcium ions? In both cases it is clear that low calcium completely inhibits morphological differentiation, since stratification is inhibited. At the level of single cells, initiation of terminal differentiation by human and rat keratinocytes definitely taken place in low calcium medium. I suggest that it occurs in mouse monolayers too, but that the terminally differentiating cells detach from the culture dish and are therefore lost on feeding. This would fit with the observation that human keratinocytes become less adhesive as part of the terminal differentiation programme and in postconfluent monolayers there is selective detachment of involucrin-positive cells (Watt and Green, 1982; Watt, 1984).

The effects of extracellular calcium ion concentration on proliferation depend on the other culture conditions. In the presence of a feeder layer, human keratinocytes seeded at clonal densities grow better in standard than low calcium medium; this is partly due to rapid degeneration of the feeders in low calcium medium (Watt, unpublished observations). Clonal growth of mouse keratinocytes in standard medium in the presence of a feeder layer has also been reported (Pera and Gorman, 1984). There are, however, reports that, in the absence of a feeder layer, human keratinocytes grow better in 0.3mM calcium that in higher or lower concentrations (Hawley-Nelson et al., 1980; Boyce and Ham, 1983).

Although human keratinocytes can initiate terminal differentiation in low calcium medium, there are undoubtedly markers of terminal differentiation which are calcium dependent. Thus, the transglutaminase that catalyses involucrin cross-linking is calcium-activated, and envelope assembly can be inhibited by EGTA and promoted by calcium ionophores (Rice and Green, 1978, 1979). In addition, the level of expression of the 195 kD envelope glycoprotein is lower in low calcium monolayers than in stratified cultures (Simon and Watt, in preparation).

What is known about intracellular calcium ion concentrations in keratinocytes? A gradient of increasing extra- and intracellular calcium ions has been reported from the basal layer to the outermost layers of mouse epidermis (Menon et al., 1985). Hennings et al. (1989) report that an increase in extracellular calcium ions causes a rise in intracellular calcium ions, and that normal and neoplastic mouse keratinocytes differ in the level of intracellular calcium ions under low calcium conditions and in response to

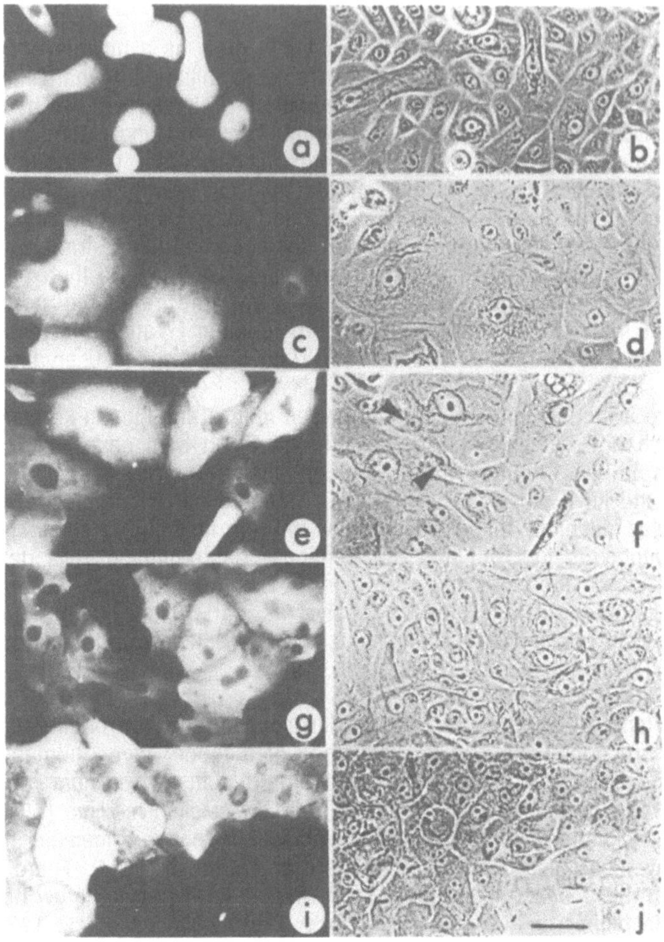

Figure 4 Immunofluorescence and phase-contrast microscopy of involucrin-positive and -negative keratinocytes grown to confluence as a monolayer in low calcium medium and switched to standard medium for A, B : 0hr; C, D : 2hr; E, F : 4hr; G, H : 8hr; I, J : 14hr. Arrows in F indicate two involucrin-negative cells which have been overlapped by neighbouring involucrin-positive cells. In (I) note sheet of involucrin-positive cells overlying the basal layer of involcurin-negative cells. Scale bar = 100μm. Reproduced from Magee et al. (1987), with permission.

changes in extracellular calcium. We have preliminary evidence that there is no difference in the resting intracellular calcium ion concentrations between involucrin-positive and -negative human keratinocytes in low or standard calcium medium (Watt et al., in preparation).

In conclusion, the extracellular calcium ion concentration can modulate morphological differentiation of keratinocytes. It can inhibit growth of mouse keratinocytes under some circumstances, but there is no strong evidence for a central role in modulating proliferation or terminal differentiation at the level of individual cells.

Retinoids Vitamin A has profound effects on epithelial differentiation: when the concentration is too low squamous metaplasia occurs and excess vitamin A causes mucous metaplasia (Fell and Mellanby, 1953; Wolbach, 1954; Wolf, 1980). In vitro correlates of these changes have been observed. Addition of retinoic acid to cultures of human keratinocytes suppresses cornified envelope assembly (Yaar et al., 1981; Green and Watt, 1982; Fig. 5) and expression of keratin 1, a keratin characteristic of terminal differentiation of keratinocytes in skin (Fuchs and Green, 1981). The enhanced expression of keratin 1 and envelopes observed in cultures at the air-liquid interface may be due to the presence of the dermal equivalent decreasing the concentration of retinoid to which the keratinocytes are exposed (Asselineau et al., 1985; Kopan et al., 1987). Excess retinoic acid leads to suprabasal expression of fibronectin receptors (Asselineau et al., 1989) and to increased cell movement, as evidenced by ridges and whorls in the cultures (Fuchs and Green, 1981).

All of these observations are consistent with a role for retinoids in determining the pathway of keratinocyte terminal differentiation between 'keratinised' and 'mucous', both at the level of individual cells and at the level of morphological differentiation. However, there is, at present, no evidence that retinoids affect commitment to terminal differentiation. Indeed the level of retinoic acid in the medium does not affect the involucrin content of the cultures, although it markedly affects the number of cells with cornified envelopes (Green and Watt, 1982; Fig. 5). Involucrin is expressed by keratinocytes during terminal differentiation even in tissues, such as cornea, which do not accumulate cornified layers (Banks-Schlegel and Green, 1981).

The molecular basis for modulation of the differentiation pathway of keratinocytes by retinoids is unclear, but three nuclear retinoic acid receptors have now been cloned, one of which is predominantly expressed in skin (Green and Chambon, 1988; Zelent et al., 1989).

Extracellular matrix So far, the different culture conditions I have described have had little effect on involucrin expression, and hence on terminal differentiation at the single cell level. Two conditions, however, which result in irreversible inhibition of DNA synthesis, followed by an increase in the proportion of cells expressing involucrin and in the proportion with cornified envelopes, are to treat cells with the phorbol ester 12-0-tetradecanoyl phorbol-13-acetate (TPA) (Parkinson and Emmerson, 1982) or to disaggregate them and suspend them in methylcellulose (Rice and Green, 1978; Rheinwald, 1979; Adams and Watt, 1989).

The ways in which suspension may induce terminal differentiation have been investigated further. By plating single involucrin-negative keratinocytes onto circular adhesive islands of a range of diameters it has been found that restricting the area of contact that a keratinocyte makes with the substratum stimulates cell cycle withdrawal and involucrin expression (Watt et al., 1988, Fig. 6). Furthermore, addition of fibronectin or antibodies to fibronectin receptors inhibits suspension-induced involucrin expression, though not cell cycle withdrawal (Adams and Watt, 1989). Taking these results together, it is possible that the level of occupancy of extracellular matrix receptors may regulate overt terminal differentiation.

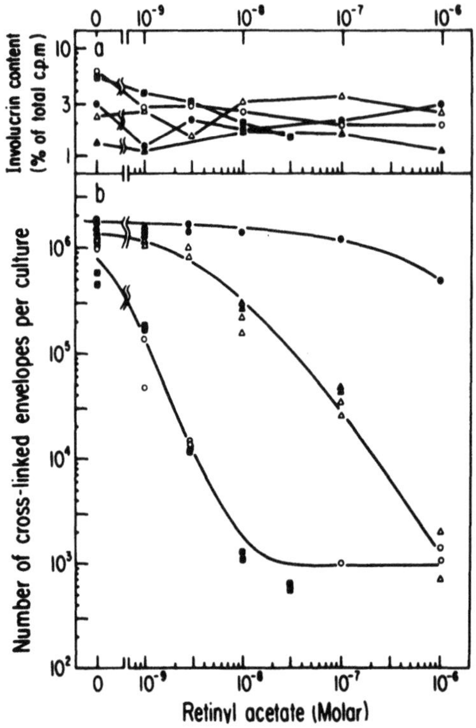

Figure 5 Effect of retinyl acetate on cornified envelope formation and involucrin content of cultured keratinocytes from different body sites. ● : epidermal; ▲ : oesophageal; △ : vaginal; ■ and ○ : conjunctival. Reproduced from Green and Watt (1982), with permission.

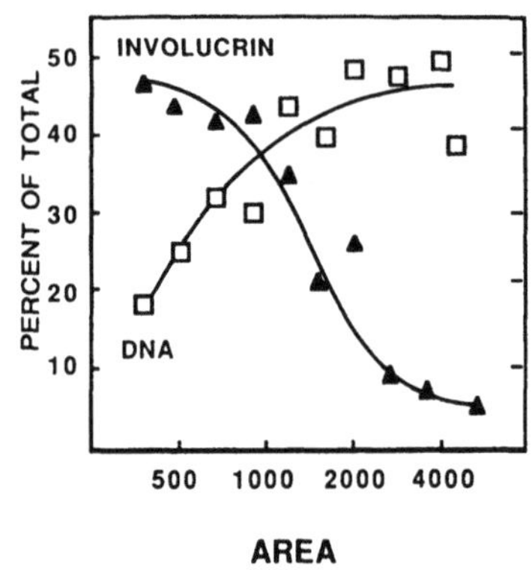

Figure 6 Effect of substratum contact area (μm^2) on proportion of keratinocytes incorporating ^{3}H-thymidine into DNA [□] or expressing involucrin (▲) 3 days after plating. Reproduced from Watt et al. (1988), with permission.

CONCLUSIONS

In summary, it is possible to recreate human epidermis in culture. Conditions which mimic the normal epidermal environment lead to improved morphological differentiation; however this is distinct from terminal differentiation at the single cell level. The extracellular calcium ion concentration modulates stratification, and hence morphological differentiation, but there is no firm evidence that it affects the commitment of individual cells to terminal differentiation. Retinoids affect the nature of the differentiation pathway following commitment. Extracellular matrix interactions can regulate overt terminal differentiation as measured by involucrin expression. However, until we can identify epidermal stem cells, the nature of the commitment events within the basal layer will remain elusive (Hall and Watt, 1989).

REFERENCES

Adams, J. C., and Watt, F. M., 1989, Fibronectin inhibits the terminal differentiation of human keratinocytes. Nature (Lond.) , 340:307.

Asselineau, D., Bernhard, B., Bailly, C., and Darmon, M., 1985, Epidermal morphogenesis and induction of the 67kD keratin polypeptide by culture of human keratinocytes at the liquid-air interface. Exp. Cell Res., 159:536.

Asselineau, D., Bernard, B. A., Bailly, C. and Darmon, M., 1989, Retinoic acid improved epidermal morphogenesis. Devel. Biol., 133:322.

Asselineau, D., Bernard, B. A., Bailly, C., Darmon, M., and Prunieras, M., 1986, Human epidermis reconstructed by culture: is it "normal"? J. Invest. Dermatol., 86:181.

Banks-Schlegel, S., and Green, H., 1980, Formation of epidermis by serially cultivated human epidermal cells transplanted as an epithelium to athymic mice. Transplantation, 29:308.

Banks-Schlegel, S., and Green, H., 1981, Involucrin synthesis and tissue assembly by keratinocytes in natural and cultured human epithelia. J. Cell Biol., 90:732.

Bell, E., Ehlrich, H. P., Buttle, D. J., and Nakatsuji, T., 1981, Living tissue formed in vitro and accepted as skin-equivalent tissue of full thickness. Science, 211:1052.

Bernstein, I. A., Ku, W. W., Kim, H. J., and Brown, R., 1987, Differentiation and stratification in the epidermis are not obligatorily coupled, in: "Psoriasis. Proceedings of the Fourth International Symposium," E. M. Farber, L. Nall, V. Morhenn, P. J. Jacobs, eds., Elsevier, New York, p78.

Boyce, S. T., and Ham, R. G., 1983, Calcium-regulated differentiation of normal human epidermal keratinocytes in chemically defined clonal culture and serum-free serial culture. J. Invest. Dermatol., 81:33S.

Compton, C. C., Gill, J. M., Bradford, D. A., Regauer, S., Gallico, G. G., and O'Connor, N. E., 1989, Skin regenerated from cultured epithelial autografts on full-thickness burn wounds from 6 days to 5 years after grafting. A light, electron microscopic and immunohistochemical study. Lab. Invest., 60:600.

De Luca, M., D'Anna, F., Bondanza, S., Franzi, A. T. and Cancedda, R., 1988, Human epithelial cells induce human melanocyte growth in vitro but only skin keratinocytes regulate its proper differentiation in the absence of dermis. J. Cell Biol., 107:1919.

Dover, R., and Watt, F. M., 1987, Measurement of the rate of epidermal terminal differentiation: expression of involucrin by S-phase keratinocytes in culture and in psoriatic plaques. J. Invest. Dermatol., 89:349.

Eckert, R. L., and Green, H., 1986, Structure and evolution of the human involucrin gene. Cell, 46:583.

Fell, H. B., and Mellanby, E., 1953, Metaplasia produced in cultures of chick ectoderm by high vitamin A. J. Physiol., 119:470.

Fuchs, E., Albers, K., and Kopan, R., 1988, Terminal differentiation in cultured human epidermal cells. Advances in Cell Culture, 6:1.

Fuchs, E., and Green, H., 1981, Regulation of terminal differentiation of cultured human keratinocytes by vitamin A. Cell, 25:617.
Green, H., and Watt, F. M., 1982, Regulation by vitamin A of envelope cross-linking in cultured keratinocytes derived from different human epithelia. Mol. Cell. Biol., 2:1115.
Green, S., and Chambon, P., 1988, Nuclear receptors enhance our understanding of transcription regulation. Trends in Genetics, 4:309.
Hall, P. A., and Watt, F. M., 1989, Stem cells: the generation and maintenance of cellular diversity. Development, 106:619.
Hawley-Nelson, P., Sullivan, J. E., Kung, M., Hennings, H., and Yuspa, S. H., 1980, Optimized conditions for the growth of human epidermal cells in culture. J. Invest. Dermatol., 75:176.
Hennings, H., and Holbrook, K. A., 1983, Calcium regulation of cell-cell contact and differentiation of epidermal cells in culture. An ultrastructural study. Exp. Cell Res., 143:127.
Hennings, H., Kruszewski, F. H., Yuspa, S. H., and Tucker, R. W., 1989, Intracellular calcium alterations in response to increased external calcium in normal and neoplastic keratinocytes. Carcinogenesis, 10:777.
Hennings, H., Michael, D., Cheng, C., Steinert, P., Holbrook, K., and Yuspa, S. H., 1980, Calcium regulation of growth and differentiation of mouse epidermal cells in culture. Cell, 19:245.
Kopan, R., Traska, G., and Fuchs, E., 1987, Retinoids as important regulators of terminal differentiation: examining keratin expression in individual epidermal cells at various stages of keratinization. J. Cell Biol., 105:427.
Mackenzie, I. C., and Fusenig, N. E., 1983, Regeneration of organized epithelial structure. J. Invest. Dermatol., 81:189S.
Magee, A. I., Lytton, N. A., and Watt, F. M., 1987, Calcium-induced changes in cytoskeleton and motility of cultured human keratinocytes. Exp. Cell Res., 172:43.
Menon, G. K., Grayson, S., and Elias, P. M., 1985, Ionic calcium reservoirs in mammalian epidermis: ultrastructural localization by ion-capture cytochemistry. J. Invest. Dermatol., 84:508.
Morrison, A. I., Keeble, S., and Watt, F. M., 1988, The peanut lectin-binding glycoproteins of human epidermal keratinocytes. Exp. Cell Res., 177:247.
Parkinson, E. K., and Emmerson, A., 1982, The effects of tumour promoters and on the multiplication and morphology of cultured human epidermal keratinocytes. Carcinogenesis, 3:525.
Pera, M. F., and Gorman, P. A., 1984, In vitro analysis of multistage epidermal carcinogenesis: development of indefinite renewal capacity and reduced growth factor requirements in colony forming keratinocytes precedes malignant transformation. Carcinogenesis, 5:671.
Potten, C. S.., and Morris, R. J., 1988, Epithelial stem cells in vivo. J. Cell Sci., Suppl. 10:45.
Pruniéras, M., Régnier, M., and Woodley, D., 1983, Methods for cultivation of keratinocytes with an air-liquid interface. J. Invest. Derm., 81:28S.
Rheinwald, J. G., 1979, The role of terminal differentiation in the finite culture lifetime of the human epidermal keratinocyte. Int. Rev. Cytol., Suppl. 10:25.
Rheinwald, J. G., 1989, Methods for clonal growth and serial cultivation of normal human epidermal keratinocytes and mesothelial cells, in: "Cell Growth and Division. A Practical Approach," R. Baserga, ed., IRL Press, Oxford, p81.
Rheinwald, J. G., and Green, H., 1975, Serial cultivation of strains of human epidermal keratinocytes: the formation of keratinizing colonies from single cells. Cell, 6:331.
Rice, R. H., and Green, H., 1978, Relation of protein synthesis and transglutaminase activity to formation of the cross-linked envelope during terminal differentiation of the cultured human epidermal keratinocyte. J. Cell Biol., 76:705.

Rice, R. H., and Green, H., 1979, Presence in human epidermal cells of a soluble protein precursor of the cross-linked envelope: activation of the cross-linking by calcium ions. Cell, 18:681.

Simon, M., and Green, H., 1984, Participation of membrane-associated proteins in the formation of the cross-linked envelope of the keratinocyte. Cell, 36:827.

Teumer, J., and Green, H., 1989, Divergent evolution of part of the involucrin gene in the hominoids: unique intragenic duplications in the gorilla and human. Proc. Natl. Acad. Sci. USA, 86:1283.

Thacher, S. M., and Rice, R. H., 1985, Keratinocyte-specific transglutaminase of cultured human epidermal cells: relation to cross-linked envelope formation and terminal differentiation. Cell, 40:685.

Watt, F. M., 1984, Selective migration of terminally differentiating cells from the basal layer of cultured human epidermis. J. Cell Biol., 98:16.

Watt, F. M., Boukamp, P., Hornung, J., and Fusenig, N. E., 1987, Effect of growth environment on spatial expression of involucrin by human epidermal keratinocytes. Arch. Dermatol. Res., 279:335.

Watt, F. M., and Green, H., 1982, Stratification and terminal differentiation of cultured epidermal cells. Nature (Lond.), 295:434.

Watt, F. M., Jordan, P. W., and O'Neill, C. H., 1988, Cell shape controls terminal differentiation of human epidermal keratinocytes. Proc. Natl. Acad. Sci. USA, 85:5576.

Watt, F. M., Mattey, D. L., and Garrod, D. R., 1984, Calcium-induced reorganization of desmosomal components in cultured human keratinocytes. J. Cell. Biol., 99:2211.

Wolbach, S. B., 1954, Effects of vitamin A deficiency and hypervitaminosis A in animals, in: "The Vitamins," W. H. Sebrell, Jr., R. S. Harris, eds. Academic Press Inc., New York. p106.

Wolf, G., 1980, Vitamin A, in: "Nutrition and the Adult, Micronutrients," R. B. Alfin-Slater, and D. Kritchevsky, eds., Plenum Publishing Corporation, New York, p.97.

Yaar, M., Stanley, J. R., and Katz, S. I., 1981, Retinoic acid delays the terminal differentiation of keratinocytes in suspension culture. J. Invest. Dermatol., 76:363.

Zelent, A., Krust, A., Petkovich, M., Kastner, P., and Chambon, P., 1989, Cloning of murine α and β retinoic acid receptors and a novel receptor γ predominantly expressed in skin. Nature, 339:714.

HUMAN KERATINOCYTE CELL LINES

Dirk Breitkreutz, Petra Boukamp, Andrea Hülsen,
Cathy Ryle, Hans-Jürgen Stark, Hans Smola,
Gabi Thiekötter, Norbert E. Fusenig

Division of Differentiation and Carcinogenesis
in Vitro, Institute of Biochemistry,
German Cancer Research Center, D-6900 Heidelberg

ABSTRACT

In the past, immortalization of human epidermal keratinocytes has been generally considered to have a massive impact on their differentiation properties. This is certainly true for many virally transformed cells. In contrast, the recently described spontaneous cell line HaCaT exhibits a virtually normal growth and differentiation pattern when transferred to an in vivo environment (cell transplants on nude mouse). These HaCaT cells also respond in vitro to external signals (e.g. levels of calcium, retinoids) similarly as normal keratinocytes. All major epidermal differentiation products are expressed (specific keratins, involucrin, filaggrin) under appropriate conditions at high levels and apparently in the right sequence. In order to provoke malignant conversion HaCaT cells were transfected with mutated cellular Ha-ras (val-12). The derived cell clones could be classified according to their growth properties in vivo into one (1) non-tumorigenic (growing as surface epithelia) and two tumorigenic groups, forming either (2) benign cysts or (3) well-differentiated squamous cell carcinomas. While the differentiation potential was retained in these cells to a varying degree, showing no clear correlation to malignant properties, all tumorigenic clones (benign and malignant) had lower growth factor requirements and maintained growth even without serum. Thus, HaCaT and derived cell lines provide readily available cell culture models and might offer useful alternatives to primary or low-passaged normal keratinocytes. According to preliminary data, some promising applications are: testing and detection of growth and differentiation factors (also drugs), synthesis of differentiation products and epidermis-derived factors (possibly at large scale) and introduction of exogenous genes or subgenomic regulatory elements.

INTRODUCTION

The site-specific differentiation and specializations in the epidermis reflect, besides inherent properties of epithelial cells, the complex mutual interactions with neighboring mesenchyme. Due to the impressive development of cell culture methodology, most aspects of normal differentiation (keratinization) can be mimicked in vitro providing appropriate environmental stimuli (Bell et al., 1983; Fusenig, 1986; Asselineau et al., 1986, 1989; Regnier et al., 1986; Kopan et al., 1987, 1989; see also E. Bell, F. Watt,

Pharmaceutical Applications of Cell and Tissue Culture to Drug Transport
Edited by G. Wilson *et al.*, Plenum Press, New York, 1991

both this volume). This in turn provides a basis for the identification of regulatory principles which presumably also act in vivo, and for the development of concepts how to modulate or to interfere with abnormal growth and differentiation processes. Although at first sight immortalization of cells seems to be a big advantage for this purpose, in the past most differentiation studies on established human cell lines have been only of very limited value. This certainly applies to most virally transformed human keratinocytes, which often might be only capable to express a truncated differentiation program at best. Based on these observations the hypothesis seemed to be verified that immortalization (presumably as a first step towards malignancy) unequivocally leads to a dramatic loss of differentiation potential (Rheinwald and Becket, 1980; Winter et al., 1980; Yuspa et al., 1985; Roop et al., 1988). After we had shown that this was not necessarily the case with murine cell lines (Breitkreutz et al., 1986), we also succeeded in developing a human keratinocyte line, HaCaT, with a virtually normal differentiation potential (Boukamp et al., 1988). That "spontaneous" transformation of the human keratinocytes, however, is a rather rare event is clearly illustrated by the fact that currently only one other cell line exists which shows apparently similar properties (Baden et al., 1987). In addition, herein we shall describe a series of cell clones derived from HaCaT cells after c-Ha-ras transfection (Fusenig et al., 1987, 1989; Boukamp, 1987; Boukamp et al., submitted) which in spite of partly different growth behavior share most differentiation properties with the parent line. The major focus will be on differentiation (in particular expression of specific keratins as early marker) and possible ways of its modulation, which, on the other hand, might have an important impact on growth control.

ORIGIN AND ESTABLISHMENT OF CELL LINES

Several attempts have been made to generate spontaneous keratinocyte cell lines from epidermis of adult human trunk skin following the protocols for establishing mouse cell lines (Fusenig et al., 1983, 1986; Breitkreutz et al., 1986). In one case we succeeded and a cell line evolved from cells growing as primary culture for five months (Fusenig et al., 1987; Boukamp et al., 1988). These keratinocytes were derived from a histologically normal skin biopsy at the far periphery of a male melanoma patient (reconfirmed by DNA mapping; Boukamp et al., 1988). The cell line was named HaCaT to indicate that cells were grown initially at low calcium concentration (0.2 mM Ca^{2+}, largely preventing differentiation, although above the concentration used for mouse cells by Hennings et al. (1980) and slightly elevated temperature (38.5°C, possibly enhancing chromosomal rearrangements).

The karyotype, when first analyzed, was hypodiploid containing two distinct marker chromosomes, but changed early to hypotetraploid coinciding with the acquisition of improved growth properties at standard cell culture conditions. At about passage 30, HaCaT cells were also transfected with constructs of mutated cellular Ha-ras (val-12) and the neomycin resistance gene for selection (with G418-containing media). Since there were no striking morphological differences, transfected clones were isolated at random and further propagated. According to their growth properties after injection into nude mice clones were grouped into three categories (Boukamp, 1987; Fusenig et al., 1987, 1989): (1) non-tumorigenic (fast regressing nodules only) and tumorigenic ones being (2) either benign (forming large differentiating cysts) or (3) malignant (well-differentiated squamous cell carcinomas, SCC). For comparison we used normal keratinocytes from both adult trunk skin and juvenile foreskin, as well as malignant SCC lines previously established from human skin tumors (Breitkreutz et al., 1981; Boukamp et al., 1982; Tilgen et al., 1983). Furthermore, two virally transformed cell lines were included in the study, one generated by transfecting adult keratinocytes with origin-defective SV40 DNA which we named HaSV (Boukamp, 1987; Fusenig et al., 1987, 1989), and the other one, HPKI, by transfecting foreskin cells with HPV16 DNA (Dürst et al., 1987). In the following, the features mainly of HaCaT cells and derived clones will be estimated using, in addition to normal cells, the other cell lines largely as reference.

GROWTH PROPERTIES AND PHENOTYPIC EXPRESSION IN CULTURE

Morphologically cultures of the parent HaCaT cell line and many of the derived ras-transfected clones very closely resembled those of normal keratinocytes. Thus, in contrast to most of the other cell lines, they were able to stratify and to form horny layers, at least at lower passage levels. Corresponding to normal cells (Hennings et al., 1980; for general review see Fusenig, 1986) stratification was dependent on the calcium level of the media as demonstrated for the benign HaCaT-ras clone II-1 (Fig. 1). Vitamin A depletion further enhanced keratinization (Fig. 1d), but this will be discussed in more detail lateron. Apart from the morphological appearance there was also no clear distinction between HaCaT cells and ras-clones in growth properties at regular media conditions (a comparison of HaCaT and HaSV cells regarding pheno- and genotype is given by Fusenig et al., 1987). However,

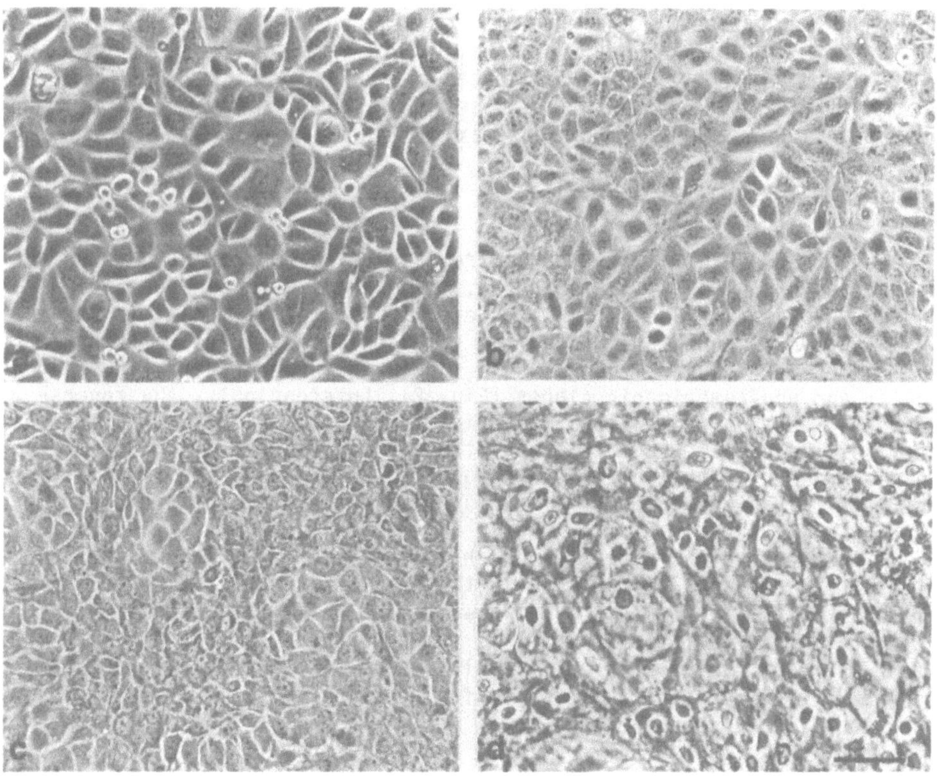

Figure 1. Influence of media composition on culture morphology (phase contrast) demonstrated for tumorigenic (benign) clone HaCaT-II-1. (a) Monolayer of highly refractory cells at low calcium (0.08 mM), (b) establishment of epithelial (desmosomal) contacts with increasing calcium (0.15 mM) and (c) onset of stratification at high calcium levels (regular media with 1.5 mM Ca^{2+}). In long-term cultures formation of keratinized squames was most pronounced in vitamin A depleted (1.5 mM Ca^{2+}) media (d), the upper layers are in focus showing some cells with small dark granules and abundant pyknotic nuclei. Bar 100 µm.

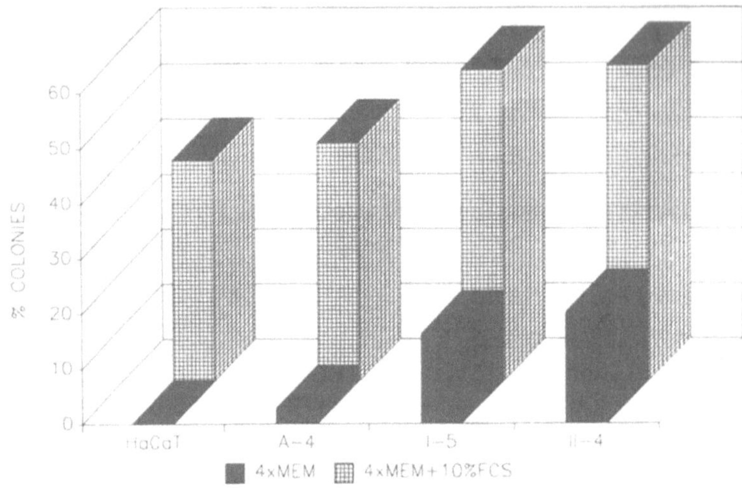

Figure 2. Comparison of colony-forming efficiency of HaCaT cells and derived clones growing with 10% fetal calf serum (FCS) or in seriumfree media. While in HaCaT cells and nontumorigenic clones (represented by clone A-4) without serum the formation of colonies was virtually completely suppressed, it was only reduced in the case of tumorigenic clones (benign group, clone I-5; malignant, II-4) generally showing similar efficiencies between both categories. The percentage indicates the number of colonies formed per 100 cells plated in the assay, mean values of three measurements are given (maximal variations in the order of 10%).

the search for minimal growth requirements revealed that both groups of the tumorigenic clones (2, benign and 3, malignant) were able to continue growth at very low serum levels (0.5%) or even in serum-free media without additional factors. Furthermore, under the same conditions these cells could still be cloned on tissue culture plastic, again in marked contrast to the parent cells and non-tumorigenic clones (Fig. 2). Interestingly, the reverse was true for soft agar growth (a "classical" transformation test for mesenchymal cells), where only HaCaT cells and the non-tumorigenic clones formed colonies. Apparently the integrated exogenous ras oncogene can exert different effects on various pathways of signal transduction in the transfected cells.

Another typical property, which is encountered during propagation is the strong adhesion of HaCaT cells to the plastic substratum. This creates the risk of counterselecting cells with high differentiation potential during passaging cultures in favor of subpopulations of less adhesive (and more aberrant) cells. Currently early passages of HaCaT cells are recloned in an attempt to avoid a steady drift in phenotype that was observed so far with continuous propagation.

Since keratins are major structural components in keratinocytes as well as indicators for type and stage of differentiation (Fuchs and Green, 1980; Franke et al., 1981; Moll et al., 1982, 1984; Sun et al., 1983; Tseng et al., 1984; ; Bowden et al., 1987; Van Muijen et al., 1987; Steinert and Roop, 1988), protein extracts were prepared from the various human cell lines and analyzed by electrophoretic separation on two-dimensional gels (as described by Stark et al. 1987). In Figure 3 the keratin patterns of cell

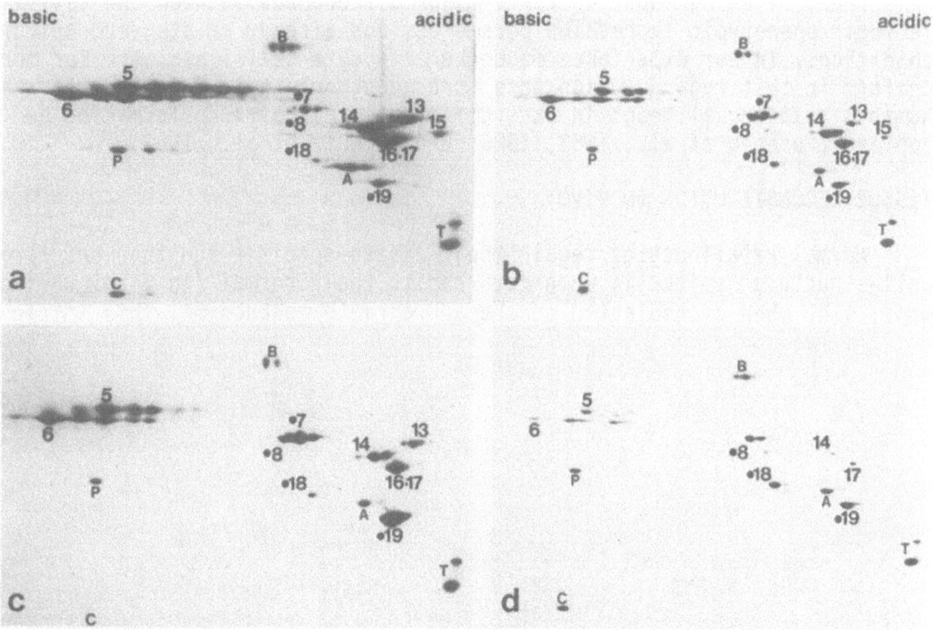

Figure 3. Keratin patterns of cultured cells on two-dimensional gels. (a) normal keratinocytes from juvenile foreskin, (b) benign HaCaT-ras clone I-7 (for pattern of parent HaCaT cells see Fig. 4a), (c, d) virally transformed cell lines, (c) HPKI (HPV16 DNA) and (d) non-tumorigenic (neo-transfected) clone of HaSV (SV40 DNA). Keratins are numbered according to Moll et al. (1982), those usually found in simple epithelia are marked (.). As internal protein standards were added: B, bovine serum albumin (68KDa, pI 6.4); P, phosphoglycerate kinase (43KDa, pI 7.4); A, α-actin (42KDa, pI 5.4); T, tropomyosin (36 KDa and 38 KDa, pI 4.9); C, carbonic anhydrase (30 KDa, pI 7.5). Separation was accomplished in the first dimension by nonequilibrium pH-gradient gel electrophoresis (NEPHGE, basic proteins on the left) and in the second by SDS polyacrylamide gel electrophoresis (SPAGE, larger proteins on top).

cultures from juvenile foreskin (a) are compared with those of several cell lines (b-d). Besides the usual "culture set" of keratins (K5, K6, K14 and K16/17) foreskin cells express varying amounts of other keratin species, most prominently K13, common in internal stratified epithelia. HaCaT cultures reaching confluence have a very similar pattern (compare Fig. 4a). On the other hand, the ras-clones (Fig. 3b, the benign clone I-7 as an example) usually express slightly higher levels of "non-epidermal" keratins (marked in this Figure: K7, K8, K18 and K19) which are usually associated with simple epithelia (SE). In HPKI cells (Fig. 3c) SE-keratins are still more prominent, most striking K19. Further remote is the keratin composition of HaSV cultures (Fig. 3d) which remains virtually unchanged in the neo- or ras-transfected variants of this cell line. First of all, in HaSV cells the level of keratins (compared to total protein) is much lower and K8, K18 and K19 become the major keratin species, generally in accord with data reported from other SV40 transformed keratinocyte lines (Taylor-Papadimitriou et al., 1982; Banks-Schlegel and Howley, 1983; Steinberg and Defendi, 1983; Bernard et al., 1985).

Thus, by using those criteria major differences between the cell lines in their phenotypic expression became obvious already at standard culture conditions. In our experience mouse keratinocyte cell lines were far more uniform in this regard, which once more questions their relevance for the human situation, although in part different modes of transformation had been applied (Fusenig et al., 1983, 1986; Breitkreutz et al., 1986).

TISSUE RECONSTITUTION IN VIVO

Normal keratinocytes regain their tissue-specific function and form well-structured epithelia when returned to their normal (in vivo) environment, i.e. when transplanted on the appropriate mesenchyme. This intrinsic

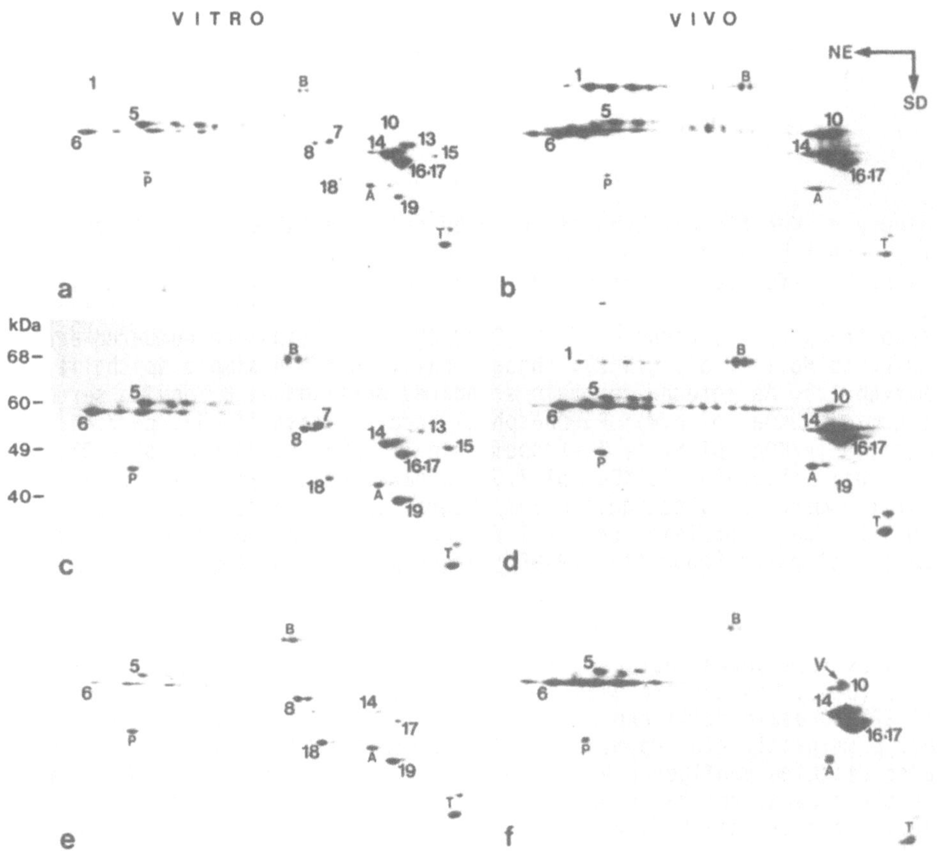

Figure 4. Comparison of keratin patterns expressed by cell lines in vitro and in vivo. (a) confluent culture of HaCaT cells and (b) corresponding transplant. (c) culture of malignant HaCaT clone II-4 and (d) transplant of the comparable clone II-3. Both clones showed also in tumors very similar keratin patterns as in transplants. (e) HaSV cells and (f) nude mouse tumor formed by a ras-transfected, malignant HaSV clone (comparable in vitro patterns). Separation (NEPHGE/SPAGE) and protein standards as in Fig. 3; V, vimentin presumably from mesenchymal components of tumor.

property has been well documented for animal and human cells (Doran et al., 1980; Banks-Schlegel and Green, 1980; Worst et al., 1982; Lavker and Sun, 1983; Fusenig et al., 1983; Breitkreutz et al., 1984; Watt et. al., 1987; Bosca et al., 1988; Boukamp et al., 1988, and submitted) and is meanwhile exploited clinically for coverage of large surface wounds mainly from severe burns (Gallico et al., 1984; Compton et al., 1989). On the other hand, transplantation has been used to test or to modulate differentiation functions in transformed keratinocytes (Wu et al., 1981; Fusenig et al., 1983, 1986; Boukamp et al., 1985, 1988, and in preparation; Breitkreutz et al., 1986). Using this approach, for example when transplanted as intact cell layer on collagen gels (Fusenig et al., 1983; for review see Fusenig, 1986), HaCaT cells formed well-structured epithelia with all differentiation markers present, detectable by immunofluorescence and protein analysis (Boukamp et al., 1988). Many of the ras-clones behaved very similarly as the parent line (histology of transplants). The malignant variants, however, while still forming surface epithelia, eventually grew invasively into the host mesenchyme and showed also a marked, though varying, loss of structural organization into distinctive cell layers.

The "suprabasal" keratins K1 and K10/11, considered as early markers in epidermal differentiation, at least to some extent also reflect the general degree of differentiation. Thus, judged by keratin analysis, the high differentiation potential of HaCaT cell transplants is shown, opposed to the pattern of a confluent culture (Fig. 4a, b). On this basis, the differentiation potential is similar in the ras-clones (compare clone II-1, below and Fig. 6d) or slightly reduced as in the case of the malignant clones II-3 and II-4 (Fig. 4c, d). Comparable levels of K1 and K10/11 were also detected in nude mouse tumors of these clones (not shown here, compare immunofluorescence below) besides a higher degree of proteolytic turnover. Similar observations had been made previously on transplants and tumors formed by the HELP series, originating from mouse keratinocytes (Breitkreutz et al., 1986), also including keratin processing as described for epidermis (Breitkreutz et al., 1984; Bowden et al., 1984, 1987). In general, most of the other human cell lines grew rather poorly as surface epithelia. As shown for a malignant ras-transfected HaSV clone (Fig. 4e, f) the keratin composition was modulated but, besides low expression of K10/11, still far from normalization. The response in the host was much alike that of our two skin carcinoma derived SCC-lines reported previously (Breitkreutz, 1981; Boukamp et al., 1982; Tilgen et al., 1983).

Immunofluorescence in addition allowed the localization of specific keratins besides other differentiation markers. This method is generally more sensitive than analyzing gels using conventional protein staining, although quantitation is still very vague on this basis. In the epithelia reconstituted by HaCaT cells several differentiation markers could be detected in distinct zones and in the right localization, and in quantities comparable to normal tissue (Boukamp et al., 1988). Using sequence-specific (Roop et al., 1983) and monoclonal (Morgan et al., 1987) antibodies against K1 and K10/11 respectively, there was a general tendency of K1 being expressed prematurely and preceding K10/11 (Ryle et al., 1989; Breitkreutz et al., in preparation). This discrepancy in coordinated expression was even more pronounced in most of the HaCaT-ras clones, where K10 synthesis mainly starts considerably later, i.e. in cross-sections far above the K1-positive cells (Fig. 5c, clone II-3), consistent with the situation in tumors (Fig. 5d, same clone). Besides this marked effect in vivo the figure also shows that suprabasal keratins can be synthesized to a fairly large extent in vitro under certain circumstances, (panel a and b, cell sheet detached from plastic and organotypic culture, respectively), this will be discussed below.

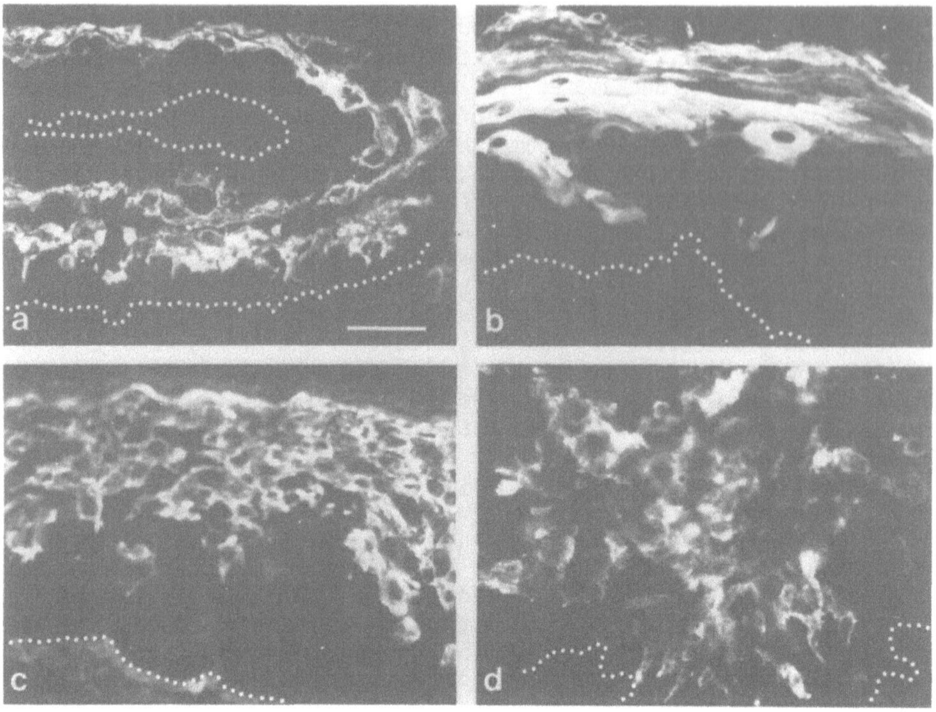

Figure 5. Localization of suprabasal keratin K10 in different epithelia formed by malignant HaCaT clone II-3 in vitro (a, b) and in vivo (c, d); indirect immunofluorescence of frozen cross-section using sequence-specific antibodies. (a) folded epithelial sheet (the lower two loops facing at the upper surface) grown at regular culture conditions (submersed) and detached by treatment with dispase, (b) organotypic culture on collagen (at air-liquid interphase), (c) cell transplant after three weeks, and (d) nude mouse tumor. In all cases the area positive for K1 was larger, involving also some basal cells, than K10-staining (seen by double fluorescence, same specimens, not shown). The lower contours of basal cells are marked (a-c), while in (d) the whole field shown was K1-positive except the small areas marked in the lower left and right corners. Bar 50 µm.

MODULATING FACTORS IN VITRO

When cultures of HaCaT cells were analyzed at different time points it became apparent that keratin synthesis was also strongly modulated by cell density (Ryle et al., 1989; Breitkreutz et al., 1989). To this extent, this was rather unexpected comparing earlier data on mouse cells (Fusenig et al., 1983; Breitkreutz et al., 1986) and other human cell lines (Breitkreutz et al., 1981; Boukamp et al., 1982; Tilgen et al., 1983). Thus, at low density, the SE-keratins K7, K8, K18 and K19, ("simple epithelial type") were much more prominent reaching the levels of the usual "culture type" pattern consisting of K5, K6, K14 and K16/K17 (the latter two not well resolved in our gel system, see Stark et al., 1987). After reaching confluence there was a transient increase of the "stratification markers" K13 and K15 (in vivo typical for internal epithelia) in HaCaT cells and most clones, and with further increasing density a significant accumulation of the "suprabasal" keratins K1 and K10/11. While this was evident by protein analysis and staining on 2D-gels, labeling experiments demonstrated this even more clearly by revealing the changes in actual synthesis. Localization of positive cells was again accomplished by indirect immunofluorescence using specific antibodies (Roop et al., 1984; Morgan et al., 1987) as previously described (Ryle et al., 1989). This disclosed that the number of strongly stained cells was raising instead of a gradual increase in staining of the whole cell population.

A more direct effect, however, was exerted by changing the retinoid concentration in the media, similarly as shown previously for normal human keratinocytes and also some cell lines (Fuchs and Green, 1981; Kim et al., 1984; Gilfix and Eckert, 1985; Kopan et al., 1987, 1989; Regnier et al., 1988; Asselineau et al., 1989). High levels of retinoic acid (added to vitamin A-free media containing delipidized serum) generally lowered overall keratin synthesis and at 10^{-6} M completely suppressed K1 and K10/11, while SE-keratins increased correspondingly (Fig. 6a; total protein extracts). Interestingly the picture looked like that of SCC- or SV40-cell lines growing at the usual vitamin A levels. The patterns normalized with decreasing doses of retinoic acid, at 10^{-8} to 10^{-9} M resembling time-matched controls in conventional media (Breitkreutz et al., 1989). In retinoid-free media (with delipidized serum only) induction of suprabasal K1 and K10/11 was optimal (Fig. 6). Synthesis of non-epidermal keratins (K7, 8, 18, 19, but also K13, 15) was largely or, in HaCaT cells, completely suppressed and the

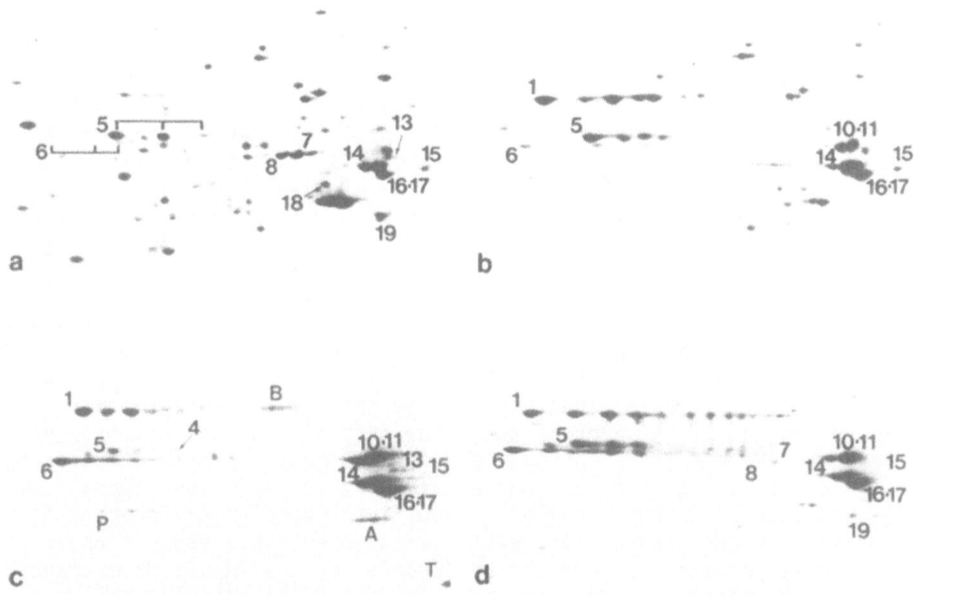

Figure 6 Modulation of keratin expression in culture by retinoic acid. (a, b) extracts of total protein from HaCaT cultures, labeled with ^{35}S-methionine (a) in the presence of 10^{-6} M and (b) in the absence of retinoic acid (both, media with delipidized serum). (c, d) cytoskeletal extracts of (e) HaCaT cell transplant and (d) culture of benign clone II-1 grown without retinoids like (b). Note that both retinoid-depleted cultures (b, d) express a very similar keratin pattern as the transplant (c). Fluorographs of labeled proteins are shown (a, b), protein staining (c, d). Separation and markers as indicated before. Reproduced (in modified form) from Breitkreutz et al. (1989), with permission.

patterns were virtually undistinguishable from those of transplants (Fig. 6c, compare Fig. 4b). Since, however, morphology was rather poor, i.e. long-term cultures looked atrophic in the absence of retinoids, this approach seems to be suitable mainly to turn on the differentiation machinery and to look for early markers (e.g. the "suprabasal" keratins).

Antibody staining on organotypic cultures (cells on collagen gels at air-liquid interphase; Fusenig et al., 1983,1986) revealed that HaCaT-ras clones (type 2 and 3 tested so far; Ryle et al. 1989; own unpublished data) stratified extensively and differentiated well, at least in the uppermost layers. HaCaT cells, on the other hand, having higher growth requirements, stratified rather poorly under these standard conditions. Preliminary data, however, indicate that morphology should substantially improve by modifying media composition or including living fibroblasts in the supporting collagen gel.

In part the effects in organotypic culture might be mediated by establishing polarity, directed transport of nutrients from "basal" to upper cells and possibly a gradient in retinoid concentration decreasing towards the surface. The generally superior quality of cell transplants, however, suggests that there must be additional systemic and local factors, also in order to allow a dynamic response to physiological requirements.

CONCLUSIONS AND FURTHER IMPLICATIONS

Of all human keratinocyte cell lines we have examined so far HaCaT cells provide an optimal benefit/effort ratio considering their differentiation properties and the relative ease of handling. This, in part, applies also to the majority of cell clones derived from HaCaT cells by transfection with cellular Ha-ras oncogene. Apparently among the cell clones there was no simple correlation between their general growth properties (non-tumorigenic, tumorigenic-benign or highly malignant) and the potential to differentiate. However, although the parent HaCaT line shares many features with normal keratinocytes, it is still not known how many genes or regulatory processes are impaired by the acquisition of immortality. In Table 1 some characteristics of HaCaT cells are listed which might be either advantageous or disadvantageous in culture models particulary in comparison to normal cells. Without question major advantages are high cell numbers which are readily available and facilitated access to relatively large amounts of specific gene products. In general, however, some caution is advised in extrapolating data from cell lines, since they certainly do not represent a true image of normal cells, even if they show only minimal deviations. On the other hand, by comparing our previous data on mouse primary cultures and cell lines (Fusenig et al., 1983; Breitkreutz et al., 1984, 1986), there are, regardless of many parallels, significant species-differences on many important aspects (concerning genotypic and phenotypic properties) thus making cells of human origin the preferable choice. Apart from transformation and mechanistic studies for example, this will certainly concern investigations on the response to regulating factors or drugs, and regulation of gene expression or specific tissue interactions. In these, like in other cases, there are some doubts about the relevance of animal models for the human situation.

Concerning another area of increasing interest HaCaT cells represent suitable targets for the introduction of foreign genes. Thus, as demonstrated above, the growth properties of HaCaT cells have been changed by transfection and stable integration of the cellular (mutated) ras-oncogene (Fusenig et al., 1987, 1989; Boukamp et al., submitted). Further, since these cells should have the necessary competence factors, they might be easily transfected to express differentiation-related genes at higher rates. This in turn could be used to examine the function of certain genes (e.g.

Table 1. Criteria for the use of HaCaT cells as an alternative for keratinocyte models

Area of Interest	Advantages	Disadvantages
Growth	unlimited, low requirements	regulation less stringent
Differentiation	all markers present in vivo, inducible in vitro	dissociation, some premature expression
Signal response	vitro: Ca^{2+}, retinoids vivo: mesenchyme	partially refractory
Stability	stable phenotype (low passages)	cytogenetic drift, variants
Gene regulation	suitable recipient: exogenous genes, subgenomic elements	possible changes in pathways ("mutants")

auto-, paracrine factors, receptors, structural components) either by overproduction or expression of their mutated forms. In transient assays using a reporter gene (such as ß-galactosidase or chloramphenicol acetyltransferase, CAT) these cells offer the possibility to scrutinize for example subgenomic regulatory elements of structural genes (preliminary experiments in collaboration with R. Nischt and C. Greenhalgh). Since these cultures can be easily manipulated in many different ways (calcium, retinoid concentration, mesenchymal and growth factors) this might become also a very useful system for the identification and isolation of the involved nuclear DNA-binding proteins.

In conclusion there will be a growing list of possible applications and approaches using HaCaT cells, which would be otherwise much more difficult or impossible to persue with normal keratinocytes.

ACKNOWLEDGEMENTS

We would like to thank Drs. I. Leigh and D. Roop for providing antisera, Drs. R. Nischt, D. Hohl and D. Greenhalgh for stimulating discussions. The skilled technical assistance of E. Böhm, A. Krischke and B. Stähle, the excellent photo work of R. Kühnl-Bontzol and the photography department, and the expert typing of the manuscript by B. Plagens is gratefully acknowledged. The work was in part supported by the Deutsche Forschungsgemeinschaft (Fu 91/2-3), the US-German Collaboration Program in Cancer Research (P.B., D.B.). C.R. was recipient of a FEBS fellowship and a guest stipend of the German Cancer Research Center.

REFERENCES

Asselineau D, Bernard BA, Bailly C, Darmon M, Pruniéras M (1986) Human epidermis reconstructed by culture: is it normal? J Invest Dermatol 86:181-186

Asselineau D, Bernard BA, Bailly C, Darmon, M (1989) Retinoic acid improves epidermal morphogenesis. Dev Biol 133:322-335

Baden HP, Kubilus J, Wolman SR, Steinberg ML, PHillips SB, Kvedar JC (1987) The NM 1 keratinocyte line is cytogenetically and biologically stable and exhibits a unique structural protein. J Invest Dermatol 89:574-579

Banks-Schlegel S, Green H (1980) Formation of epidermis by serially cultivated human epidermal cells transplanted as an epithelium to athymic mice. Transplantation 29:308-313

Banks-Schlegel SP, Howley PM (1983) Differentiation of human epidermal cells transformed by SV40. J Cell Biol 96:330-337

Bell E, Sher S, Hull B, Merrill C, Rosen S, Chamson A, Asselineau D, Dubertret L, Coulomb, B, Lapiere C, Nusgens B, Neveux Y (1983) The reconstitution of living skin. J Invest Dermatol 81:2s-10s

Bernard BA, Robinson SM, Semat A, Darmon M (1985) Reexpression of fetal characters in Simian virus 40-transformed human keratinocytes. Cancer Res 45:1707-1716

Bosca AR, Tinois E, Faure M, Kanitakis J, Roche P, Thivolet J (1988) Epithelial differentiation of human skin equivalents after grafting onto nude mice. J Invest Dermatol 91:136-141

Boukamp P, Tilgen W, Dzarlieva RT, Breitkreutz D, Haag D, Riehl RK, Bohnert A, Fusenig NE (1982) Phenotypic and genotypic characteristics of a cell line from a squamous cell carcinoma of human skin. J Natl Cancer Inst 68:415-127

Boukamp P, Rupniak HTR, Fusenig NE (1985) Environmental modulation of the expression of differentiation and malignancy in six human squamous cell carcinoma cell lines. Cancer Res 45:5582-5592

Boukamp P (1987) Studies on carcinogenesis in human epidermal cells. Establishment and characterization of primary cultures and immortal cell lines and malignant transformation by the cellular Harvey-ras oncogene. Thesis, University of Kaiserslautern, FRG

Boukamp P, Petrusevska RT, Breitkreutz D, Hornung J, Markham A, Fusenig NE (1988) Normal keratinization in a spontaneously immortalized aneuploid keratinocyte cell line. J Cell Biol 106:761-771

Bowden PE, Quinlan RA, Breitkreutz D, Fusenig NE (1984) Proteloytic modification of acidic and basic keratins during terminal differentiation of mouse and human epidermis. Eur J Biochem 142: 29-36

Bowden PE, Stark H-J, Breitkreutz D, Fusenig NE (1987) Expression and modification of keratins during terminal differentiation of mammalian epidermis. In: Moscona AA, Sawyer R (eds) Current Topics in Developmental Biology, Vol. 22. Academic Press, pp 35-68

Breitkreutz D, Tilgen W, Boukamp P, Fusenig NE (1981) Correlation of prekeratin peptides and ultrastructure in epithelial cells of human skin tumors in vivo and in vitro. Anticancer Res 1:323-328

Breitkreutz D, Bohnert A, Herrmann E, Bowden PE, Boukamp P, Fusenig NE (1984) Differentiation specific functions in cultures and transplanted mouse keratinocytes: Environmental influences on ultrastructure and keratin expression. Differentiation 26:154-169

Breitkreutz D, Hornung J, Poehlmann J, Brown-Bierman L, Bohnert A, Bowden PE Fusenig NE (1986) Environmental induction of differentiation-specific keratins in malignant mouse keratinocyte lines. Eur J Cell Biol 42:255-267

Breitkreutz D, Boukamp P, Stark H-J, Ryle C, Fusenig NE (1989) Response of established keratinocyte lines to modulators of epidermal differentiation. In: Reichert U, Shroot B (eds) Pharmacology and the Skin. Vol 3, Karger Basel, pp 8-14

Compton C, Gill JM, Bradford DA (1989) Skin regenerated from cultured epithelial autografts on acute burn wounds: A light electron microscopic and immunohistochemical study. Lab Invest 60:600-612

Doran TI, Vidrich A, Sun T-T (1980) Intrinsic and extrinsic regulation of the differentiation of skin, corneal and esophageal epithelial cells. Cell 22: 17-25

Dürst M, Dzarlieva R, Boukamp P, Fusenig NE, Gissmann L (1987) Molecular and cytogenetic analysis of immortalized human primary keratinocytes obtained after transfection with human papillomavirus type 16 DNA. Oncogene 1:251-256
Franke WW, Schiller DL, Moll R, Winter S, Schmid E, Engelbrecht I, Denk M, Krepler R, Platzer B (1981) Diversity of cytokeratins. Differentiation specific expression of cytokeratin polypeptides in epithelial cells and tissues. J Mol Biol 153:933-959
Fuchs E, Green H (1980) Changes in keratin gene expression during terminal differentiation of the keratinocyte. Cell 19:1033-1042
Fuchs E, Green H (1981) Regulation of terminal differentiation of cultured human keratinocytes by vitamin A. Cell 25:617-625
Fusenig NE, Breitkreutz D, Dzarlieva RT, Boukamp P, Bohnert A, Tilgen W (1983) Growth and differentiation characteristics of transformed keratinocytes from mouse and human skin in vitro and in vivo. J Invest Dermatol 81:168s-175s
Fusenig NE (1986) Mammalian epidermal cells in culture. In: Bereiter-Hahn I, Matoltsy AG, Richards KS (eds) Biology of the Integument, Part 2 Vertebrates. Springer Verlag, Berlin, Heidelberg, pp 409-442
Fusenig NE, Boukamp P, Breitkreutz D, Karjetta S, Petrusevska RT (1987) Oncogenes and malignant transformation of human keratinocytes. In: Anticarcinogenesis and radiation protection. P. Cerutti (eds). Plenum Publishing Corp., New York, p. 227-231
Fusenig NE, Boukamp P, Breitkreutz D, Hülsen A, Karjetta S, Stanbridge E (1989) Transformation of human skin epithelial cells in vitro: Concepts and stages of transformation. In: Chadwick K (ed) Cell transformation systems relevant to radiation induced cancer in man. Adam Hilger IOP Publishing, Bristol; in press.
Gallico GG, O'Connor NE, Compton CC, Kehinde O, Green H (1984) Permanent coverage of large burn wounds with autologous cultured human epithelium. N Engl J Med 311:488
Gilfix BM, Eckert RL (1985) Coordinate control by vitamin A of keratin gene expression in human keratinocytes. J Biol Chem 260:14026-14029
Hennings HD, Michael C, Cheng P, Steinert K, Holbrook K, Yuspa, SH (1980) Calcium regulation of growth and differentiation of mouse epidermal cells in culture. Cell 19:245-254
Kim KH, Schwartz F and Fuchs E (1984) Differences in keratin synthesis between normal epithelial cells and squamous cell carcinoma are mediated by vitamin A. Proc Natl Acad Sci USA 81:4280-4284
Kopan R, Traska G, Fuchs E (1987) Retinoids as important regulators of terminal differentiation: examining keratin expression in individual epidermal cells at various stages of keratinization. J Cell Biol 105: 427-440
Kopan R, Fuchs E (1989) The use of retinoic acid to probe the relation between hyperproliferation-associated keratins and cell proliferation in normal and malignant epidermal cells. J Cell Biol 109:295-307
Lavker RM, Sun T-T (1983) Rapid modulation of keratinocyte differentiation by the external environment. J Invest Dermatol 80:228-237
Moll R, Franke WW, Schiller DL, Geiger B, Krepler R (1982) The catalog of human cytokeratins: Patterns of expression in normal epithelia, tumors and cultured cells. Cell 31:11-24
Moll R, Moll I, Franke WW (1984) Differences of expression of cytokeratin polypeptides in various epithelial skin tumors. Arch Dermatol Res 276: 349-363
Morgan PR, Shirlaw PJ, Johnson NW, Leigh IM, Lane EB (1987) Potential applications of anti-keratin antibodies on oral diagnosis. J Oral Pathol 16: 212-222
Regnier M, Schweizer J, Michel S, Bailly C, Pruniéras M (1986a) Expression of high molecular weight (67K) keratin in human keratinocytes cultured on dead de-epidermized dermis. Exp Cell Res 165:63-72

Regnier M, Desbes C, Bailly C, Darmon M (1988) Differentiation of normal and tumoral keratinocytes cultured on dermis: Reconstruction of either normal or tumoral architecture. In Vitro 24:625-632

Rheinwald JG, Beckett MA (1980) Defective terminal differentiation in culture as a consistent and selectable character of malignant human keratinocytes. Cell 22:629-632

Roop DR, Cheng CK, Titterington L, Meyers CA, Stanley JR, Steinert PM, Yuspa SH (1984) Synthetic peptides corresponding to keratin subunits elicit highly specific antibodies. J Biol Chem 259:8037-8040

Roop DR, Krieg TM, Mehrel T, Cheng CK, Yuspa SH (1988) Transcriptional control of high molecular weight keratin gene expression in multistage mouse skin carcinogenesis. Cancer Res 48:3245-3252

Ryle C, Breitkreutz D, Stark HJ, Leigh I, Steinert P, Roop, Fusenig NE (1989) Density-dependent modulation of synthesis of keratins 1 and 10 in the human keratinocyte line HaCaT and in ras-transfected tumorigenic clones Differentiation 40:42-54

Stark HJ, Breitkreutz D, Limat A, Bowden P, Fusenig NE (1987) Keratins of the human hair follicle: "Hyperproliferative" keratins consistently expressed in outer root sheath cells in vivo and in vitro. Differentiation 35:236-248

Steinberg ML, Defendi V (1983) Transformation and immortalization of human keratinocytes by SV40. J Invest Dermatol 81:131s-136s

Steinert PM, Roop DR (1988) Molecular and cellular biology of intermediate filaments. Ann Rev Biochem 57:593-625

Sun T-T, Eichner R, Nelson WG, Tseng SCG, Weiss RA, Jarvinen M, Woodcock-Mitchell J (1983) Keratin classes: molecular markers for different types of epithelial differentiation. J Invest Dermatol 81:109s-115s

Taylor-Papadimitriou J, Purkis JP, Lane EB, McKay IA, Chang, SE (1982) Effects of SV 40 transformation on the cytoskeleton and behavioural properties of human keratinocytes. Cell Differ 11:169-180

Tilgen, W, Boukamp P, Breitkreutz D, Dzarlieva RT, Engstner M, Haag D, Fusenig NE (1983) Preservation of morphological, functional, and karyotypic traits during long-term culture and in vivo passage of two human skin squamous cell carcinomas. Cancer Res 43:5995-6011

Tseng SCG, Hatchell D, Tierney N, Huang J-W, Sun T-T (1984) Expression of specific keratin markers by rabbit corneal, conjunctival and esophageal epithelia during vitamin A deficiency. J Cell Biol 99:2279-2286

Van Muijen GNP, Warnaar SO, Ponec M (1987) Differentiation-related changes of cytokeratin expression in cultured keratinocytes and in fetal, newborn, and adult epidermis. Exp Cell Res 171:331-345.

Watt FM, Boukamp P, Hornung J, Fusenig NE (1987) Effect of growth environment on spatial expression of involucrin by human epidermal keratinocytes. Arch Dermatol Res 279:335-340

Winter H, Schweizer J, Goerttler K (1980) Keratins as markers of malignancy in mouse epidermal tumours. Carcinogenesis 1:391-398

Worst PKM, Mackenzie IC, Fusenig NE (1982) Reformation of organized epidermal structure by transplantation of suspensions and cultures of epidermal and dermal cells. Cell Tissue Res 225:65-77

Wu YJ, Rheinwald JG (1981) A new small (40 kD) keratin filament protein made by some cultured human squamous cell carcinomas. Cell 25:627-635

Yuspa SH, Kilkenny AE, Stanley J, Lichti U (1985) Keratinocytes blocked in phorbol ester-responsive early stage of terminal differentiation by sarcoma viruses. Nature 314:459-462

REGULATION OF HAEMOPOIETIC CELL DEVELOPMENT

Nydia G. Testa, Thomas M. Dexter, and Terence D. Allen

Paterson Institute for Cancer Research
Christie Hospital & Holt Radium Institute
Manchester, UK

HAEMOPOIESIS

The constant supply of blood cells required to maintain a normal steady state by replacing the cells lost through senescence is of the order of 4×10^{11} cells per day in a normal human adult, and is even larger in response to haemopoietic stress like blood loss or infection. All blood cells are derived from stem cells in the bone marrow tissue in the adult. The stem cells are characterized by a) their capacity to reconstitute the haemopoietic system of individuals who had received a potentially lethal dose of irradiation, b) their capacity of self-renewal by which they can regenerate their own population after serious and repeated depletion caused by cytotoxic injury, and c) their multipotentiality, i.e. they give rise to all the myeloid and lymphoid cell types.

These properties imply an extensive capacity for proliferation and differentiation, which has been demonstrated by serial transplantation of bone marrow in mice (Harrison and Astle, 1982). The transplanted stem cells were functional for a time equivalent to several life spans of the donor mice, even when subjected to the strong proliferation stress induced by the transplantation procedure. By differentiation, stem cells give rise to committed progenitor cells limited in their differentiation capacity to one or two cell lineages, but which proliferate actively to give rise to the more mature morphologically recognisable cells that comprise the majority of the cells in the bone marrow, and which will eventually give rise to the mature, functional end-cells found in the blood stream. The development of clonal assays has allowed the characterisation of these primitive cell populations, and the understanding of the hierarchical structure of the haemopoietic tissue illustrated in Figure 1.

Although stem cells can strictly be assayed only by measuring their capacity to repopulate the bone marrow, an in vivo clonal assay in mice detects a cell population, CFU-S (colony-forming units in spleen) which overlaps partially with the stem cell population (recently reviewed by Lord and Testa, 1988).

In other species the most primitive haemopoietic cell detected by a clonal assay is the Mix-CFC, colony forming cell, with the capacity to give rise to several lineages. Mix-CFC conserve at least some self-reproduction capacity, and overlap partially with the CFU-S population, as well as with

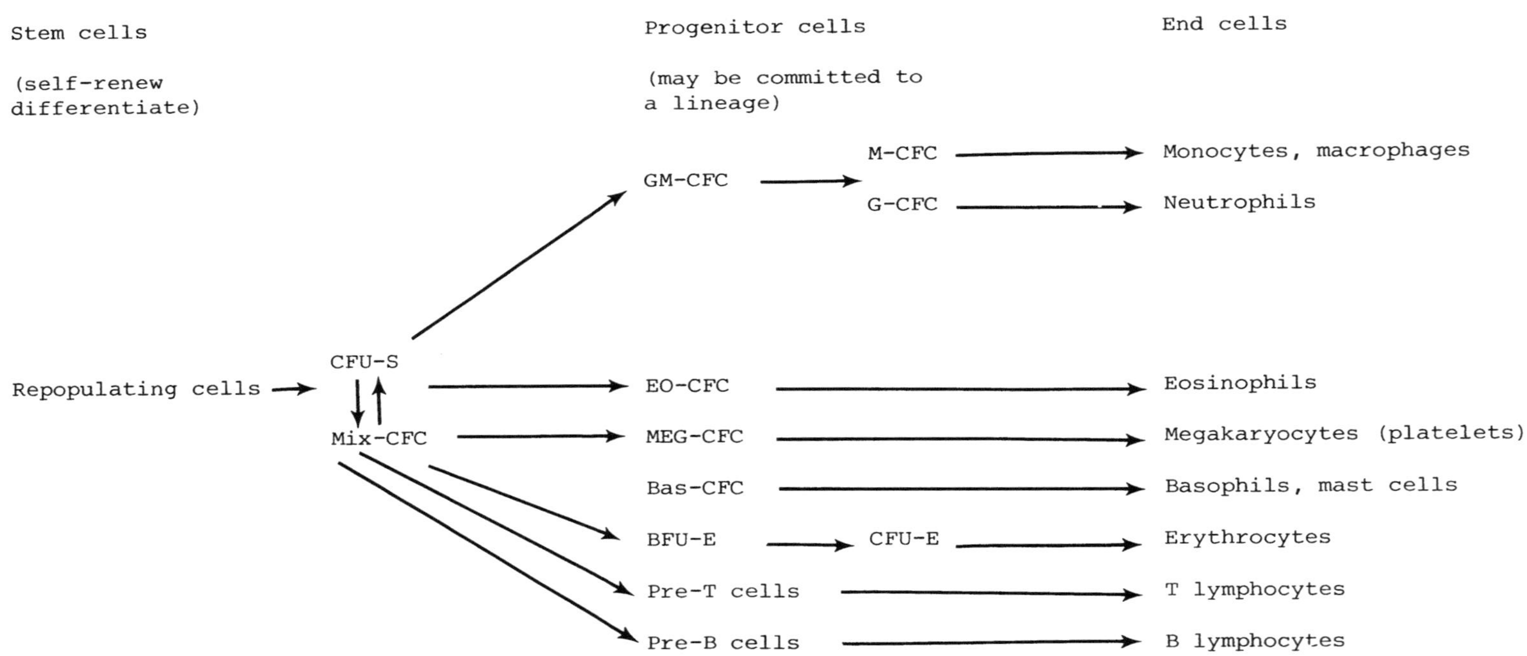

FIGURE 1. The Structure of the haemopoietic tissue

more mature progenitor cells (Metcalf et al, 1979, Fauser and Messner, 1979) Humphries et al 1981). The latter, usually called colony forming cells (CFC) or CFU (colony forming units) are characterised by a prefix denoting the lineage their progeny beongs to: Meg: megakaryocyte, E: erythroid, M: monocyte-macrophage, Eo: eosinophil, Ba: basophil. Early erythroid progenitors are caled BFU-E (burst-forming units), a name which describes the particular appearance of the erythroid colonies. Technical, extensive reviews on the methodology used to grow these progenitors have been published (Metcalf, 1977, Testa 1985).

REGULATION OF CELL PRODUCTION

Haemopoietic growth factors

The conditions in which CFC are grown in vitro select the populations that are going to develop and give rise to a progeny of mature cells. They also select, as will be seen below, which lineage of development progenitors will follow (Table 1). These effects are mediated through factors which are now available in pure, recombinant form. The four major haemopoietic growth factors (generally called CSFs, or colony stimulating factors) have considerable overlap in their actions on haemopoietic cells. Thus, Il3 stimulates the growth of Mix-CFC (Metcalf, 1984) and increases the survival of CFU-S in vitro (Spivak et al, 1985). Therefore, it appears that Il3 acts on primitive multipotential cells as well as on the more mature committed progenitors. After stimulation with Il-3, Mix-CFC produce neutrophils, macrophages, eosinophils, basophils and erythrocytes. Production of erythrocytes also requires stimulation with erythropoietin, a growth factor which stimulates cells in the erythroid lineage, and also may stimulate megakaryocytes (reviewed by Goldvasser et al, 1989). GM-CSF stimulates preferentially progenitors of granulocytes and macrophages, but may also stimulate others when used at high concentrations (Table 1). From these data, it is apparent that Il-3 and GM-CSF show considerable overlap in their target cells. The latter, however, does not appear to promote the development of basophils (reviewed by Gough and Nicola, 1989). Although G-CSF and M-CSF act on committed progenitors to promote almost exclusively the development of neutrophilic granulcytes and macrophages respectively, when acting together or with other factors may act synergistically and stimulate other, more immature progenitors, like Mix-CFC. The cell lineages produced, however, will be only those promoted by these factors when acting along ie. neutrophilic granulocytes and macrophages. (Hayworth et al, 1988). The lineages promoted by other combinations of factors will also be related to those produced by each acting in isolation. Thus multi-potential cells and committed progenitors can respond to several growth factors. This implies that they express simultaneously the receptors specific for each growth factor and that the fine regulation needed for the adequate production of the different blood cell types required in normal steady state and in very diverse pathological circumstances may be achieved by modulating the balance of growth regulators to which cells are exposed in the bone marrow. This concerns not only the haemopoietic growth factors, but also a variety of other bioregulators (like Interleukins 1, 4, 6, 7) which, although concerned mainly with immune processes, also modulate the production of haemopoietic cells. The clearest example, and the only one known to have direct colony-stimulating activity for myeloid cells, is Il-5, which stimulates differentiation of Eo-CFC (reviewed by Sanderson, 1989)

The haemopoietic microenvironment

Contrary to a belief held for a long time, the bone marrow is a highly structured tissue, with discrete environmental domains in which distinct populations of stem and progenitor cells are found in different proportions in different areas of the marrow, and are exposed to the influence of

TABLE 1 Human haemopoietic growth factors

Factor	Target colony forming cells	Major Types of Cells Produced
Interleukin-3 (IL-3, also called Multi-CSF)	Mix , GM,GM,Ba Eo,BFU-E, Meg	Neutrophils monocyte-macrophages, eosinophils, basophils and mast cells; erythrocytes (if erythropoietin also present).
Granulocyte-macrophage colony stimulating factor (GM-CSF)	Mix(±), GM,G,M Eo(±),BFU-E(±) Meg(±)	Neutrophilic granulocytes, monocyte-macrophages, eosinophils, megakaryocytes ; erythrocytes; (as above)
Granulocyte colony stimulating factor (G-CSF)	GM,G	Neutrophilic granulocytes
Macrophage colony stimulating factor (M-CSF also called CSF1)	GM,M	Macrophages
Interleukin-5 (Il-5)	Eo	Eosinophils
Erythropoietin	E,Meg(±)	Erythrocytes

regulatory cells which are not themselves distributed randomly through the tissue (Testa et al, 1985, Lord and Testa, 1988). These domains may promote the preferential development of specific haemopoietic lineages, a role that may be largely mediated through growth factors. The CSFs are produced by several stromal cell types present in the marrow, as well as by T cells and macrophages. Thus fibroblasts can produce M-CSF, endothelial cells produce GM- and G-CSF, activated T lymphocytes can produce Il-3 and GM-CSF and macrophages can produce M-CSF (reviewed by Metcalf, 1984). It is known that growth factors may bind selectively to components of the extracellular matrix (Gordon et al 1987, Roberts et al 1987). Significantly, it has been demonstrated that one of these molecules, heparan sulphate, needs to be bound to (stromal?) cell membranes to have the ability to bind and present Il-3 or GM-CSF (and probably other factors) to the target cells (Roberts et al, 1987). This phenomenon offers a possible mechanism which allows high concentrations of specific factors in discrete sites of the marrow, and may explain the observation that active haemopoiesis in long term bone marrow cultures requires direct contact between a regulatory stroma and stem cells (Dexter, 1982).

Long term bone marrow cultures

In long term bone marrow cultures, active haemopoiesis is dependant on the presence of a complex layer of stromal cells. Haemopoiesis takes place in the absence of added growth factors, and in the practical absence of serum (Eliason, 1984). If a permeable membrane, or a layer of agar separate the haemopoietic cells from the regulatory stroma, haemopoiesis is

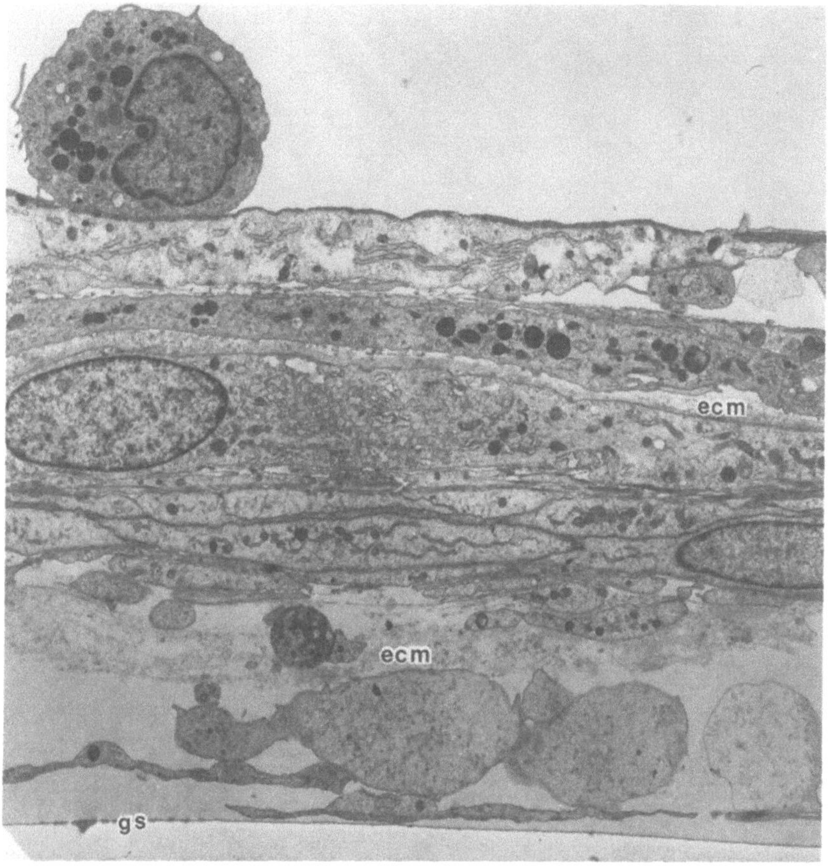

FIGURE 2. Transmission electron micrograph of a section through a human bone marrow culture cut at right angles to the growing surface (gs, dark line at bottom of micrograph). This plane of section shows the multilayered nature of the stromal cells in this non-haemopoietic area of the culture. The stromal fibroblastoid cells continue to proliferate throughout the lifetime of the culture, and may generate areas of adherent multilayers twenty to thirty cells in depth. The adherent layer is also permeated with extracellular maxtrix (ecm).

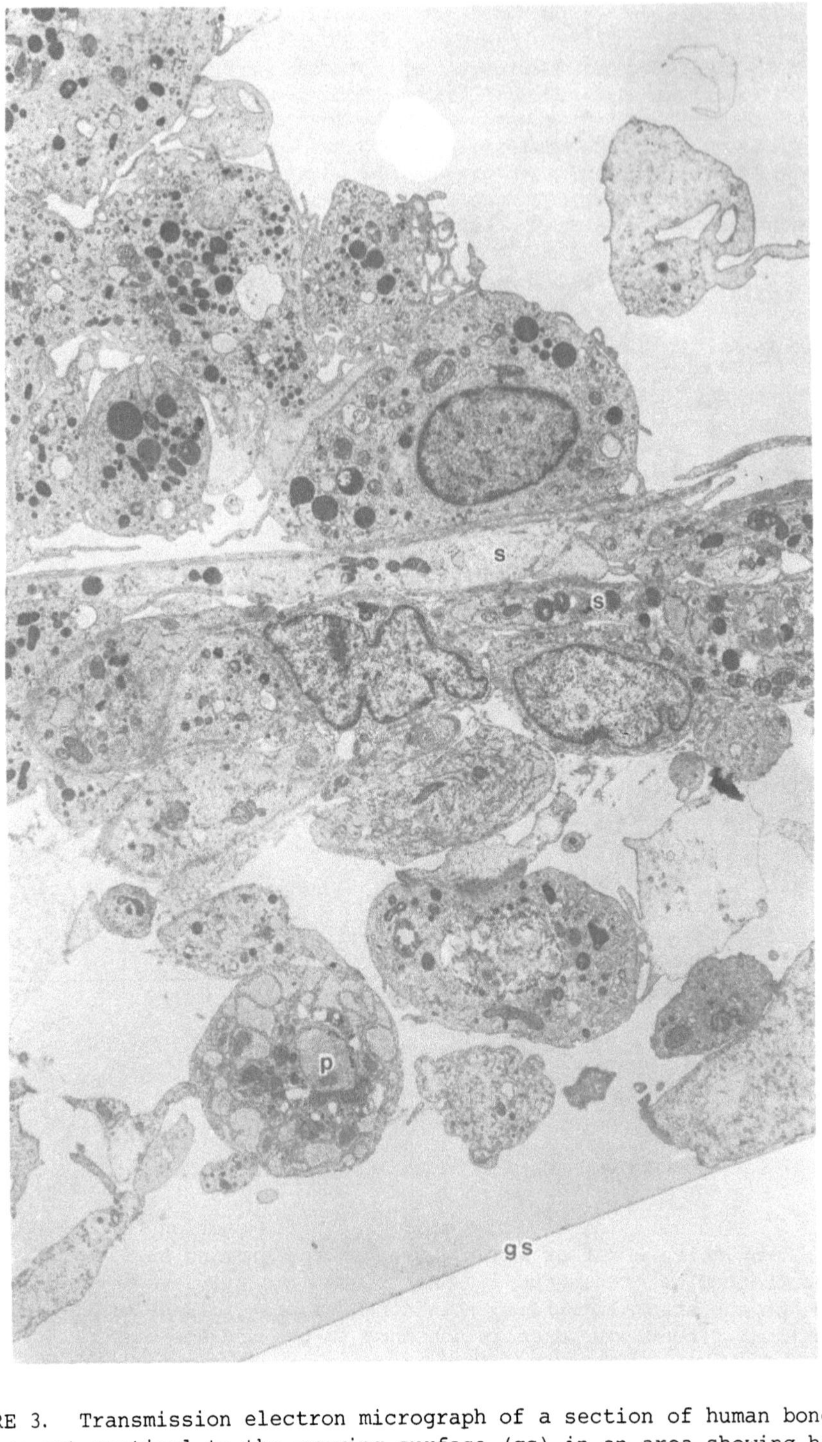

FIGURE 3. Transmission electron micrograph of a section of human bone marrow culture cut vertical to the growing surface (gs) in an area showing haematopoietic activity. The majority of cells in the field are at various stages of myeloid differentiation, with the more mature cells generally distributed towards the upper regions of the multilayer. This field is partitioned by a sheet of stromal cells in the centre (S), overlying and enclosing a mixture of myeloid precursors, and emphasising the microenvironmental nature of haemopoietic sites. The more mature myeloid cells appear above, at the surface of the adherent layer. An early plasmocyte (P) is also apparent close to the growing surface (gs).

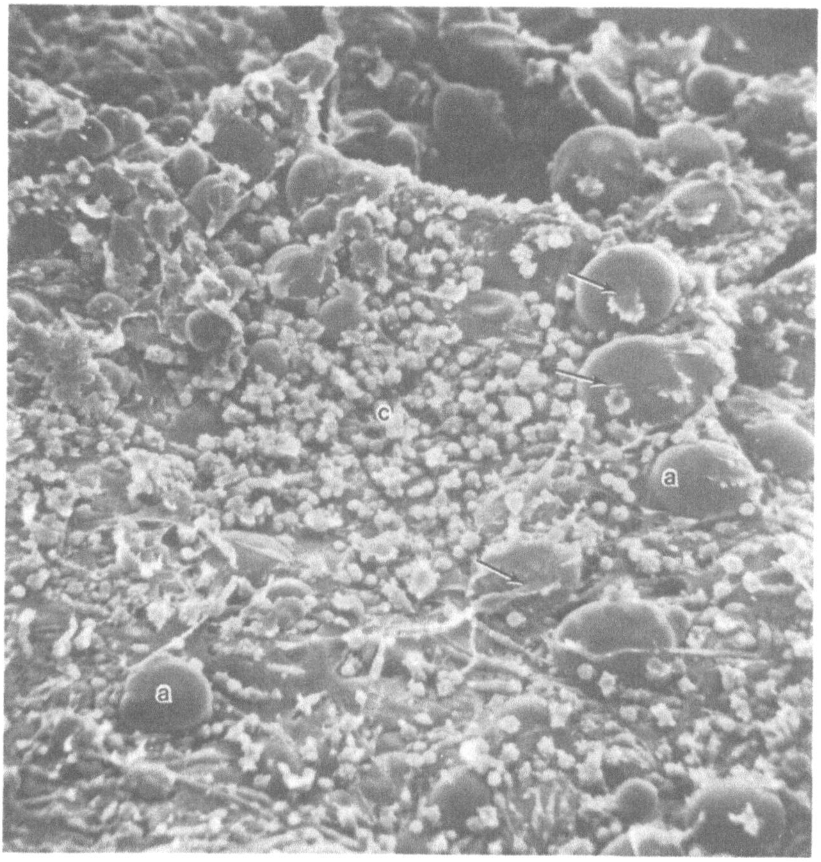

FIGURE 4. Scanning electron micrograph of an area of murine long term bone marrow culture, again displaying the three dimensional aspect of the more highly populated regions of the adherent layer. In this region, there has been spontaneous detachment of the adherent layer in the foreground which became fixed in a folded back configuration, exposing an 'underside' view of the densely packed developing granulocytes and monocytes in a cobblestone region (c). The large spherical cells are mature adipocytes (a), and several flattened profiles (arrowed), show cells which have become detached from direct attachment to the growing surface.

not maintained. This implies that soluble factors which act on stem cells are not released into the growth medium. Factors produced by the stromal cells probably remain bound to cell membrane-associated molecules such as heparan sulphate. This may hold for the CSFs, as well as for yet unknown factors regulating the stem cell population.

There is evidence for the role of the microenvironment in vivo (Weiss, 1976, Testa et al, 1985, Lord and Testa, 1988) as well as in vitro in long term bone marrow cultures (Dexter, 1982, Allen and Dexter, 1983). In these cultures, haemopoiesis occurs within well developed, multi-layered stromal cells (Figures 2,3). In active haemopoietic cultures up to 30 or so cell layers (stromal and haemopoietic) can be counted and considerable amounts of extracellular matrix are seen. Large numbers of haemopoietic cells (Figure 4) are continuously produced in these cultures for up to 12-16 weeks for human cells, and for more than a year for murine cells (Coutinho et al, 1989, Dexter, 1982). That stem cells remain functional in these cultures has been indicated by their capacity to reconstitute the haemopoietic system of experimental animals (Dexter and Spooncer 1978). These results allowed the use of long-term bone marrow cultures as a source of haemopoietic cells for autologous bone marrow transplants in patients without suitable compatible donors (Chang et al 1986, Testa et al, 1987, Coutinho et al 1989). The results to date allow optimism, with a rate of regeneration after transplantation similar to that reported for autologous transplants of freshly harvested bone marrow, but with the added advantage that the cultures allow biological "purging" of leukaemic cells. However, to date neither experimental data, nor data from the transplanted patients suggest that stem cells have either increased in numbers, or changed their functional properties to make the capacity to regenerate in the haemopoietic tissue in the long term, any better than freshly harvested normal bone marrow stem cells. This is true using the standard long term cultures, or after the methodological modifications tried in several laboratories. However, expansion in the more mature progenitor cells and their progeny may perhaps be used to improve the short term regeneration of the tissue after intensive chemotherapy for malignancy. Whether this will be more effective than reinfusing the high numbers of primitive cells which can be harvested from peripheral blood after treatment with growth factors (Gianni et al, 1989) remains to be determined.

CONCLUSIONS

The availability of clonal cultures has allowed the study of pure growth factors on pure cell populations of haemopoietic progenitor cells. This has led to the detailed knowledge of the structure of the haemopoietic tissue summarised here, and has allowed investigations of the sensitivity of different cell populations to modulators of cell production and differentiation and also to pharmacological agents.

Our knowledge of the microenvironmental regulation of haemopoiesis can now be enlarged by studying the ways in which cell behaviour can be modified. The roles of a) cell-cell interactions (for example between factor-producing or factor-binding cells and haemopoietic cells), b) interactions between cells and molecules in the extracellular matrix and c) interactions between cells and soluble factors can now be analysed, and eventually the processes involved could be manipulated. The physiological relevance of the methodology available has already been demonstrated not only in experimental in vivo studies, but in the case of long term bone marrow cultures, by establishing their clinical usefulness as a source of transplantable bone marrow.

ACKNOWLEDGEMENTS

The authors are supported by the Cancer Research Campaign, Great Britain.

REFERENCES

Allen, T. and Dexter, T.M., 1983. Long term bone marrow culture: an ultrastructural review, Scanning electron microsc. 4, 1951.

Chang, J., Coutinho, L., Morgenstern, G., Scarffe, J.H., Deakin, D., Harrison, C., Testa, N.G. and Dexter, T.M. 1986 Marrow from AML in relapse and grown in long term culture can be used to reconstitute the haemopoietic system by autologous bone marrow transplantation, The Lancet 1, 293.

Coutinho, L.H., Testa, N.G., Chang, J., Morgenstern, G., Harrison, C., Dexter, T.M. 1989 The use of cultured bone marrow cells in autologous transplantation, Leukemia, in press.

Dexter, T.M. (1982) Stroma cell associated haemopoiesis. J. Cell Physiol., suppl. 1, 87.

Dexter, T.M. and Spooncer, E. 1978. Loss of immunoreactivity in long term bone marrow culture, Nature, 275: 135.

Eliason, J.F., 1984. Long term production of hemopoietic progenitors in cultures containing low levels of serum, Exp. Hematol. 12: 559.

Fausser, A.A. and Messner, H.A. 1979. Identification of megakaryocytes, macrophages and eosinophils in colonies of human bone marrow containing neutrophilic granulocytes and erythroblasts, Blood 53, 1023.

Gianni, A.M., Bregni, M., Stern, A.C., Siena, S., Tarella, C., Pileri, A and Bonadona, G., 1989. Granulocyte-macrophage colony-stimulating factor to harvest circulating haemopoietic stem cells for autotransplantation, The Lancet ii, 580.

Goldvasser, E., Beru, N. and Smith, D. 1989. Erythropoietin in: Cellular and molecular biology of colony stimulating factors, Dexter, T.M., Garland,J. and Testa, N.G. eds. M. Dekker, New York p.257.

Gordon, M.V., Riley, G.P., Watt, S.M., Greaves, M.F. 1987. Compartmentalization of a haematopoietic growth factor (GM-CSF) by glycosaminoglycans in the bone marrow microenvironment, Nature 326, 403.

Gough, N.M. and Nicola, N. 1989. Granulocyte-macrophage colony-stimulating factor in: Cellular and molecular biology of colony stimulating factors, Dexter, T.M., Garland, J. and Testa, N.G. eds. M. Dekker, NY p.111.

Harrison, D. and Astle, C.H. 1982. Loss of stem cell repopulating ability on transplantation; Effects of donor age, cell number and transplantation procedure, J. Exp. Med 156, 1767.

Heyworth, C.M., Ponting I.L.O. and Dexter T.M. 1988. The response of haemopoietic cells to growth factors: developmental implication of synergistic interactions, J. Cell Sci. 91, 239.

Humphries, R.K., Evans, A.C. and Eaves, C.J. 1981. Self-renewal of hemopoietic stem cells during mixed colony formation in vitro, Proc.Natl. Acad. Sci. USA, 3629.

Lord, B.I. and Testa, N.G. 1988 The hemopoietic system: structure and regulation. in: Hematopoiesis, long term effects of chemotherapy and radiation. Testa, N.G. and Gale, R.P. eds. M. Dekker, New York p.1.

Metcalf, D. 1977 Haemopoietic colonies. Springer Verlag, Berlin.

Metcalf, D. 1984. The haemopoietic colony stimulating factors. Elsevier. Amsterdam.

Metcalf, D. 1988. The molecular control of blood cells. Harvard University Press. Cambridge, Mass.

Metcalf, D., Johnson, G.R., Burges, A.W. 1979 Colony formation in agar by multipotential hemopoietic cells J. Cell Physiol. 98, 401.

Nicola, N. 1989 Granulocyte colony stimulating factor in cellular and molecular biology of colony stimulating factors. Dexter, T.M., Garland, J. and Testa, N.G., eds. M. Dekker, New York, p.77.

Roberts, R.A., Gallaher, J.T., Spooncer, E., Allen, T.D., Bloomfield, F. and Dexter, T.M. 1988. Heparan sulphate bound growth factors: a mechanism for stromal cell mediated haemopoiesis.. Nature 332, 376.

Sanderson, C.J. 1989 Eosinophil differentiation factor (Interleukin 5) in: Cellular and MOlecular biology of colony stimulating factors Dexter, T.M., Garland, J. and Testa, N.G. eds. M. Dekker New York p.231.

Spivak, J.L., Smith, R.R. and Ihle, J.N. 1985. Interleukin 3 promotes the in vitro proliferation of murine pluripotent stem cells. J. Clin. Invest. 76: 1613.

Testa, N.G. 1985. Clonal assays for haemopoietic and lymphoid cells in vitro. In: Cell clones: manual of mammalian cell techniques. Potten, C.S., Hendry, J.H. eds.

Testa, N.G., Hendry, J. and Molineux, G. 1985. Long-term bone marrow damage in experimental systems and in patients after radiation or chemotherapy. Anticancer Res.5, 101.

Testa, N.G., Coutinho, L., Chang J., MOrgenstern, G., Scarffe, J.H. and Dexter, T.M. 1987. The use of cultured bone marrow cells for autologous transplantation in patients with acute myeloblastic leukaemia. Haematol. Blood Trans. 31, 75.

Testa, N.G. and Dexter, T.M. 1989. Haemopoietic growth factors: their role in cell development and their clinical use, Cytotechnology, in press.

Weiss, L., 1976. The hematopoietic environment of the bone marrow: an ultrastructural study of the stroma in rats. Anat. Rec. 186, 161.

MULTILINEAGE HEMATOPOIETIC EXPRESSION IN A THREE-DIMENSIONAL LONG TERM BONE MARROW CULTURE SYSTEM

Brian A. Naughton and Gail K. Naughton*

Medical Laboratory Sciences Dept., Hunter College School of Health Sciences, New York, New York 10010; Marrow-Tech Incorporated*, Elmsford, New York 10523

INTRODUCTION

The initial attempts to grow bone marrow cells in semi-solid or liquid medium promoted only the terminal differentiation of hematopoietic stem cells in the absence of self-renewal, with cultures becoming hematopoietically unproductive after 1-2 weeks of culture. Subsequent work in the murine system by Dexter and co-workers demonstrated that the sustained growth of hematopoietic cells in liquid culture was possible if these cells were plated onto a pre-established monolayer of bone marrow stromal cells [3]. These cells synthesize the matrix necessary to support hematopoiesis and contribute trophic/regulatory factors to the hematopoietic microenvironment. Inclusive of this group are fibroblasts, adipocytes, reticular adventitial cells, macrophages, and endothelia [4]. The Dexter long-term bone marrow culture (LTBMC) system sustains the self renewal of murine pluripotential stem cells (CFU-S) although the lineage restricted progenitors become predominantly myeloid in character after the first several weeks of culture [3,5]. This system can be modulated to favor the growth of lymphoid cells [6,7] or erythroid progenitors [8], but single cultures do not produce substantial numbers of each hematologic cell type concurrently.

LTBMC of human [9] and non-human primate [10] cells has been achieved using modifications of the Dexter technique. Although the presence of myeloid progenitors (CFU-C) was reported in a human LTBMC of 20 weeks duration [9] and in a pro-simian culture system for at least 25 weeks [9], the hematopoietic potential steadily decreased with time. In this regard, studies of the adherent zone in human monolayer-type LTBMC revealed the persistence of both myeloid (CFU-C) and erythroid (BFU-E) progenitors, but their numbers diminished by approximately 90% between weeks 2 and 7 of culture. The present report describes hematopoiesis in a human LTBMC established on suspended nylon filtration screen templates. The resulting three-dimensional matrix supported the concurrent expression of multiple hematopoietic lineages for the entire 12 week experimental period.

MATERIALS AND METHODS

Specimens Bone marrow was aspirated from the posterior iliac crest of seven hematologically normal adult volunteers after informed consent was obtained. Specimens were collected into heparinized tubes and suspended in 8 mL of RPMI 1640 medium which was conditioned with 10% fetal bovine serum (FBS) and 5-10% horse serum (HS) and supplemented with 5 x 10^{-8}M hydrocortisone, fungizone, and streptomycin. The cell clumps were physically disaggregated and divided into aliquots of 5-10 x 10^{6} (for establishment of the stromal matrix) and 5-10 x 10^{5} (for cryopreservation and eventual hematopoietic cell seeding) nucleated cells.

Preparation of the three-dimensional matrix Nylon filtration screen (#3-210/36, Tetko, Inc., N.Y.) was used as a template for stromal cell growth. This consisted of fibers which were 90 um in diameter and assembled into a square weave pattern with sieve openings of 210 um. 8mm x 45mm pieces of screen were soaked in 0.1 M acetic acid for 30 min and incubated in FBS for 1-2 hr to enhance cellular attachment. The screens were placed in Lab Tek slide chambers (Nunc Inc., Naperville, Il) and inoculated with 5 x 10^{6} fresh bone marrow cells, After 1-2 hr of incubation (5% CO_2, $35^{\circ}C$, humidity>90%) the screens were transferred into 25 cm^2 Corning culture flasks and gently floated in an additional 5mL of medium. Adherence and growth were monitored using inverted phase contrast microscopy and scanning electron microscopy (SEM). After 7-13 days, bone marrow stromal cells completely enveloped each nylon strand and stretched across the sieve openings. The second inoculum was applied when approximately 70% of the mesh openings were "bridged" with stromal cells.

Cryopreservation Aliquots of 5-10 x 10^{5} fresh bone marrow cells were cryopreserved until the stromal matrix was suitable for inoculation. The cells were suspended in RPMI 1640 conditioned with 10% FBS and 5-10% HS and placed on crushed ice. The cryoprotectants dimethylsulfoxide and glycerol were added to final concentrations of 10% and 5% respectively, just prior to the initiation of a programmed freezing cycle (CryoMed 1010 Controller). The initial freezing rate was $-1^{0}C$ from $+4^{0}C$ to $-40^{0}C$ and thereafter was $-10^{0}C$/min until $-90^{0}C$ was achieved. Specimens were stored in the vapor phase of liquid nitrogen until needed.

Inoculation of hematopoietic cells Cryopreserved specimens were rapidly thawed in a $37^{0}C$ water bath, placed into sterile tubes containing 5-7 mL of pre-warmed complete medium, centrifuged at 600 x g, and the pellet resuspended in 1 mL volumes. After an additional wash, cell viability was determined using the trypan blue exclusion method. Nylon screen stromal cell cultures were transferred to Lab Tek dishes and were inoculated with 5-10 x 10^{5} viable cells in 1 mL. After 2 hr of incubation (5% CO_2, $37^{0}C$, >90% humidity) the cultures were gently floated in $25cm^2$ Corning flasks containing 5 mL of medium.

Methods of Evaluation

1. **Cell Counts:** Total cell counts and cytospin preparations

were made using the spent media removed at each feeding (4-5 day intervals). Cell counts were performed using the hemocytometer method. Cytospins were stained with Wright's Giemsa and differential counts were performed on random fields. Differential counts were performed also on cytospin preparations of cells which were dislodged from the adherent zone following treatment with a 1:1 mixture of collagenase (Type I, Sigma, St. Louis, Mo.) and trypsin (10 ug/mL) in Ca^{++}/ Mg^{++} free HBSS and mild agitation over a 3-5 hour period.

2. **Progenitor Cells Assays**: Cells of the adherent zone were harvested at various intervals of LTBMC and assayed for CFU-GM and CFU-GEMM in semi-solid methylcellulose (MC) cultures[12]. Briefly, cells were cultured at a density of 1 x 10^5/mL in 35mm culture dishes containing Iscove's Modified Dulbecco's Medium supplemented with 0.8% MC, 30% FBS, 5 x 10^{-5}M mercaptoethanol and 0.3% bovine serum albumin along with 1.0 U recombinant colony stimulating factor (CSF) and 1.0 U erythropoietin (Amgen, Thousand Oaks, CA.) Triplicate plates for each time interval were cultured for 11-14 d (5% CO_2, 37^{0}C). Plates were read at 40X magnification with an inverted microscope and classified by conventional methods for colony size and presence of hemoglobinized cells [13]. Erythroid progenitors were assayed using the plasma clot system [14]. Adherent cells from LTBMC were plated at a cell density of 5 x 10^5/mL and clotting was initiated by the addition of 0.1 mL of NCTC-109 (Microbiological Associates, Bethesda, Md) containing 1 U of grade 1 bovine thrombin (Sigma Chemical Co., St. Louis, Mo). The clotting mixture was dispensed in 0.1 mL aliquots into 0.2 mL microtiter culture wells (Linbro Scientific Co., New Haven, Conn) and incubated at 5% CO_2 in high humidity. The culture system contained 2 I.U./mL erythropoietin. Plasma clots were incubated for up to 14 days and were fixed and stained according to the method of McLeod et al (14). Erythroid colonies in 3-6 clots were counted and the results were expressed as the number of erythroid colonies per 10^5 cells plated. CFU-E were enumerated on day 7 and BFU-E were counted on days 12-14.

3. **Cytofluorographic analysis**: At various intervals of LTBMC, single cell suspensions were prepared by incubating adherent zone cells with collagenase and trypsin followed by extensive washing. Cells were then incubated in HBSS without Ca^{++} or Mg^{++} for 45-60 min and reacted with the following monoclonal antibodies which were conjugated to fluorescein isothiocyanate (FITC): Mo-1, T 3, B 1, Plt-1, and MY-9 (Coulter Immunology, Hialeah, Fla.). The fluorescence wavelength of the EPICS C flow cytometer was set to 488 nm. Right angle and forward light scatter histograms were used to set bitmaps. Analysis of log fluorescence of each sample was performed by setting windows to exclude the fluorescence of bone marrow cells labelled with mouse IgM- or IgG-FITC alone.

4. **Scanning Electron Microscopy**: At various intervals of LTBMC, nylon screens were cut into approximately four equal parts, fixed in 3% gluteraldehyde-phosphate buffer solution, washed, dehydrated in acetone, and placed in a critical point

dryer. Specimens were coated with 60% gold and 40% palladium and viewed with an Amray SEM.

The results are represented as mean values ± 1 standard error of the mean. Levels of significance (P) were calculated using Students' t test.

RESULTS

Bone marrow stromal cells adhered to nylon meshes which were acid treated and incubated with FBS and completely surrounded each nylon strand by 2-4 days after inoculation. Stromal elements which were plated onto acid and FBS-treated nylon screens then began to stretch across the sieve spaces and by 7-13 days a minimum of 70% of the mesh openings were bridged by their processes. Hematopoietic cells of the second marrow inoculum seeded in the natural interstices formed by the developing stromal cells and colonization, evident in all cultures by 3-6 days, were sustained for 12 weeks (Fig.1) Hematopoietic colonies were typically "sandwiched" between stromal cell layers but a limited amount of surface colonization was noted also. As can be observed in a low power scanning electron micrograph, (Fig. 2) cell growth in the 210 um sieve space occurred in a three-dimensional pattern. It was common to observe myeloid colonies proliferating in close apposition to erythroid and other colonies (Fig. 3). Analysis of cytospin preparations of adherent zone cells revealed the presence of hematopoietic blasts of all lineages (Fig. 4). MC

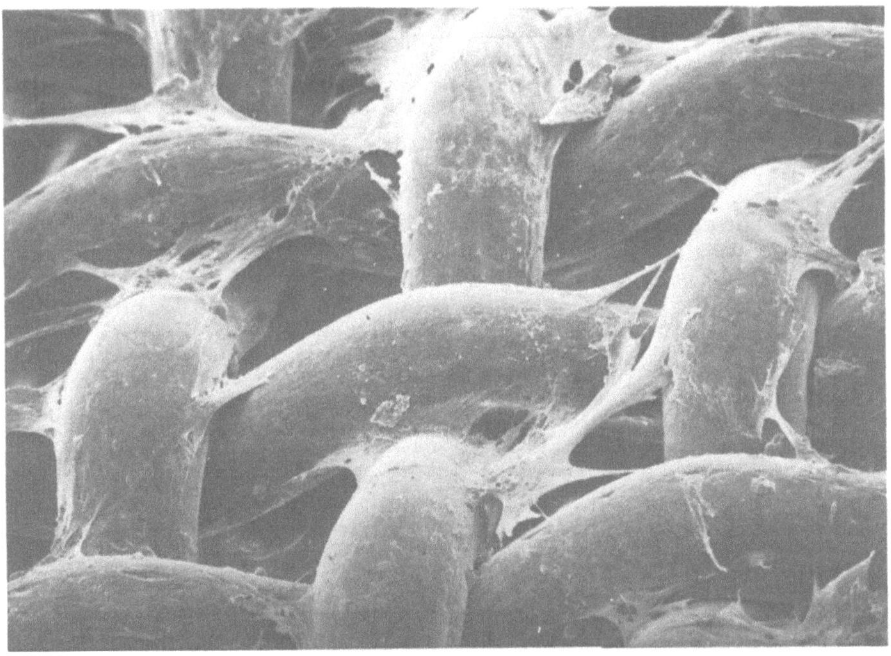

Figure 1. SEM of a suspended nylon screen culture 8 days following the first inoculation. Stromal cells completely surround each nylon filament and span the sieve spaces. 550X

assays for hematopoietic progenitors in the adherent zones at 2, 6, 8, and 10 weeks of LTBMC confirmed these findings (Fig. 5). Although mean numbers of BFU-E dropped by approximately 15% by the second week of culture and 45% by the tenth week, the levels of CFU-E and CFU-GEMM remained uniform over the same period and CFU-GM numbers exceeded input levels at 2 and 6 wks of culture ($P<0.05$) (Fig. 5). Absolute counts of all progenitors were substantially higher than input levels at all time periods studied (Fig. 5). Cytofluorographic analysis of the adherent zones at 2, 7, and 10.5 weeks confirmed the presence of early myeloid cells (MY-9), B (B-1) and T (T-3) lymphocytes, granulocytes/macrophages (Mo-1), and megakaryocytes/platelets (Plt-1) (Table I). Although the cell population of the adherent zone was highly variable, multi-lineage hematopoietic expression was evident. Hematologic cells released into the non-adherent zone of the suspended LTBMC consisted primarily of mature and end-stage precursors of the myeloid, erythroid, and lymphoid lines (Fig. 6). In addition, most of the erythrocytes observed were anucleate. Detached stromal cells and clusters of blasts were also observed in the "spent" medium of these cultures. The blasts probably originated from surface colonies which became detached when the mesh was manipulated. It is unlikely that their appearance in the medium indicates an egress from the adherent zone, where they are firmly ensconced. The major non-adherent stromal elements were fibroblasts and macrophages. Stromal cells were found routinely in the non-adherent zones of these cultures (Fig. 6), perhaps due to the sloughing off of surface cells as hematopoiesis occurred within the matrix. Figure 7 is a composite sketch depicting the orientation of cells within the three-dimensional matrix.

Figure 2. SEM of a nylon screen culture at 8.5 weeks after the second inoculation showing the relationship between hemato-poietic colonies and stromal processes.

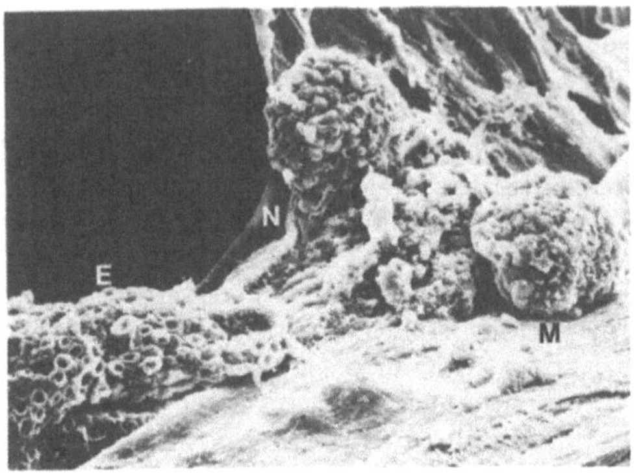

Figure 3. SEM depicting myeloid (M) and erythroid (E) colonies closely apposed. 30 day culture. 1560 X.

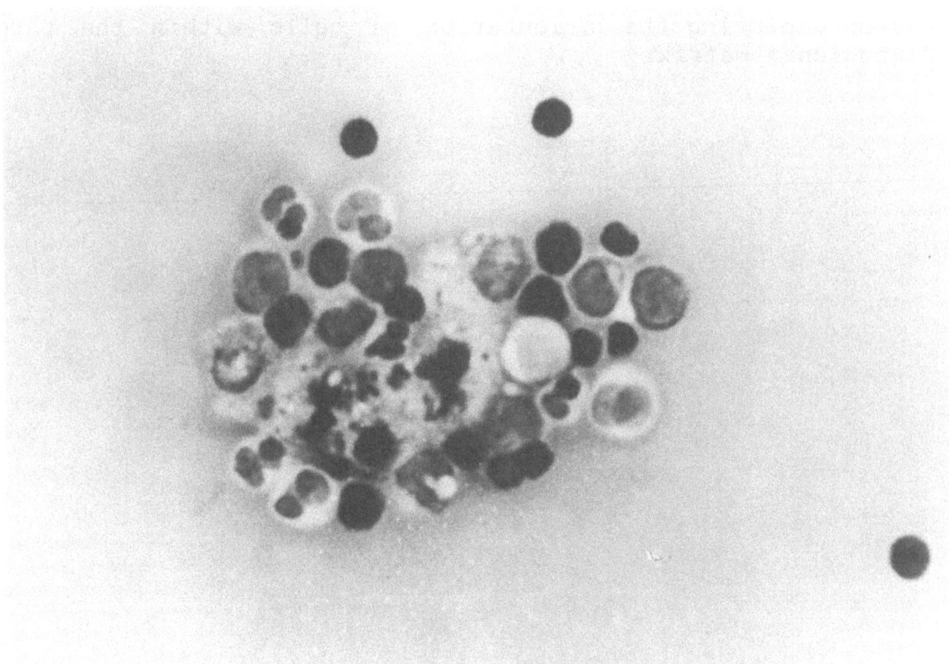

Figure 4. Cytospin preparation of enzyme-dissociated adherent zone showing cells representative of several hematologic lineages. Wright-Giemsa stain. 440X.

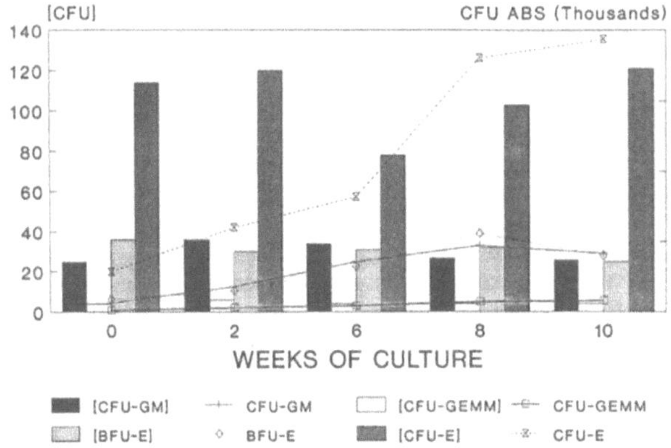

Figure 5. Mean absolute numbers (line drawing) and concentration per 10^5 viable cells (bar graph) of hematopoietic progenitor-derived colonies generated from cells removed via enzyme treatment of adherent zones of nylon screen cultures at various intervals after second (hematopoietic cell) inoculation. Cell numbers removed from the mesh were 10^6 at 2 wks, 2.1 x 10^6 at 6 wks, 3.5 x 10^6 at 8 wks, and 3.2 x 10^6 at 10 wks. These counts reflect a 70-80% depopulation of adherent cells.

Table I. Mean percent reactivity (± 1 standard error of the mean) of uncultured bone marrow and cells from 2, 7, and 10.5 wks of culture with various FITC-labelled monoclonal antibodies.

	fresh marrow	2 wk culture	7 wk culture	10.5 wk culture
B 1	11.96 ± 1.13	10.83 ± 1.56	7.46 ± 1.02	14.38 ± 1.81
T 3	9.90 ± 0.64	16.64 ± 2.41	12.18 ± 1.59	12.99 ± 1.90
MY 9	1.46 ± 0.54	3.60 ± 0.29	3.79 ± 0.51	3.38 ± 0.21
Plt-1	8.72 ± 1.83	3.88 ± 0.98	7.61 ± 1.04	11.09 ± 3.26
Mo-1	15.60 ± 0.84	22.09 ± 2.17	17.88 ± 1.95	24.15 ± 2.74

Results reflect the difference between full scale accumulation and the fluorescence of the Mouse IgG_1-FITC control.

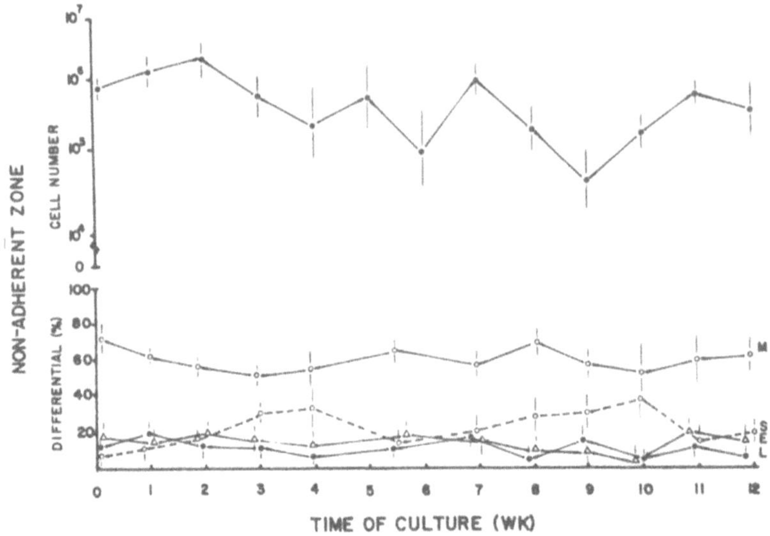

Figure 6. Cellular content of the non-adherent zone at various intervals of culture. E=erythroid, M=myeloid, L=lymphoid, S=stromal.

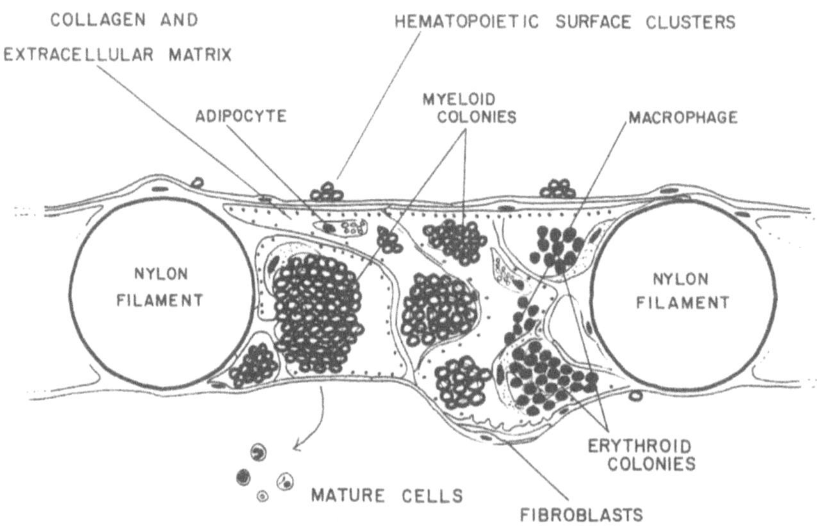

Figure 7. Schematic of the three-dimensional culture.

DISCUSSION

LTBMC systems established on monolayer templates support the self-renewal of murine hematopoietic stem cells for several months [3,5] but the potential of this system to support hematopoiesis in primate LTBMC is considerably lower. In this regard, myeloid progenitors (CFU-C) have been reported to persist in human [9] and non-human primate [10] monolayer-type LTBMC but the proliferative capacity of these cells steadily diminishes with time. Erythroid progenitors suffer a similar fate, but decline at a faster rate [11]. The cellular content of the non-adherent zone of monolayer-type LTBMC assumes a predominantly myeloid character after several weeks [3,5,16] with increased numbers of macrophages appearing in the spent medium [16]. These cultures may be modulated via the addition of exogenous regulatory factors such as erythropoietin [8] or by varying the constitution of the stroma monolayer [16,27]. However, after extended periods they generally support the growth and terminal differentiation of only one or two hematopoietic cell lines [5,8].

Hematopoiesis in LTBMC systems appears to be dependent upon regulatory and microenvironmental factors provided by the stromal support cells [18]. In this regard, qualitative differences have been observed in stroma capable of supporting myelopoiesis and erythropoiesis. Alkaline phosphatase positive stromal cells have been associated with myeloid development whereas erythropoiesis was found to occur in areas populated by acid phosphatase positive cells [17]. Thus, the ability of the fetal liver to support erythropoiesis and the propensity of the stroma from this organ to promote erythroid maturation in LTBMC [19] suggests a regulatory role of the stromal elements. Fibroblasts and other stromal cells were reported to synthesize colony stimulating factors [19,20] and contribute glycosaminoglycans and proteoglycans to the microenvironment[21]. Stromal cells can be be stimulated to synthesize hematotrophic factors by radiation [20] or following exposure to the products of other cells, such as macrophages [22].

The nylon screen template utilized in the present study provides a greater area for cell attachment and proliferation than a two dimensional surface. Despite pretreatment with a number of agents to maximize the extent and uniformity of cell attachment [23], stromal cell seeding efficiency rarely exceeds 40% [24]. Hematopoietic cell seeding (second inoculation) however, is much more efficient [23,24] and colonization is observed between the stromal cell processes and on the outermost surfaces of the nylon screen cultures within 3 days of inoculation. Immature hematopoietic cells are held tenaciously by the adherent zone whereas the smaller surface colonies are loosely attached, appearing at times to be part of the non-adherent zone. These structures are routinely observed in monolayer-type LTBMC and have been termed the pseudoadherent layer [11]. No progenitor cell assays were perfomred on cells of the non-adherent zones of nylon screen LTBMC since these contained primarily late stage precursor and mature hematologic cells. Studies of the adherent zones revealed that the numbers of CFU-GM, BFU-E, CFU-E, and CFU-GEMM were sustained for the 12 week experimental period (Table I). Furthermore, the numbers of CFU-GM significantly (p<0.0.5) exceeded input levels at 2 and 6 weeks of culture. Kinetic studies of monolayer-type human LTBMC demonstrate that BFU-E and CFU-GM of the non-adherent zone actively cycle whereas their counterparts in the adherent layer alternate between mitotic quiescence and cell division [25]. The failure to retain cells with high proliferative potential in the adherent zone may contribute to the progressive decline in progenitor cells which is noted in these monolayer cultures. The hematopoietic potential of LTBMC systems appears to be dependent upon the ability of stromal cells to elaborate growth/regulatory factors and provide physical support for developing blood cells. This, in turn, may be related to the proliferative activity of the support cells. In an earlier study, we obser-ed a significant decline in hematopoiesis on nylon screen LTBMC which were suspended in flasks containing fully confluent monolayers of rat bone marrow stromal cells [24]. Overcrowding and loss of secretory activity by stromal elements in monolayer-type LTBMC may contribute to the progressive decline in hematopoiesis which is noted in these cultures. Consistent with this hypothesis is the observation that declining human LTBMC which were "recharged" with fresh marrow failed to sustain significant growth [26]. A recent report indicates that irradiation of the stroma of declining LTBMC may rejuvenate its ability to support hematopoiesis [27]. Stromal cells in nylon screen cultures do not become confluent as quickly as in monolayer-derived systems and the ratio of stromal to hematopoietic cell numbers is relatively uniform over several weeks of culture [23,24]. However, in our experience, stromal cell overgrowth of nylon screen cultures inhibits the release of maturing cells from the adherent zone and diminishes hematopoiesis [24]. Demidepopulation of adherent cells with Ca^{++}/Mg^{++} free washes and/or brief enzyme treatment can prolong the "life" of some of these cultures by detaching the outermost stromal cell layers, albeit at the expense of the surface colonies (Naughton _et al,_ unpublished observat-ions).

Most of the LTBMC studies which have been published to date utilize plastic culture flasks as a substrate. In addition, implanted Millipore diffusion chamber filters [28] and cellulose ester membranes [29] have been reported to support hematopoiesis for extended periods _in vivo_. The morphology and differentiation patterns of cells growing in three-dimensional matrices such as cellulose sponge [30] and collagen gels [31,32] appears to more closely resemble the _in vivo_ condition than monolayer-derived cultures. Hematopoietic cells of nylon screen LTBMC are interspersed between the processes of the stromal cells rather than attaching uniformly to one side of a cell that is affixed to a non-permeable surface. This three-dimensional structure maximizes cell-cell interaction, enhances diffusion, and apparently inhibits the egress of immature hematopoietic cells from the adherent zone. The 210 um mesh opening provides sufficient volume for the establishment of microenvironments conducive to the expression of several hematopoietic lineages. Cytofluorographic analysis of the adherent zone using monoclonal anitbodies reactive against the various cell lines revealed the presence of early myeloid precursors, mature granulocytes, B and T lymphocytes, megakaryocytes/platelets, and macrophages/monocytes (Table II). The question as to the precise origin of the lymphocytes found in these cultures has not been resolved; reports by others indicate that these cells are relatively short lived in LTBMC [3,7,33]. However, maintenance of Thy-1 positive lymphocytes in LTBMC has been described previously [34]. In addition, studies of monolayer-type LTBMC revealed the continued proliferation of T lymphocyte-CFU [35,36]. Although terminal differentiation to mature T_3+ lymphocytes failed to occur within the marrow culture system, TL-CFU generated in the adherent zones were capable of differentiating into phenotypically mature T cells with the proper stimulation [35]. In addition, progenitors produced in LTBMC were reported to generate mature T cells which displayed helper activity for cytolytic T cell precursors. Preliminary studies of suspended nylon screen culture of marrow depleted of B and T lymphocytes by erythrocyte rosetting after treatment with 2-aminoethylisothorium bromide (AET)[37] revealed the re-expression of the CD4 phenotype and the B cell leukocyte common antigen after 2 weeks. (Naughton et al, unpublished observations). Additional experiments are in progress to determine the precise range of lymphoid differentiation in three dimensional culture and answer the question whether these cells originate de novo or are a residuum of the initial marrow inoculum. Suspended nylon screen bone marrow cultures offer a physiological model for the study of normal and abnormal hematopoiesis. Some of the potential pharmaceutical applications include the following:

1. **toxicity testing**: Since elements of all of the hematopoietic lineages are expressed simultaneously, this system could be utilized to determine the effects of chemotherapy or industrial agents on various cell lines.
2. **evaluation of bone marrow after purging**: Unlike the suspension cultures which were developed to study leukemic clones [38], monolayer-derived LTBMC appear to select for the normal progenitor cell population at the expense of the leukemic clones in some patients [39], possibly by promoting the differentiation of these abnormal cells [40]. Preliminary studies with AML marrow grown in suspended nylon screen LTBMC revealed the persistence of morphologically abnormal blast cells and Auer rods for 3.5 weeks, with little evidence of terminal differentiation (Naughton et al, unpublished observations). Since three-dimensional cultures appear to support both normal and abnormal hematopoiesis, they may prove to be a useful tool for evaluating the efficacy of a bone marrow purging procedure.
3. **studying and testing hematopoietic growth factors**
4. **producing stem cells for transplants**: With further development, adherent zone cells of suspended nylon screen cultures could be used as adjunctive therapy for bone marrow transplant recipients and possibly for patients with a variety of cytopenic conditions.
5. **adaptability for the growth of other tissues**: This three-dimensional system has also been used to grow skin cells, hepatocytes, and endocrine pancreatic cells using stromal cells derived from these organs.

ACKNOWLEDGEMENTS

The authors wish to thank Glen L. Feye and Ernest Campany for their technical assistance.

REFERENCES

1. Bradley, TR, Metcalf, D: The growth of mouse bone marrow cells in vitro. Australian J Exptl Biol Med Sci 44:287, 1966.

2. Pike, BL, Robinson, WA: Human bone marrow colony growth in agar gel. J Cell Physiol 76:77, 1970.
3. Dexter, TM, Allen, TD, Lajtha, LG: Conditions controlling the proliferation of haematopoietic stem cells in vitro. J Cell Physiol 91:335, 1977.
4. Weiss, L: The hematopoietic microenvironment of the bone marrow: an ultrastructural study of the stroma in rats. Anat rec 186:161, 1974.
5. Dexter, TM, Wright, EG, Krizsa, F, Lajtha, LG: Regulation of haemopoietic stem cell proliferation in long-term bone marrow cultures. Biomedicine 27:344, 1977.
6. Schrader, JW, Schrader, S:In vitro studies on lymphocyte differentiation. I. Long-term in vitro culture of cells giving rise to functional lymphocytes in irradiated mice. J Exptl Med. 148:823, 1978.
7. Whitlock, CA, Witte, ON: Long-term culture of B lymphocytes and their precursors from murine bone marrow. Proc Nat Acad Sci 79:3608, 1982.
8. Eliason, JF, NG, Dexter, TM: Erythropoietin-stimulated erythropoiesis in long-term bone marrow culture. Nature 281:382, 1979.
9. Gartner, S, Kaplan, S; Long-term culture of human bone marrow cells. Proc Natl Acad Sci 77: 4756, 1980.
10. Moore, MAS, Sheridan, AP: Pluripotential stem cell replication in continuous human, prosimian, and murine bone marrow culture. Blood Cells 5:297, 1979.
11. Coulombel, L, Eaves, AC, Eaves, CJ: Enzymatic treatment of long-term human marrow cultures reveals the preferential location of primitive hemopoietic progenitors in the adherent layer. Bloood 62:291, 1983.
12. Iscove, NM, Sieber, F, Winterhalxer, KH: Erythroid colony formation in cultures of mouse and human bone marrow. J Cell Physiol 83:309, 1974.
13. Fauser, AA, Messner, A: Identification of megakaryocytes, macrophages, and eosinophils in colonies of human bone marrow containing neutrophilic granulocytes and erythrocytes. Blood 53:1023, 1979.
14. Clarke, BJ, Housman, HE: Characterization of an erythroid precursor cell of high proliferative capacity in normal peripheral blood. Proc Nat Acad Sci 74:3879, 1977.
15. McLeod, DC, Shreeve, MM, Axelrad, AA: Improved plasma clot culture system for productiion of erythrocytic colonies in vitro: Quantitative assay method for CFU-E. Blood 44:517, 1974.
16. Slovick, FT, Abboud, CN, Brennan, JK, Lichtman, MA: Survival of granulocytic progenitors in the nonadherent and adherent compartments of human long-term marrow cultures. Exp Hematol 12:327, 1984.
17. Westen, H. Bainton, DF: Association of alkaline phosphatase positive reticulum cells in bone marrow with granulocytic precursors. J Exptl Med 150:919, 1979.
18. Campbell, AD, Wicha, MS: Extracellular matrix and the hematopoietic microenvironment. J Lab Clin Med 112:140, 1988.
19. Slaper-Cortenbach, I, Ploemacher, R, Lowenberg, B: Different effects of human bone marrow and fetal liver stromal cells on erythropoiesis in long-term bone marrow culture. Blood 69:135, 1987.
20. Bagby, GCK, McCall, E, Layman, DL: Regulation of colony stimulating activity productiion. Interaction of fibroblasts, mononuclear phagocytes, and lactoferrin. J Clin Invest 71:340, 1983.

21. del Rosso, M, Cappaletti, R, Dini, G, Fibbi, G, Vannucchi, S, Chiarugi, V, Guazelli, C: Involvement of glycosaminoglycans in detachment of early myeloid precursors from bone marrow stromal cells. Biochem Biophys Acta 676:129, 1981.
22. Naparstek, E, Donnelly, T, Kase, K, Greenberger, JS: Biological effects of in vitro x-irradiation of murine long-term bone marrow cultures on the production of granulocyte-macrophage colony-stimulating factors. Exp Hematol 13:701, 1985.
23. Naughton, BA, Naughton, GK: Hematopoiesis on nylon mesh templates. I. Long term culture of rat bone marrow cells. J Medicine 18:219, 1987.
24. Naughton, BA, Naughton, GK: Hematopoiesis on nylon mesh templates: comparative long-term bone culture and the influence of stromal support cells. In:Molecular and Cellular Controls in Hematopoiesis, D, Orlic, editor, Annals of the New York Academy of Science, vol 554:pp 140-155, 1989.
25. Cashman, J, Eaves, AC, Eaves, CJ: Regulated proliferation of primitive hematopoietic progenitor cells in long-term human marrow culture. Blood 66:1002, 1985.
26. Moore, MAS, Broxmeyer, HE, Sheridan, APC, Meyers, PA, Jacobsen, N, Winchester, RJ: Continuous human bone marrow culture: Ia antigen characterizations of probable pluripotential stem cells. Blood 55:682, 1980.
27. Meagher, RC, Salvado, AJ, Wright, DG: An analysis of the multilineage production of human hematopoietic progenitors in long-term bone marrow culture: Evidence that reactive oxygen intermediates derived from mature phagocytic cells have a role in limiting progenitor cell self-renewal. Blood 72:273,1988.
28. Benestad, HB: Formation of granulocytes and macrophages in diffusion chamber cultures of mouse blood leukocytes, Scand J Haematol 7:279, 1970.
29. Seki, M: Hematopeoitic colony formation in a macrophage layer provided by intraperitoneal injection of cellulose acetate membrane. Transplantation 16:544, 1973.
30. Leighton, J, Mark, R, Rush, G: Patterns of three dimensional growth in collagen coated cellulose sponge:carcinomas and embryonic tissues. Cancer Rie 28:286, 1968.
31. Douglas, WHJ, McAteer, JA, DEll'Orco, RT, Phelps, D:The formation of histiotypic structures from monodispersed rat lung cells cultured on three-dimensional sustrate. In Vitro 12:373, 1976.
32. Yang, J, Elias, JJ, Petrakis, NL, Wellings, SR, Nandi, s: Effects of hormones and growth factors on human mammary epithelial cells in collagen gel culture. Cancer Res 41: 1021, 1981.
33. Jones-Villeneuve, E. Phillips, RA: Potentials for lymphoid differentiation by cells from long-term cultures of bone marrow. Exp Hematol 8:65, 1980.
34. Jones-Villeneuve, EV, Rusthovoen, JJ, Miller, RG, PHillips, RA: Differentiation of Thy-1 bearing cells from progenitors in long-term bone marrow cultures. J Immunol 124:597, 1980.
35. Touw, I, Lowenberg, B: Production of T lymphocyte colony-forming units in human long-term bone marrow cultures. Blood 64:656, 1984.
36. Nabholz, M, MacDonal, HR: Cytolytic T lymphoctyes. Ann Rev Immunol 1:273, 1983.

37. Madsen, M, Johnsen, HE, Wendelboe, Hansen, P, Christiansen, SE: Isolation of hunger T and B lymphocytes by E-rosette gradient centifugation. Characterization of the idolated subpopulations. J.Immunol Methods 33:323-336, 1980.
38. Nara, N, McCulloch, EA: The proliferation in suspension of progenitors of the blasts in acute myeloblastic leukemia. Blood 65:1484-1493, 1985.
39. Coulombel, L, Kalousek, DK, Eaves, CJ, Gupta, CM, Eaves, AC: Long-term marrow culture reveals chromosomally normal hematopoietic progenitor cells in patients with Philadelphia chromosome-positive chronic myelogenous leukemia. N Engl J Med 308:1493-1498, 1983.
40. Iland, HJ, Croaker, GM, Repka, E, Radloff, TJ, Vincent, PC: Long-term bone marrow culture induces terminal differentiation of human myeloid cells. Exp Hematol 15:1109-1114, 1987.

ISOLATION, CULTURING AND IN VITRO ACTIVATION OF LIVER MACROPHAGES

Gerrit L. Scherphof, Toos Daemen and
Johannes T.P. Derksen

Laboratory of Physiological Chemistry
State University Groningen, Bloemsingel 10
9712 KZ Groningen, The Netherlands

INTRODUCTION

Systemic administration of particulate drug carriers such as liposomes generally leads to rapid accumulation of the carrier in cells of the mononuclear phagocyte system (MPS) (Scherphof et al., 1983). The mechanism of entry is most likely to be of endocytic nature (Scherphof et al., 1985), implying that the carrier-drug complex will be subject to intralysosomal processing. For a drug to be released from the lysosomal compartment in an active or activatable form it will have to be dissociated from the carrier and pass the lysosomal membrane when, as will most often be the case, its action is required in the cytosol. The great majority of the MPS is concentrated in the liver and the spleen, the organs which usually are responsible for the bulk of the removal of the drug carrier from the circulation. In order to be able to study the intracellular fate of such drug carriers as a function of various carrier parameters such as size and composition without being bothered by the involvement, either directly or indirectly, of other cell types, it might be advantageous to perform such studies in vitro with isolated cells. Since the liver macrophage population comprises the largest population of MPS cells in any one organ and as such contributes most extensively to the elimination of particulate drug carriers from the blood, we set out to investigate in detail the uptake and intracellular processing of one of the most versatile drug carriers, the liposome, by isolated liver macrophages (Kupffer cells) in monolayer culture. Also the action of a biologically active compound, the immunomodulator muramyl dipeptide (MDP), encapsulated in liposomes, was studied in this system.

METHODS

Isolation and culturing of liver macrophages

Macrophages can be isolated from liver in good yield and high viability by means of pronase-digestion of the parenchymal cells (the hepatocytes), followed by density-gradient and counter-current centrifugation of the remaining non-parenchymal cell fraction as described by Knook and Sleyster (1976) and as modified by Dijkstra et al. (1984a). Briefly, the liver is perfused in situ with Gey's balanced salt solution (GBSS) and then for 3 min with 0.2% pronase. Small pieces of the excised organ are then further incubated at 37° in the same pronase solution for 45-60 min in the presence of DNAse and the remaining (non-parenchymal) cells are spun down and resuspended. The cell suspension is further purified on a metrizamide concentration gradient which yields a non-parenchymal cell preparation, virtually free of hepatocytes and debris. By means of counter-flow centrifugation the non-parenchymal cells are fractionated into endothelial cells and macrophages (Kupffer cells). Sub-fractionation of the Kupffer cell preparation is achieved by stepwise increasing the counter flow rate, yielding the desired number of sub-fractions, separated according to size (Derksen et al., 1988). The Kupffer cells or sub-fractions thereof are collected by centrifugation and resuspended in culture medium containing 20% fetal calf serum. Aliquots of this suspension are seeded on appropriate culture plates and allowed to attach and to recover from the isolation procedure during 24 hours.

Cytotoxicity assay

Macrophage-mediated tumor cell lysis was assayed by quantifying the release of radioactivity from target tumor cells which had been labeled by incorporation of ^{3}H-deoxythymidine (dThd). To this end the cells werd incubated for 20 h with ^{3}H-dThd, washed free from non-incorporated radioactivity and cold-pulsed for 3-4 h to deplete intracellular pools of ^{3}H-dThd. The labeled cells were harvested by trypsinization and added in the indicated amounts to immunomodulator-activated macrophages. Typically, after 48 h the culture medium was collected and assayed for radioactivity, corrected for radioactivity released during an incubation with control macrophages. Percent specific cytolysis was calculated as 100 times the ratio of this corrected amount of radioactivity and that present initially in the added tumor cells, also corrected for the amount released during incubation with control macrophages. Cytostatic activity was determined by measuring the rate of ^{3}H-dThd incorporation into target tumor cells after a 24-h incubation with control or activated macrophages.

Liposome preparation

Liposomes of the indicated compositions were prepared by polycarbonate-filter extrusion (MLV) or reverse-phase evaporation (REV) according to established procedures (Olson et al., 1979). Non-encapsulated test or marker solutes were removed by gel filtration. IgG-coated liposomes were prepared by covalent linkage of bovine or rabbit immunoglobulins, randomly thiolated by means of the heterobifunctional reagent N-succinimidyl-S-acetylthioacetate (SATA), to liposomes containing the phospholipid derivative maleidophenylbutyrylphosphatidylethanolamine (MPB-PE) (Derksen et al., 1985). Thus, a stable, non-reducible thioether linkage is obtained.

RESULTS

The use of different liposome markers

Rat-liver macrophages obtained as described above were shown to be capable of binding and endocytosing liposomes in vitro as of about one day after their isolation, as indicated by the rates at which a radiolabeled and metabolically stable liposome marker (^{3}H-inulin encapsulated in the aqueous space) became associated with the cells under these conditions (Dijkstra et al., 1984a). Within 2-3 days after isolation maximal uptake values were obtained. Both binding (4 °C) and uptake (37 °C) were saturable, while uptake was energy-dependent. Biodegradable liposome

TABLE 1

KUPFFER CELL SUB-FRACTIONS OBTAINED BY FRACTIONATED COUNTER-FLOW CENTRIFUGATION

Fraction	Elutriation flow rate (ml/min)	Cell size class* diameter, μm)	Recovery (%) (mean± S.D.; n=16)	Peroxidase positive cells(%)	Cellular protein (pg/cell) (mean±S.D.; n=16)
A	25.0	9.2-10.2	49.4±3.4	85	53.4± 5.4
B	30.0	10.2-11.2	26.4±2.8	86	75.7± 8.0
C	35.0	11.2-12.0	11.0±1.6	88	97.6±10.0
D	40.0	12.0-13.0	8.3±1.4	95	120.1±11.0
E	46.5	13.0-14.0	5.0±1.4	95	141.9±11.0

* Calculated from the nomogram supplied by the manufacturer of the elutriation centrifuge (Beckman Inc.)

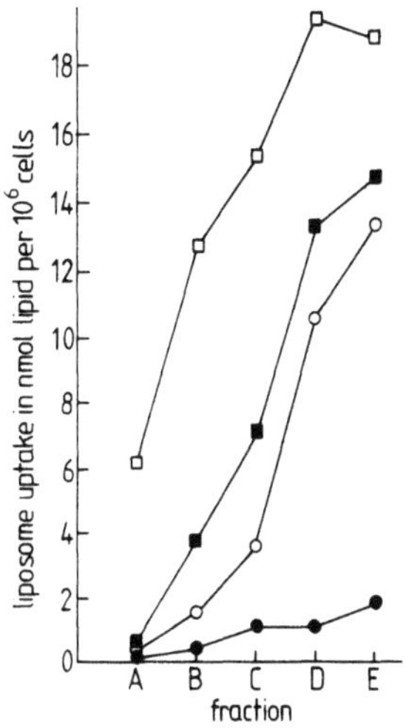

FIGURE 1

In vivo uptake of [^{3}H]-cholesteryl hexadecylether containing liposomes by subfractions of Kupffer cells. Multilamellar liposomes composed of egg-PC:cholesterol:DCP in a 4:5:1 molar ratio and containing [^{3}H]-CE were injected into rats via the lateral tail vein at the given doses. After 18 h macrophage subfractions A-E, of increasing cell diameter, were isolated (see Methods and Table 1). Aliquots were taken for cell counts and radioactivity measurements.
Given is the mean uptake in nmol of lipid per 10^6 cells after the injection of 5 μmol (□), 1 μmol (■), 0.5 μmol (O) and 0.1 μmol (●) of lipid. Duplicate experiments agreed within 10%.
CE, cholesterol ester; PC, phosphatidylcholine; DCP dicetylphosphate.

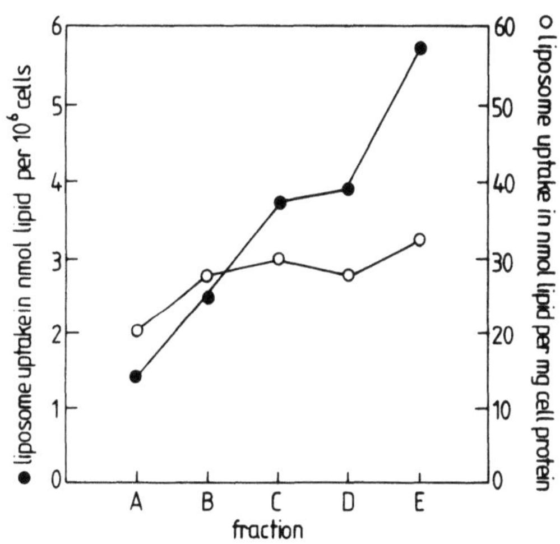

FIGURE 2

In vitro uptake of [^{3}H]-cholesteryl hexadecylether-containing liposomes by Kupffer cell subfractions. Multilamellar liposomes composed of egg-PC:cholesterol: DCP (in a 4:5:1 molar ratio) contained [^{3}H]-CE. Fifty nmol of liposomal lipid were added per well to macrophage subfractions A through E ($10\text{-}30 \times 10^4$ cells), cultured in monolayer and incubated at 37°C in medium containing 10% fetal calf serum. After a 4 h incubation the wells were washed and lysed with 0.5 M NaOH. Aliquots were taken for the determination of protein and radioactivity content. Given is the mean uptake in nmol lipid per 10^6 initially plated macrophages (solid symbols) or the mean uptake in nmol of lipid per mg cellular protein (open symbols) of triplicate determinations agreeing within 5%.
Abbreviations, see Fig. 1.

markers such as phosphatidyl-^{14}C-choline or ^{125}I-labeled albumine were found to be readily degraded in the lysosomal compartment as indicated by the chloroquin and NH_4Cl sensitivity of the degradation pathway (Dijkstra et al., 1984b). Water-soluble break-down products of the phosphatidylcholine as well as of ^{14}C-choline-labeled sphingomyelin were efficiently reutilized by the cells for cellular phosphatidylcholine synthesis, i.e. only little label was released from the cells (Dijkstra et al., 1985a). In contrast, water-soluble break-down products of the radiolabeled albumine were readily released from the cells, thus offering a useful method to assess the rate of intracellular liposome degradation (Dijkstra et al., 1985b).

An even more convenient way to assess liposome breakdown is offered by the simultaneous use of the degradable marker cholesteryl-^{14}C-oleate and the non-degradable ether analogue ^{3}H-cholesteryl hexadecylether (Derksen et al., 1987). As the ^{14}C-oleate formed during hydrolytic degradation of the liposomes is readily released by the cells in presence of serum (albumine serves as an acceptor) whereas the internalized cholesterylether remains firmly cell-associated, the rise in ^{3}H/^{14}C ratio of the cells provides an easily accessible parameter of liposome degradation. This approach has also been applied in vivo to study the effect of liposome composition on the rate of degradation (Roerdink et al., 1986;1989); a complication that strongly influences the interpretation of the observations in this case is the efficient incorporation of the radiolabeled fatty acid released from the Kupffer cells into the lipids of the parenchymal cells. Determination of the isotopic ratio of a chloroform extract of whole liver will therefore result in a grossly underestimated degree of liposome degradation. In order to obtain more reliable values it will be necessary to isolate the Kupffer cell fraction of the liver and determine its isotopic ratio. Obviously, the use of cultured Kupffer cells made it possible to become aware of this potential pitfall.

Kupffer cell heterogeneity

Liver macrophages can be differentiated according to size by means of stepwise increasing the counter flow rate of the elutriation centrifuge as described in the Methods section. In this way an arbitrary number of sub-fractions can be obtained as presented in Table 1. The 5 sub-fractions elutriated between 22.5 and 46.5 ml/min and varied in average diameter from about 9.5 to 14 nm. 80-95% of the cells in each fraction stained positive for endogenous peroxidase, the fractions containing the larger cells showing the highest enrichment in peroxidase-positive cells. These larger cells constitute a relatively small proportion of the total Kupffer cell population, while the smaller cells (fractions A and B) represent almost half of the total population. During in vivo experiments we observed that these small cells participate to only low extents in the process of liposome elimination from the circulation. This was found both for uncoated (Daemen et al., 1989d) as well as for IgG-coated liposomes (Derksen et al., 1988). The relative contribution of small and large cells to liposome clearance was dependent on the liposome dose. Upon administration of increasing amounts of liposomal lipid the participation of the smaller cells appeared to become more important (Fig. 1). In vitro this behaviour could only partly be reproduced. While at one day after isolation of the Kupffer cells there was still an increased uptake of liposomes per cell with increasing cell size (Fig. 2), uptake per mg of protein, which takes into consideration differences in cell size, already shows a levelling off of the uptake differences. After two days in culture the differences in uptake tend to disappear altogether or even turn into the reverse, the larger cells appearing to lose their capacity to take up liposomes, while the smaller cells seem to have increased their uptake capacity (Derksen et al., 1988), as shown in Fig. 3 for IgG-coated liposomes. The F_c-receptor specificity of binding and uptake is demonstrated by the inhibitory effect of aggregated immunoglobulin at both temperatures and in all subfractions. Thus, it appears that for studies on Kupffer cell sub-fractions extrapolations from the in vivo to the in vitro situation should be made with caution; although the uptake capacity of the total population may appear to remain unchanged, substantial changes may occur within the subfractions during the culture period. Similar changes may occur with respect to other parameters such as degradation capacity, i.e. hydrolytic activity, or the capacity to become activated (see Table 2).

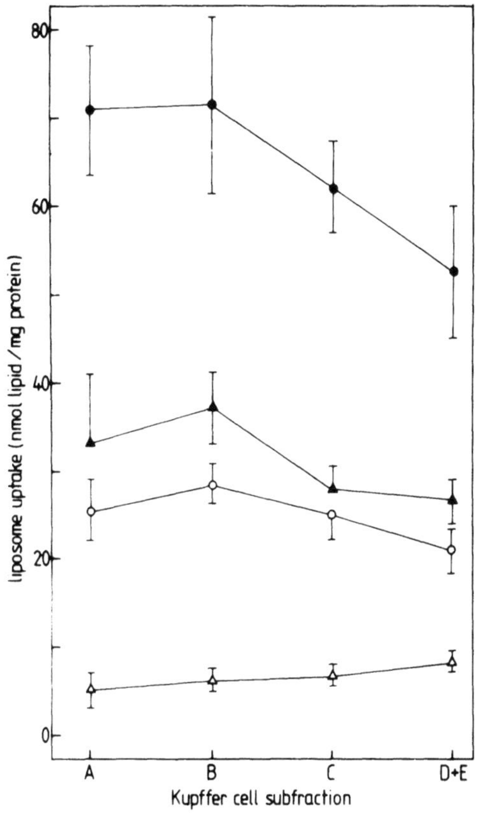

FIGURE 3

Rabbit Ig-coated liposome uptake by different size classes of Kupffer cells in vitro. After 2 days in culture Kupffer cell sub-fractions were incubated with [^{3}H]cholesteryl hexadecylether-labeled rabbit Ig-coated liposomes (50 nmol total liposomal lipid/ml) at 37°C (circles) or 4°C (triangles) in the absence (filled symbols) or presence (open symbols) of 1 mg/ml aggregated human immunoglobulin. After 4 h of incubation, Kupffer cells were harvested and cell-associated ^{3}H radioactivity was determined. Data represent means of four incubations; bars, SD.

A general comment with respect to extrapolation from the in vitro to the in vivo situation refers to the intrinsically different conditions which the cells experience in terms of medium-volume to cell-number ratio. In the intact liver some 10^8 Kupffer cells are perifused with 6-7 ml of plasma, whereas in the in vitro situation typically 10^6 cells are covered with 0.5 ml of medium, while on microtiter plates the amount of medium per cell is even considerably larger. It is likely that such differences are of influence on parameters such as liposome uptake. For example, upon in vivo administration of a dose of 5 µmol liposomal lipid at least 50% of it will be taken up by the 10^8 Kupffer cells of the liver, i.e. 25 nmol of lipid per 10^6 cells; this is approximately one order of magnitude higher than the maximal amounts of uptake found in vitro with similar liposome preparations. Obviously, this consideration is applicable to all in vitro systems of tissue cells.

TABLE 2

SUSCEPTIBILITY OF DIFFERENT TUMOR CELL LINES AND NORMAL CELL LINES TO MACROPHAGE-MEDIATED CYTOLYSIS AND CYTOSTASIS

Macrophage[a] treatment (µg MDP/ml)	dpm released in supernatant (% specific cytolysis)[b]				
	B16	C26	P815	LLC	Fb (cell type[c])
liposomal MDP					
(1.0)	1082±74(41)	891±68(81)	1021±42(97)	240± 9(4)	ND[e]
(0.5)	929±54(31)	821±68(71)	953±50(88)	240±20(4)	2721± 97(4)
free MDP					
(50.0)	841±99(26)	761±28(64)	972±88(91)	269± 9(8)	2752±147(5)
	dpm incorporated (% proliferation inhibition)[d]				
	B16	C26	P815	LLC	Fb (cell type[c])
liposomal MDP					
(1.0)	220±79(98)	99± 8(94)	ND	136± 3(81)	ND
(0.5)	962±21(89)	109± 7(93)	ND	214±41(78)	164±21(96)
free MDP					
(50.0)	535±23(94)	134±13(91)	ND	159±15(75)	183±18(95)

a. Per well 25×10^4 liver macrophages were incubated with medium, free MDP or liposome-encapsulated MDP (1 ng of MDP per nmol of liposomal lipid).
b. After 4 h the wells were washed and 10^4 [^{3}H]dThd-labeled target cells were added per well. After another 48 h, ^{3}H release into the supernatant was determined in triplicate experiments. Numbers in parentheses, percentage of specific cytolysis; see Methods.
c. B16, murine B16-B6 melanoma cells; C26, murine colon carcinoma cells; P815, murine mastocytoma cells; LLC, monkey kidney cells; Fb, human fibroblasts.
d. After an incubation of 4 h with the immunomodulators the wells were washed and 10^4 target cells were added per well. After 24 h of coculture [^{3}H]dThd was added to the wells for another 20 h. At this time, the cultures were washed 3 times with PBS, and the cells were lysed with NaOH. Radioactivity of the lysate was determined in triplicate. Numbers in parentheses, percentage of proliferation inhibition; see Methods.
e. ND, not determined.

The use of liposomes to concentrate a low-molecular weight immunomodulator such as muramyl dipeptide into macrophages was initiated by Fidler (Fidler, 1980; Fidler et al., 1981). He showed that liposome encapsulation of this compound resulted in a several orders of magnitude increase in the efficacy with which it could activate alveolar lung macrophages, both in vitro and in vivo, to a tumorcytotoxic state. In animal models it was shown that treatment of tumor-bearing animals with established lung metastases with liposome-encapsulated MDP resulted in efficient eradication of the metastatic growth. Since hepatic metastases originating from primary colo-rectal carcinoma are a major cause of death by cancer, as no efficient radio- or chemotherapeutic treatment is as yet available, the activation of liver macrophages by means of liposomal immunomodulators would seem a meaningful approach to combat this disease. With isolated Kupffer cells in monolayer culture we demonstrated that these cells do possess the potential to attain the tumorcytotoxic state upon incubation with MDP encapsulated in liposomes (Daemen et al., 1986). Table 2 shows the cytolytic as well as the cytostatic effect of Kupffer cells treated with free and liposome-encapsulated MDP on a number of different tumor cell lines and two non-tumor cell lines. It is shown that only the tumor cells are susceptible to

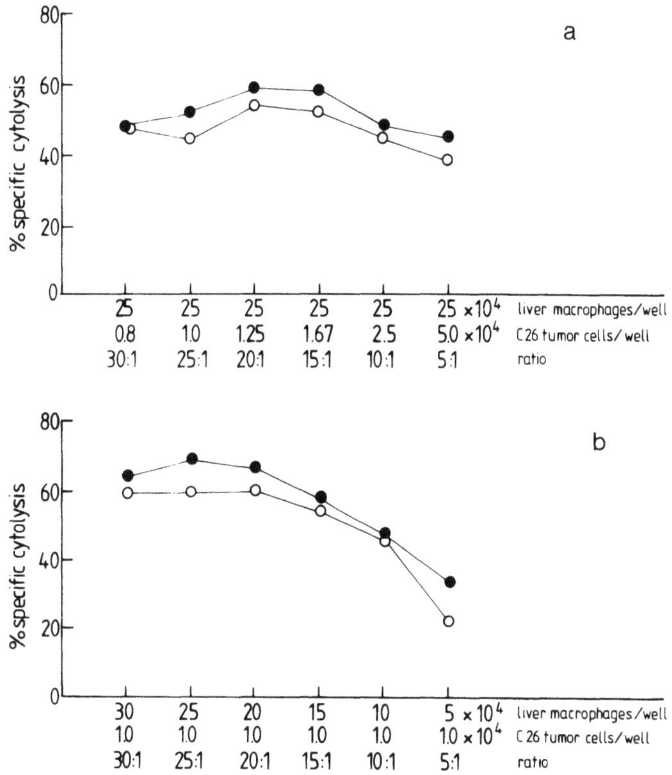

FIGURE 4

The effect of the macrophage to tumor cell ratio on macrophage-mediated cytolysis. Macrophages were incubated with medium (control), with free MDP (50 μg/ml, -O-) or with liposome-encapsulated MDP (0.5 µmol of lipid containing 0.5 µg of MDP per ml; -●-). C26 colon adenocarcinoma cells were used as target cells.

a. The ratio was changed by varying the number of tumor cells added to a constant number of macrophages (25×10^4 per well) plated.
b. The ratio was changed by varying the plating density of the macrophages and adding a fixed number of C26 cells (10^4 per well).

The experiment was carried out in triplicate. Plotted are the mean percentages of specific cytolysis (see Methods). Bars, SD.

macrophage-induced cell lysis; on the other hand, both normal and tumor cells are susceptible to the cytostatic action of the macrophages. This would argue in favor of the view that the activated macrophages do not specifically recognize the tumor cells but rather that the normal cells, in contrast to the tumor cells, are capable of repairing the damage inflicted by the macrophages before it becomes lethal (Stewart et al., 1988).

In a separate set of experiments we found that also for liver macrophages the encapsulation within liposomes resulted in an increase in immunostimulating potency of the MDP of several hundred-fold. It seems most likely that the (major) target site of the MDP molecule lies intracellularly and that the encapsulation into liposomes allows a much more efficient way of transport of the MDP into the cell by virtue of the high endocytic uptake of liposomes by these cells. In contrast, activation of Kupffer cells by lipopolysaccharide (LPS) or Lipid A was several hundred-fold reduced by incorporating it into liposomes (Daemen et al., 1989a). This suggests that the target site of these immunostimulants is at the cell surface of the macrophages and that the association with liposomes prevents the recognition of LPS or Lipid A by the cell-surface binding site. Association of these agents with liposomes would lead to endocytic uptake and transfer to the lysosomal system, where they would be subject to hydrolytic enzymes and inactivation. This view was supported by our observation that incubation of LPS or Lipid A with lysosomal enzymes results in inactivation of both compounds, while MDP, upon incubation with lysosomal enzymes, fully retains its immunomodulating activity. Thus, the uptake of MDP through the endocytic route, as a result of liposome encapsulation, would not lead to its inactivation. Obviously, these results could only be obtained by virtue of the use of isolated cells. On the other hand, also in this case substantial discrepancies exist between the in vitro and the in vivo system. For example, in order to obtain maximal cytotoxicity in vitro the macrophages have to form a confluent monolayer (Fig. 4), suggesting that the macrophages have optimal cytotoxic activity under conditions when several cells can simultaneously attack one tumor cell (Daemen et al., 1989b). Such a situation will never be attained in vivo, where the Kupffer cells are spread through the liver as single cells, relatively far apart. In vivo, however, we will have to take into consideration a quite different phenomenon, which does not play a role in vitro. That is the phenomenon that, upon a single injection with liposomal MDP, the macrophage population in the liver transiently doubles its size within 24 hours (Daemen et al., 1989c). This increase in cell numbers could be ascribed to both local proliferation of pre-existing cells and to extrahepatic recruitment, presumably from circulating monocytes or bone marrow. The macrophage population is gradually restored to control levels after about three days.

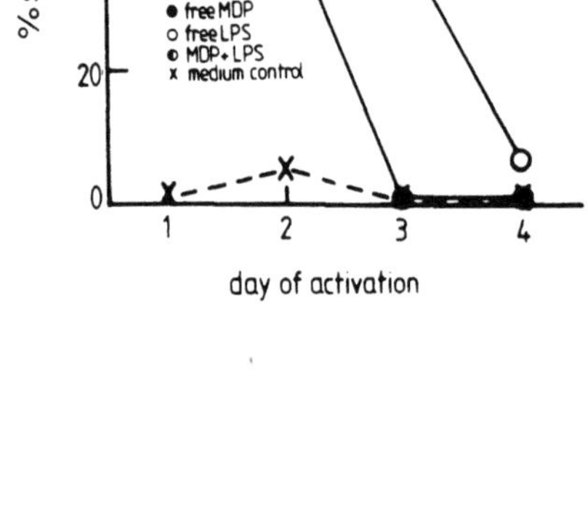

FIGURE 5

Activation of the total liver macrophage population, 1 to 4 days after initiation of in vitro culture.

Macrophages were isolated and plated in 96-well microtiter plates. On day 1, 2, 3 or 4 after isolation the macrophages were incubated with medium (control; x), MDP (50 μg/ml; ●), LPS (100 ng/ml; O) or a combination of MDP and LPS (50 μg/ml of MDP, 100 ng/ml of LPS; ◐). After a 4 h incubation, the immunomodulators were removed and 10^4, [^{3}H]dThd-labeled C26 colon adenocarcinoma cells were added per well. After 48 h of coculture ^{3}H release into the supernatant was determined. The experiment was carried out in triplicate. Plotted are the mean percentages of specific cytolysis (see Methods). Standard deviations were less than 10% of the mean.

Reactivation after loss of cytotoxic activity

From an immunotherapeutic point of view it is important to know how long the liver macrophages, once activated, retain their cytotoxic potency. In addition, it would be of value to know whether the cells, after having lost their cytotoxicity, can be reactivated to a tumoricidal state. Such experiments are difficult to carry out in vivo; therefore, also in this case we set out to investigate this problem with the isolated cells in primary monolayer culture. First of all, we determined how long after isolation the cells were still susceptible to activation. It turned out that they lost the capacity to become activated by MDP or LPS within 2 and 3 days, respectively, while a combination of these two immunomodulators was still able to elicit cytotoxicity 4 days after isolation (Fig. 5). For individual Kupffer cell subfractions, we found similar results, i.e. the MDP/LPS combination produced the highest cytotoxicities in all subfractions and were even able to do so on day 4 after isolation. In addition, it was found that the largest cells lost their capacity to become activated more rapidly than the smaller cells. After activation by MDP or a combination of MDP and LPS all three fractions had lost most of their cytotoxic capacity when, after 24 h, the target tumor cells were added. Activation by liposome-encapsulated MDP, however, resulted in cytotoxicity which lasted for at least one day in the smaller cells; the largest cells had lost, also under these conditions, their cytotoxic activity. All subfractions were refractory to reactivation by MDP two days after they were activated for the first time with this agent. LPS and MDP+LPS resulted in reduced levels of cytotoxicity of the small and intermediate size cells upon a second stimulation two days after the first, while the largest cells had completely lost their response to a second stimulus. Recently, these results were confirmed for the in vivo situation (Hoedemakers et al., unpublished), by determining the capacity of Kupffer cells to become activated in vitro after they had been activated in vivo with liposomal MDP two days before. In this connection, it has to be noted that the macrophage population that is isolated from the liver of a rat that has received a dose of liposomal MDP two days before still contains a substantial proportion of newly arrived cells (see above). It is conceivable that the low levels of cytotoxicity observed after a second (in vitro) stimulation are due to the presence of these new cells which were presumably not activated by the first (in vivo) stimulus.

CONCLUDING REMARKS

Primary monolayer cultures of rat liver macrophages (Kupffer cells) can serve as a convenient model to study the properties of these cells with respect to liposome uptake and processing and their tumorcytotoxic characteristics following treatment with immunomodulating agents. Qualitatively, several characteristics that these cells display in vivo are retained under in vitro culture conditions. Quantitatively, however, substantial differences may exist or be provoked as a result of the in vitro conditions. As one example may serve the 10-fold difference in liposome uptake capacity, presumably due to the unfavorable ratio of medium volume to number of cells under in vitro conditions. Another example is the shift of endocytotic capacity within the Kupffer cell population from the larger cells to the smaller cells during the culture period. Similar shifts in cytotoxic potency may occur between the subfractions under long-term culture conditions. Therefore, a word of caution would be appropriate for those who would wish to extrapolate in vitro observations on tissue macrophages in monolayer culture to the in vivo situation. The answer to the question: "Can one obtain meaningful information from macrophage cultures?" should therefore be: "That depends on the particular question one asks!"

ACKNOWLEDGMENTS

The authors wish to thank Aletta Veninga, Henriette Morselt, Jan Wijbenga and Bert Dontje for expert technical assistance and Lineke Klap for secretarial help. The work described was financially supported by the Netherlands Organization for the Combat of Cancer, Koningin Wilhelmina Fonds (Grant no. GUKC 83-010) and the Netherlands Organization for Scientific Research (NWO) under the auspices of the Foundation for Medical and Health Research (Medigon, grant no. 135360). MDP was generously supplied by Ciba-Geigy, Basel.

REFERENCES

Daemen, T., Veninga, A., Roerdink, F.H. and Scherphof, G.L. (1986) In vitro activation of rat liver macrophages to tumoricidal activity by free or liposome-encapsulated muramyl dipeptide. Cancer Res. 46, 4330-4335.

Daemen, T., Veninga, A., Dijkstra, J. and Scherphof, G. (1989a) Differential effects of liposome incorporation on macrophage-activating potencies of LPS, Lipid A and MDP; differences in susceptibility to lysosomal enzymes. J. Immunol. 142, 2469-2474.

Daemen, T., Veninga, A., Roerdink, F.H. and Scherphof, G.L. (1989b) Conditions controlling tumor cytotoxicity of rat liver macrophages mediated by liposomal muramyl dipeptide. Biochim. Biophys. Acta 991, 145-151.

Daemen, T., Koudstaal, J., Scherphof, G.L. and Hardonk, M.J. (1989c) Population kinetics of rat liver macrophages after intravenous administration of liposome-encapsulated MDP, in: Cells of the Hepatic Sinusoid, Vol. 2 (E. Wisse, D.L. Knook and K. Decker, eds.) Kupffer Cell Foundation, Rijswijk, The Netherlands, pp. 400-405.

Daemen, T., Veninga, A., Roerdink, F.H. and Scherphof, G.L. (1989d) Endocytic and tumoricidal heterogeneity of rat liver macrophage populations; implications for drug targeting. Select. Cancer Therap., in press.

Derksen, J.T.P. and Scherphof, G.L. (1985) An improved method for the covalent coupling of proteins to liposomes. Biochim. Biophys. Acta 814, 151-155.

Derksen, J.T.P., Morselt, H.W.M. and Scherphof, G.L. (1987) Processing of different liposome markers after in vitro uptake of immunoglobulin-coated liposomes by rat liver macrophages. Biochim. Biophys. Acta 931, 33-40.

Derksen, J.T.P., Morselt, H.W.M. and Scherphof, G.L. (1988) Uptake and processing of immunoglobulin-coated liposomes by subpopulations of rat liver macrophages. Biochim. Biophys. Acta 971, 127-136.

Dijkstra, J., Van Galen, W.J.M., Hulstaert, C.E., Kalicharan, D., Roerdink, F.H. and Scherphof, G.L. (1984a) Interaction of liposomes with Kupffer cells in vitro. Exp. Cell Res. 150, 161-176.

Dijkstra, J., Van Galen, M. and Scherphof, G. (1984b) Effects of ammoniumchloride and chloroquine on endocytic uptake of liposomes by Kupffer cells in vitro. Biochim. Biophys. Acta 804, 58-67.

Dijkstra, J., Van Galen, M., Regts, J. and Scherphof, G. (1985a) Uptake and processing of liposomal phospholipids by Kupffer cells in vitro. Eur. J. Biochem. 148, 391-397.

Dijkstra, J., Van Galen, M. and Scherphof, G. (1985b) Effects of (dihydro)cytochalasin B, colchicin, monensin and trifluoperazine on uptake and processing of liposomes by Kupffer cells in culture. Biochim. Biophys. Acta 845, 34-42.

Fidler, I.J., Sone, S., Fogler, W.E. and Barnes, Z.L. (1981) Eradication of spontaneous metastases and activation of alveolar macrophages by intravenous injection of liposomes containing muramyl dipeptide. Proc. Natl. Acad. Sci. USA 78, 1680-1684.

Fidler, I.J., (1980) Therapy of spontaneous metastases by intravenous injection of liposomes containing lymphokines. Science 208, 1469-1471.

Knook, D.L. and Sleyster, E.Ch. (1976) Separation of Kupffer and endothelial cells of the rat liver by centrifugal elutriation. Exp. Cell Res. 99, 444-449.

Olson, F., Hunt, C.A., Szoka, F.C., Vail, W.J. and Papahadjopoulos, D. (1979) Preparation of liposomes of defined size distribution by extrusion through polycarbonate membranes. Biochim. Biophys. Acta 557, 9-23.

Roerdink, F., Regts, J. and Scherphof, G. (1986) Effect of lipid composition on the uptake and intracellular degradation of liposomes by Kupffer cells, in: Cells of the hepatic sinusoid (A. Kirn, D.L. Knook and E. Wisse, eds.) Kupffer Cell Foundation, Rijswijk, pp. 131-136.

Roerdink, F.H., Regts, J., Handel, T., Sullivan, S.M., Baldeschwieler, J.D. and Scherphof, G.L. (1989) Effect of cholesterol on the uptake and intracellular degradation of liposomes by liver and spleen; a combined biochemical and gamma-ray perturbed angular correlation study. Biochim. Biophys. Acta 980, 234-240.

Scherphof, G., Roerdink, F., Dijkstra, J., Ellens, H., De Zanger, R. and Wisse, E. (1983) Uptake of liposomes by rat and mouse hepatocytes and Kupffer cells. Biol. Cell 47, 47-58.

Scherphof, G.L., Dijkstra, J., Spanjer, H.H., Derksen, J.T.P. and Roerdink, F.H. (1985) Uptake and intracellular processing of targeted and non-targeted liposomes by rat Kupffer cells in vivo and in vitro. Ann. N.Y. Acad. Sci. 446, 368-384.

Stewart, C.C., Stevenson, A.P. and Hibbs, J. (1988) Effector mechanisms for macrophage-induced cytostasis and cytolysis of tumor cells, in: Macrophages and Cancer (G. Heppner and A.M. Fulton, eds.) CRC Press, pp. 40-59.

IN VITRO EVALUATION OF MACROPHAGE MEDIATED HOST DEFENSES AGAINST NEOPLASTIC DISEASE

R. Kirsh[1,2] and P. J. Bugelski[3]

Departments of Drug Delivery[2] and Experimental Pathology[3]
SmithKline & French Laboratories
King of Prussia, Pa. 19406

Introduction

The ability of malignant tumors to disseminate to organs distant from the site of the primary lesions is the principal reason for failure in cancer therapy.(1) There are several reasons for this high failure rate. First, at the time of initial diagnosis, metastatic lesions may be too small to detect using currently available diagnostic methodologies; second, the anatomic location of the metastatic lesion may limit the dose of antineoplastic drug that can reach the site without excessive toxicity; and third, the heterogeniety of tumor cell phenotypes within a single lesion is sufficiently diverse, that the responsiveness of cells from one lesion to an antineoplastic modality can differ from both the primary lesion and other metastases.(2-4) This latter issue suggests that the development of an effective therapy for metastatic disease approaches that will circumvent the problems of cellular heterogenicity.

Within the past several years, considerable information has been obtained indicating that cells of the mononuclear phagocyte series play a major role in the recognition and destruction of tumor cells in situ and that amplification of macrophage-mediated tumoricidal activity could offer a potentially useful modality for the therapeutic management of cancer, including metastatic disease.(5-8)

Macrophages rendered tumoricidal through interaction with a variety of synthetic or naturally occurring agents acquire the capability of selectively recognizing tumor cells regardless of their degree of phenotypic diversity.(9) In addition, macrophage mediated tumoricidal activity appears to be devoid of the problems of generation of resistance which is often observed with cytotoxic drug therapy and specific immunotherapeutic modalities.(9-11) For these reasons, a significant effort is currently being undertaken in both academic and industrial laboratories to identify novel agents that selectively enhance macrophage-mediated tumoricidal activity in vivo.(10)

In this article, we will discuss the methodologies involved in the culture of rodent macrophages and human monocytes; activation to the tumoricidal state, and evaluation of macrophage-mediated tumoricidal activity in vitro.

In Vitro Culture Of Rodent And Human Mononuclear Phagocytes

Preparation of Murine and Rat Peritoneal Macrophages: Specific pathogen-free mice or rats are injected intraperitoneally (i.p.) with 2 ml or 5 ml thioglycollate broth respectively, and killed (ether euthanasia) 5 days later. The animals are swabbed with iodine solution and with 70% ethanol, and pinned to a dissecting board. The abdominal skin is reflected and 5 ml sterile Ca^{2+}-,Mg^{2+}-free Hanks' BSS (HBSS) is injected i.p. (18 gauge needle). Following gentle massage, the peritoneal fluids are aspirated (18 gauge needle) and the cell suspension is centrifuged (600 x g, 5 min) and resuspended in serum-free Eagle's DMEM. The cells are plated into 120-mm culture dishes (10^7 cells per dish) and the cultures washed 40 min later, to remove all nonadherent cells. Rapid adherence to glass or plastic in serum-free medium is selective for macrophages. The adherent macrophages are gently scraped off with a rubber policeman.

Rodent Alveolar Macrophages: Experimental animals are anesthetized by i.p. injection of sodium pentobarbital, thorax and abdomen swabbed with iodine solution and 80% ethanol. Skin covering the thorax is carefully reflected and the salivary gland removed exposing the trachea.

Carefully insert syringe needle into the trachea at the cricoid cartilage and lavage lungs with prewarmed HBSS using 5 ml/rat and 1 ml/mouse. Following lavage step, alveolar macrophages should be purified as described for peritoneal macrophages. Yield of alveolar macrophages should be approximately 5 x 10^6/rat and 3 x 10^5/mouse.

Human peripheral blood monocytes: Peripheral blood monocytes are isolated from one day old platlet pheresis residues obtained from the Red Cross.[12] Briefly, residues are washed with HBSS and mononuclear cells (lymphocytes and monocytes) are isolated by density gradient centrifugation on Ficoll-Hypaque (Lymphoprep® Accurate Chemical Co). The mononuclear cell fraction collected at the interface is further subfractionated into lymphocytes and monocytes by centrifugation over isoosmotic 46% Percoll (Pharmacia Inc., Upsalla) at 550 x g for 45 minutes. Monocytes are collected at the interface. The purified monocyte fraction is then washed two times with HBSS and concentration adjusted to 3-4 x 10^6 cells/ml. Following this method, monocyte purity is routinely greater than 95% based on morphology and phagocytic activity.

II. STRATEGIES FOR MACROPHAGE ACTIVATION IN CANCER THERAPY

A. The Activated Macrophage

Considerable attention has been devoted to the role of activated macrophages in host defense against invading bacteria, parasites, and tumor cells. The term "activated macrophage" is an operational definition and is used differently by various authors to describe acquisition of a variety of functions that are not exhibited by resident tissue macrophages. Examples of the use of the term activation can be found referring to activated macrophages as cells that show oxidative metabolism but lack microbicidal or tumoricidal activity, whereas other authors, including ourselves, limit the use of the term to macrophages that exhibit both microbicidal and tumoricidal properties. Acquisition of tumoricidal properties may be accompanied by other phenotypic alterations including increased phagocytosis, secretion of neutral proteases and acid hydrolases, synthesis and release of arachadonic acid metabolites, expression of an altered ectoenzyme profile, the ability to suppress natural killer (NK)-cell activation, and an enhanced ability to kill intracellular microorganisms.[13-15] In this review, macrophages that display a biochemical and/or physiological profile different from resident tissue macrophages, but which do not express tumoricidal activity, will be referred to as stimulated macrophages.

Despite this useful general definition, the term "macrophage activation" if often the source of a great deal of confusion and misunderstanding. Of the many reasons for this confusion, the following deserve emphasis:

- Macrophages have on their surfaces at least 30 distinct receptors and are capable of the secretion of over 75 different substances. The expression or production of most of these is alterable by stimulation or activation.[16-18] Preoccupation with the measurement of many of these parameters after the in vitro stimulation of macrophages with defined stimuli does not reflect what is likely to be a complex regulatory process in vivo.[13]

- No definitive evidence exists for the presence of macrophage populations with distinct differentiation patterns.[19,21] Despite this fact, macrophage populations are quite heterogeneous in regard to expression of various phenotypic and functional markers.[20,21]

- Significant differences exist in regard to phenotypic and functional parameters of macrophage activation among different species and even among different strains within a species.[16,22-24] Most distressing is the observation that without standardization of animal vendors, mouse strains, tumor target types, reagent sources, and other variables, quite different results can be obtained in regard to assessments of macrophage function using mouse macrophages.[25]

- In vivo studies regarding macrophage activation by cytokines have been minimal.[26,27] Because of this there is a general lack of understanding of the dose, route of administration, and kinetic requirements for macrophage activation in vivo using cytokines.

- Macrophages obtained from different anatomical sites can also vary considerably in their functional capabilities.[28-31] Data obtained with one population cannot be extrapolated to another in all cases.

B. Pathways of Macrophage Activation

Macrophages can be rendered tumoricidal by a wide variety of naturally occurring and synthetic agents. Currently, there are two major classes of macrophage-activating agents that appear attractive as potential therapeutic candidates: microbial cell wall components and structural analogs related to these agents and various lymphokines. Macrophages can be activated by agents that either act directly or by agents that induce other cells to release mediators that evoke macrophage activation.

The direct activation of macrophages is induced by interaction with agents such as bacterial endotoxin (lipopolysaccharide, PLS), lentinin, glucan, picibanil, and diverse bacteria such as Mycobacteria, Propionibacteria, Nocardia, or subunits isolated from the cell walls of these microorganisms.[15,32-42]

Most of the microbial cell-wall derived materials are unsuitable for clinical use due to significant toxicities. In addition, detailed structural characterization of many of the components responsible for macrophage activation by these materials has yet to be accomplished. One notable exception is the water soluble, low molecular weight (MW 459) synthetic dipeptide, N-acetyl-muramyl-l-alany-D-isoglutamine, referred to as muramyl dipeptide (MDP). MDP is the smallest structural unit capable of inducing all of the adjuvant activities of mycobacteria in a water in oil emulsion (complete Freund's adjuvant).[42,44] MDP has been shown to stimulate a wide variety of macrophage functions in vitro including secretion of prostaglandins and collagenase, generation of superoxide anion, release of interleukin-1 (IL-1), and to augment macrophage-mediated bacteriocidal and tumoricidal activities.[43-46] In vivo, MDP has been shown to protect experimental animals against lethal infection by a range of bacteria,[47-54] parasites,[53] fungi,[47,50] and viruses.[55,56] However, MDP, although far less toxic than intact mycobacteria, is much less effective than the intact bacteria in activating macrophages to the tumoricidal state in vivo. Its efficacy in vivo is limited by its extremely rapid clearance (90% of MDP injected intravenously can be detected in the urine within two hours).[45] Even when administered in high doses, it fails to stimulate significant macrophage-mediated antitumor activity.

Indirect macrophage activation can be achieved by lymphokines released by antigen-or mitogen-stimulated lymphocytes, which interact with specific receptors on the macrophage surface.[57] This activity is referred to by the predictable designation as macrophage activating factor (MAF), but the exact identity of the mediator(s) involved and whether different subpopulations of macrophages are activated by different mediators in lymphokine preparations is still unclear.[58-60] Current limitations in lymphokine production and purification technology dictate that the lymphokine preparations containing MAF activity are crude and vary markedly in their biological activities. In particular, the crucial question of whether there is a single, specific lymphokine that merits classification as MAF, or whether such activity can be attributed exclusively to other lymphokines such as gamma-interferon (IFN-γ), is currently an area of intense investigation. MAF activity and IFN-γ share many physicochemical and biological characteristics including co-purification by a variety of chromatographic procedures, similar chemical stabilities, and abilities to induce an oxidative burst and expression of microbicidal and tumoricidal properties.[62-65] Conversely, several groups have now demonstrated MAF activity in culture supernatants from T cell lines that contain no detectable IFN-γ activity in association with materials whose physicochemical and immunological properties are distinct from IFN-γ.[65-68] The biological effects of these MAFs have some overlap with the effects of IFN-γ with respect to shared induction of macrophage-mediated cytotoxicity. However, several significant differences can be cited. First, interferon-induced macrophage-mediated cytotoxicity requires the presence of a second signal, supplied by bacterial endotoxin, for complete expression of the tumoricidal phenotype,[65] whereas MAF-induced tumoricidal activity is independent of second signals.[66] Second, MAF does not induce the expression of Fc receptors or Class I or II histocompatibility antigens and does not have direct antiviral activity, whereas interferon treatment induces these effects.[69] However, formal proof of the existence of a specific non-IFN lymphokine with macrophage activation activity must await cloning of the gene and rigorous chemical analysis of the gene product.

1. Cytokine-Mediated Macrophage Activation

Because of the potential importance to immunotherapeutic applications, much recent work has focused on the ability of different cytokines to induce activation. The following list describes briefly what is known about various classes of agents in addition to interferon and their ability to directly or indirectly activate macrophages either alone or in combination with other agents.

a. Interleukins

- IL-1 has been reported to directly effect macrophage activation by promoting monocyte-mediated tumor cytotoxicity.[70] It has also been reported to induce tumor necrosis factor (TNF) in human monocytes and may be an autocrine system in mouse macrophages.[71] It may also indirectly contribute to macrophage activation by inducing cytokine production in other cells.[72]

- Interleukin-2 (IL-2) has recently been described as directly activating human monocytes for tumor cytolysis.[73] It has been reported not to be active in inducing reactive oxygen metabolite secretion and microbicidal activity in monocytes.[74] Interestingly, IL-2 may also activate autoreactive killer cells that are cytotoxic to human monocytes.[75]

- Interleukin-4 (IL-4) has recently been described to activate mouse macrophages for tumor cytolysis.[76]

b. Tumor Necrosis Factors (TNF)

- TNF-α induces its own production and monocyte-mediated cytotoxicity by human monocytes.[71] Again it was reported not to induce oxidative metabolism and microbicidal activity by itself[74] but has been reported to synergize with IFN-γ for induction of both tumoricidal and microbicidal activities in mouse macrophages.[77] TNF-α also induces cytokine production by monocytes, macrophages, and fibroblasts with the potential for inducing macrophage activation, namely IL-1 and granulocyte-macrophage colony-stimulating factor (GM-CSF).[78,79]

- TNF-β (lymphotoxin) snyergizes with IFN-γ for induction of tumoricidal and microbicidal properties in mouse macrophages.[77]

c. Colony-Stimulating Factors (CSF)

- GM-CSF can induce tumoricidal activities in both granylocytes and macrophages as well as microbicidal activity and enhanced production of reactive oxygen intermediates in murine macrophages and human monocytes.[80,81]

- Colony-stimulating factor-1 (CSF-1) induces tumoricidal activity and molecules with the potential for inducing activation of macrophages.[82,93]

C. Markers of the Stages of Macrophage Activation

Most models of the stages of activation of macrophages emphasize the progressive acquisition of the ability to perform complex functions.[16,21,23] These models also tend to emphasize the difference between resident tissue macrophages and inflammatory macrophages either stimulated with various potential activators or not. How the peripheral blood monocyte fits into this scheme has not been addressed to a large extent. Recent work suggests that murine monocytes closely resemble macrophages taken from sties of inflammation, and are distinct from resident peritoneal macrophages.[16,84] Many studies in the murine system have emphasized the importance of the inflammatory macrophage in regard to responsiveness to signals for induction of tumoricidal activity.[23-31] Others demonstrate clearly that resident macrophages are capable of being activated to both the microbicidal[85] as well as tumoricidal states.[85]

Macrophages taken from sites of inflammation are readily distinguished from resident macrophages based on several functional characteristics including increased phagocytosis and pinocytosis, increased secretion of lysosomal enzymes, increased expression of several receptors, decreased expression of certain ectoenzymes, and many others.[16,17] These types of changes in functional and biochemical markers are also associated with primed and activated macrophage populations.[23] Primed macrophages can be defined as macrophages that are prepared to complete a complex function such as microbicidal or tumoricidal functions, but require an additional signal. Activated macrophages are capable of mediating such functions.[16] Since many of the functional and biochemical changes seen in inflammatory macrophages are also seen in primed and activated cells many of these parameters are not useful markers for the tumoricidal or microbicidal phenotype.

At present no marker exists for the activated phenotype which is absolutely predictive of microbicidal or tumoricidal activity. The following, however, represent recent descriptions of markers that are associated with the activated phenotype:

- Selective augmentation of S-adenosylmethionine in murine macrophages after activation by IFN-γ, as well as a decreased level of RNA synthesis and alteration in ribosomal RNA accumulation.[87,88]

- Activation and translocation of protein kinase C during activation.[89]

- Elevated selective recognition and capture of tumor target cells by activated macrophages and the potential involvement of lymphocyte function associated 1 (LFA-1) antigen in this process.[90]

- Induction of specific receptors for IL-2 on both human peripheral blood monocytes and alveolar macrophages.[91]

- Protein changes associated with the stages of activation have been identified by two different groups. Specific changes in two proteins are associated with the acquisition of tumoricidal activity (p47b and p71/73).[91] Similarly, a 120 kDa protein has been identified to be expressed upon activation.[92]

Mechanism of Macrophage Mediated Tumoricidal Activity In Vitro

Tumor cell populations are heterogeneous with regard to many phenotypic characteristics; however, their sensitivity to lysis by appropriately activated macrophages appears characteristic of most neoplastic cells. The mechanism by which activated macrophages selectively recognize and destroy neoplastic target cells is poorly understood. Early reports on the mechanism of macrophage-mediated tumoricidal activity utilizing activated murine macrophages and a range of syngeneric, allogeneic or xenogeneic tumor types characterized the cytotoxic mechanism as requiring direct cell to cell contact.[93] Recently, these observations have been extended to human peripheral blood monocytes activated in vitro with liposome-encapsulated MDP.[94,95] These observations of direct cell-to-cell contact required for macrophage-mediated cytolycis with a range of tumor cell types, suggest the presence of common recognition molecule(s). However, the biochemical structures involved in macrophage cellular and biochemical binding -tumor cell and cytolysis have not yet been conclusively identified.

The cellular and biochemical events involved in macrophage binding and lysis of tumor cells are poorly understood. Macrophage-mediated tumor cell cytolysis appears independent of various tumor cell properties such as altered histocompatibility antigens, tumor-associated antigens, drug sensitivities, cell cycle kinetics, and invasiveness or metastatic potential.[9] Although the exact mechanism by which macrophages recognize and destroy neoplastic cells has not yet been elucidated, the studies mentioned above suggest that the tumor cell properties recognized by activated macrophages are intimately associated with neoplastic transformation. Furthermore, in contrast to the rapid emergence of tumor cell variants that are resistant to chemotherapy, radiation, hyperthermia, or T lymphocyte or natural killer (NK) cell-mediated cytolysis, the generation of variants resistant to macrophage-mediated recognition and/or cytolysis appears an extremely rare event.[9]

The apparent rarity of tumor cell resistance to killing by activated macrophages observed in studies of cultured cells appears to be equally valid for tumor cell populations in vivo. For example, tumor cells harvested from metastases have been found to be fully susceptible to lysis by activated macrophages indicating that progressive growth of the metastatic lesions results from an excessive tumor burden rather than the emergence of tumor cell variants resistant to killing by macrophages. Collectively, these data suggest that understanding the macrophage activation process and designing approaches for the activation of tumoricidal macrophages in vivo could provide a strategy for adjunct therapy of metastatic disease, i.e., one this is capable of circumventing the problem of tumor cell heterogenicity.

References

1. Poste, G. and Fidler, I.J., The pathogenesis of cancer metastasis. Nature (London), 283, 139, 1980.

2. Poste, G. and Greig, R., On the genesis and regulation of cellular heterogeneity in malignant tumors, Invasion and Metastasis, 2, 137, 1982.

3. Kerbel, R.S., Implications of immunological heterogeneity of tumors, Nature (London), 280, 358, 1979.

4. Owens, A.H., Coffey, D.S., and Baylin, S.B., Eds., in Tumor Cell Heterogeneity: Origins and Implications. Academic Press, New York, 1982.

5. Eccles, S.A., Macrophages and cancer, in Immunological Aspects of Cancer, Castro, J.E., Ed., University Park Press, Baltimore, 1978, 123.

6. Chirigos, M. A., Mitchell, M., Mastrangelo, M.J., and Krim, M., Eds., Modulation of Cellular Immunity in Cancer by Immune Modifiers, Raven Press, New York, 1981.

7. Witz, I.P. and Hanna, M.G., Jr., Eds., In situ expression of tumor ummunicty, Contemp. Top. Immunobiol., 10, 1, 1980.

8. Mantovani, A., Giavazzi, R., Polentaniti, N., Spreafico, F., and Garanttini, S., Divergent effects of macrophage toxins on growth primary tumors and lung metastases in mice, Int. J. Cancer, 25, 617, 1980.

9. Fidler, I.J., and Poste, G., Macrophage-mediated destruction of malignant tumor cells and new sstrategies for the therapy of metabolic disease, Springer Semin. Immunopathol., 5. 161, 1982.

10. Poste, G., and Kirsh, R. (1982). Liposome-encapsulated macrophage activation agents and active non-specific immunotherapy of neoplastic disease. In "Cell Function and Differentiation" G. A. Koyunoglou, A.E. Evangelopoulos, J. Georgatsos, G. Palarologos, A. Trakatellis, and C. P. Tsiganos, eds., 309-319, Alan R. Liss, New York.

11. Trope, C., Different susceptibilities of tumor cell subpopulations to cytotoxic agents, in Design of Models for Testing Cancer Chemotherapeutic Agents, Fidler, I.J. and White, R. J., Eds., Van Nostrand, New York, 1982,64.

12. Colotta, R., Peri, G., Villà, A. and Mantovani Rapid killing of actinoiycen D treated tumor cells by human mononuclear cells. J. Immunol., 132, 936, 1984.

13. North, R.J., The concept of activated macrophages, J. Ummunol., 121, 806, 1978.

14. Hanna, N., Role of natural killer cells in control of cancer metastasis, Cancer Metastasis Rev., 1, 45, 1982.

15. Alexander, P. and Evans, R., Endotoxin and double stranded RNA render macrophages cytotoxic, Nature (London). 232, 76, 1971.

16. Adams, D.O. and hamilton, T.A., The cell biology of macrophage activation, Annu. Rev. Immunol., 2, 283, 1984.

17. Nathan, C.F. and Cohn, Z.A., Cellular components of inflammation: monocytes and macrophages, in Textbook of Pheumatology, Kelley, W., Harris, E., Ruddy, S., and Sledge, Cl., Eds., W. B. Saunders, Philadelphia, 1985, 144.

18. Takemura, R. and Werb, Z., Secretory products of macrophages and their physiological functions, Am. J. Physiol. 246, Cl. 1984.

19. Gordon, S. and hirsch, S., Differntiation antigens and macrophage heterogencity, in Macrophages and Natural Killer Cells, Norman, S. and Sorkin, E., Eds., Plenum Press, New Yoark, 1982, 391.

20. Walker, W.S., Functional heterogencity of macrophages, in Immunobiology of the Macrophage, Nelson, D., Ed., Academic Press, Orlando, Fla. 1986, 91.

21. Adams, D.O. and Marino, P.A., Activation of mononuclear phagocytes for destruction of tumor cells as a model for sutdy of macrophage development, in Contemporary Topics in Hematology-Onocology, Gordon. A., Stiber, R., and LoBue, J., Eds., Plenum Press, New York, 1984, 69.

22. Fischer, D.G., Hubbard, W.J., and Koren, H.S., Tumor cell killing by freshly isolated peripheral blood monocytes, Cell. Immunol., 58, 426, 1981.

23. Johnson, W.J., Marino, P.A., Schrieber, R.D., and Adams, D.O., Sequential activation of murine mononuclear phagocytes for tumor cyutolysis: differential expresion of markers by macrohages in the several stages of development, J. Ummunol., 131, 1038, 1983.

24. Johnson, W.J., DiMartino, M.J., and Hanna, J., Macrophage activation in rat models of inflammation and athritis: determination of markers of stages of activation, Cell. Immunol., 103,54, 1986.

25. Pace, J.L., Varesio, J., Russell, S.W., and Blasi, E., The strain of mouse and assay conditions influence whether nuLFN-γ primes or activates macrophages for tumor killing, J. Leukocyte Biol., 37, 475, 1985.

26. Murray, H.W., Spitainy, G.L., and Nathan, C.F., Activation of mouse peritoneal macrophages in vitro and in vivo by interferon-γ, J. Immunol., 138, 491, 1987.

27. Black, C.M., Catterall, J.R., and Remington, J.S., In vivo and in vitro activation of alveolar macrophages by recombinant interferon-γ, J. Immunol., 138, 491, 1987.

28. Lehrer, R.I., Ferrari, L.G., Patterson-Delafield, J., and Sorrell, T., Fungicidal activity of rabbit alveolar and peritoneal macrophages against Candida albicans, Infect. Immun., 28, 1001, 1980.

29. Collins, F.M., Niederbuhl, C.J., and Campbell, S.G., Bactericidal activity of alveolar and peritoneal macrophages exposed in vitro to three strains of Pasteurella multocida, Immun., 39, 779, 1983.

30. Schaffner, A., Douglas, H., Braude, A>I>, and Davis, C.E., Killing of Aspergillus spores depends on the abatomical source of the macrophage, Infect. Immun. 42, 1109, 1983.

31. Black, C.M., Beaman, B.L., Donovan, R.M., and Goldstein, E., Intracellular acid phosphatase content and ability of different macrophage populations to kill Nocardia asteroides, Infect. Immun., 47, 375, 1985.

32. Cleveland, R.P., Meltzer, M.S. and Zbar, B., Tumor cytotoxicity in vitro by macrophagtes from mice infected with Mycobacterium bovis strain BCG, J. Natl. Cancer Inst., 52, 1887, 1974.

33. Droller, M.J. and Remington, J.S., A role for the macrophage in the in vivo and invitro resistance to murine bladder tumor cell growth, Cancer Res., 35, 49, 1975.

34. Kaplan, A.M., Morahan, P.S., and Regilson, W., Induction of macrophage mediated tumor cell cytotoxicity by by pyran copolymer, J. Natl. Cancer Inst., 52, 1919, 1974.

35. Scott, M.T., In vitro cortisone sensitivity of nonspecific antitumor activity of Corynebacterium parvum activated mouse peritoneal macrophages. J. Natl. Cancer Inst., 54, 789, 1975.

36. Sone, S. and Fidler, I.J., In situ activation of tumoricidal properties in rat alveolar macrophages and rejection of experimental lung metastases by intravenous injections of Nocardia rubra cell wall skeleton, Cancer Immunol. Immunother., 12, 203, 1982.

37. Chihara, G., Maeda, Y>Y>, Hamuro, J., Sasaki, T., and Fukuoka, F., Inhibition of mouse sarcoma 180 by the polysaccharide from Lentinus erodes, Nature (London), 222, 687. 1969.

38. Chirigos, M.A., Ed., Immune modulation and control of neoplasia by adjuvant therapy, Progress in Cancer Research and Therapy, Vol. 7, Raven Press, New York, 1978.

39. Fenichel, R. and Chirigos, M.A., Eds., Immune Modulating Agents and Their Mechanisms, Vol. 25, Marcel Dekker, New york, 1985.

40. Halper, B., Corynebacterium parvum: Applications in Experimental and Clinical Onocology, Plenum Press, New Yor, 1975.

41. Hersh, E., Chirigos, M.A., and Mastrangelo, A., Eds., Augmenting agents in cancer therapy, Progress in Cancer Research and Therapy, Vol. 16, Raven Press, New York, 1981.

42. Mihich, E., Ed., Immunological Approaches to Cancer Therapeutics, John Wiley & Sons, New York, 1982.

43. Lederer, E., Synthetic immunostimulants derived from the bacterial cell wall, J. Med. Chem., 23, 819, 1980.

44. Chedid, L. and Audibert, F., New approaches for control of infections using synthetic or semi-synthetic constructs containing MDP, Springer Semin. Immunopathol., 8, 401, 1985.

45. Parant, M., Parant, F., Chedid, L., Yapo, A., J.F., and Lederer, L., Fate of the synthetic immunoadjuvant muramyl depeptide (^{14}C-labelled) in the mouse, Int. J. Immunopharmacol., 1, 35, 1979.

46. Chedid, L., Parant, M., Parant, F., Lefrancier, P., Choay, J., and Lederer, E., Enchancement of nonspecific immunity to Klebsiella pneumoniae infection by a synthetic immunoadjuvant (N-acetyl-muramyl-L-alanyl-D-isoglutamine) and several analogs, proc. Natl. Acad. Sci. U.S.A., 74, 2089, 1977.

47. Dietrich, F.M., Sackmann, W., Zak, O., and Dukor, P., Synthetic muramyl dipeptide immunostimulants: protective effects and increased efficacy antibiotics in experimental bacterial and fungal infections in mice, in Current Chemotherapy and Infectious Disease, Vol. 2, Nelson, J.D. and Grassi, C., Ed., Am. Soc. Microbiol., Washington, D.C., 1980, 1730.

48. Fraser-Smith, E.B. and Matthews, T.R., Protective effects of muramyl dipeptide analogs against infections of Pseudomonas aeruginosa or Candida albicans in mice, Infect. Immunol., 34, 676, 1981.

49. Humphries, R.C., Heniki, P.R., Ferraresi, R.W., and Krahenbuhl, J.L., Effects of treatment with muramyl dipeptide and certain of its analogs on resistance to Listeria monocytogenes in mice, Infect. Immunol., 30, 462, 1980.

50. Matthers, T.R., and Fraser-Smith, E.B., Protective effect of muramyl dipeptide and analogs against Pseudomonas aeruginosa and Candida albicans infections in mice, Current Chemotherapy of Infectious Disease, Vol. 2, Am. Soc. Microbiol., Washington, D.C., 1980, 1734.

51. Onozuka, K., Saito-Taki, T., and Nakano, M., Effect of muramyl dipeptide analog on Salmonella enteritidis infection in beige mice with Chediak-Higashi syndrome, Microbiol. Immunol., 28, 1211, 1984.

52. Parant, M., Parant, F., and Chedid, L., Enhancement of the neonate's nonspecific immunity to Klebsiella infection by muramyl dipeptide, a synthetic immunoadjuvant, Proc. Natl. Acad. Sci. U.S.A., 75, 3395, 1978.

53. Kierszenbaum, F. and Ferraresi, R.W., Enhancement of host resistance against Trypanosoma cruzi infection by the immunoregulatory agent muramyl dipeptide, Infect. Immunol., 25, 273, 1979.

54. Krahenbuhl, J.L., and Humphries, R.C., Effects of treatment with muramyl dipeptide on resistance to Mycobacterium leprae and Mycobacterium marinum infection in mice, Immunopharmacology, 5, 329, 1983.

55. Dietrich, F.M., Lukas, B., and Schmidt-Rupin, K.H., MTP-PE (synthetic muramyl peptide): prophylactic and therapeutic effects in experimental viral infections, Commun. 13th Int. Cong. Chemother., Vienna, August 28, 1983.

56. Koff, W.C., Fidler, I.J., Showalter, S.D., Chakrabarty, M.K., Hampar, B., Ceccorulli, L.M., and Kleinerman, E.S., Human monocytes activated by immunomodulators in liposomes lyse herpesvirus-infected but not normal cells, Science, 224, 1007, 1984.

57. Poste, G., Kirsh, R., and Fidler, I.J., Cell surface receptors for lymphokines, Cell. Immunol. 44, 71, 1979.

58. Andreesen, R., Bross, K.J., Osterholz, J., and Emmrich, F., Human macrophage maturation and heterogeneity: analysis with a newly generated set of monoclonal antibodies to differentiation antigens, Blood, 67(5), 1257, 1986.

59. Shen, H.H., Talle, M.A., Goldstein, G., and Chess, L., Functional subsets of human monocytes defined by monoclonal antibodies. A distinct subset of monocytes contains the cells capable of inducing the autologous mixed lymphocytes culture, J., Immunol., 130, 698, 1983.

60. Zembala, M., Uracz, W., Ruggiero, I., Mytar, B., and Pryjma, J., Isolation and functional characterization of FcR+ and FcR- human monocyte subsets, J. Immunol., 133, 1293, 1984.

61. Schreiber, R.D., Pace, J.L., Russell, S.W., Altman, A., and Katz, D.H., Macrophage-activating factor produced by a T cell hybridoma: physiochemical and biosynthetic resemblance of γ-intergeron. J. Immunol., 131, 826, 1983.

62. Le, L., Prensky, W., Yip, Y.K., Chang, Z., Hoffman, T., Stevenson, H.C., Balazs, I., Sadlik, J.R., and Vilcek, J., Activation of human monocyte cytotoxicity by natural and recombinant immune interferon, J. Immunol., 131, 2821, 1984.

63. Schreiber, R.D., Pace, J.L., Russell, S.W., Altman, A., and Katz, D.H. Macrophage activating factor produced by a T cell hybridoma: physiochemical and biosynthetic resemblance to γ-interferon, I. Immunol., 131, 826, 1083.

64. Roberts, W.K. and Vasil, A., Evidence for the identity of murine gamma interferon and macrophage activating factor, J. Interferon Res., 2, 519, 1982.

65. Kleinerman, E.D., Zicht, R., Sarin, P.S., Gallo, R.C., and Fidler, J. Constitutive production and release of a lymphokine with macrophage activating factor activity distinct from γ-interferon by a human T cell leikemia virus-positive cell line, Cancer Res., rr, 4470, 1984.

66. Meltzer, M.S., Gilbreath, M., Nacy, C.A., and Schreiber, R.D., Macrophage activation factor from EL-4 cells distinct from murine gamma interferon (IFN), Fed. Proc. Fed. Am. Soc. Biol., 44, (Abstr.), 1697, 1985.

67. Crawford, R., Hoover, D., Finbloom, D., Gilbrath, M., Nacy, C., and Meltzer, M., Physicochemical properties of a human lymphokine (LK) distinct from gamma interferon (IFN) that activates monocytes to kill Leishmania donovani, Fed. proc. Fed. Am. Soc. Exp. Bio., 44 (Abstr.) 1697, 1985.

68. Lee, J.C., Rebar, L., Young, P., Ruscetti, R.W., Hanna, N., and Poste, G., Identification and characterization of a human T cell line-derived lymphokine with MAF-like activity distinct from interferon-γ, J. Immunol., 136, 1322, 1986.

69. Lee, J.C., Badger, A.M., Johnson, W.J., Sung, C.P., and Horan, P., Human non-gamma interferon macrophage activating actor (MT-2/MAF) elicits multiple and cross-species effects on macrophage activationa, Abstr. Sixth Int. Congr. Immunol., 1986.

70. Onozaki, K., Matsushima, K., Kleinerman, E.D., Saito, T., and Oppenheim, J.J., Role of IL-1 in promoting human monocyte-mediated tumor cytotocivity, J. Ummunol., 135, 314, 1985.

71. Philip, R. and Epstein, L.B., TNF as immunomodulator and mediator of monocyte cytotoxicity induced by itself, IFN-γ and IL-1, Nature (London) 323, 86, 1986.

72. Billiam, A., Damme, J.V., Opdenakker, G., Fibbe, W.E., Falkenburg, J.H.F., and Content, J., IL-1 as a cytokine inducer, Immunobiology, 172, 323, 1986.

73. Malkovsky, M., Loveland, B., North, M., Asherton, G.L., Gao, L., Ward, P., and Fiers, W., Recombinant IL-2 directly augments the cytotocivity of human monocytes, Nature (London), 325, 262, 1987.

74. Nathan, C.F>, Prendergast, T.J., Wiebe, M.E., Stanley, E.R., Platzer, E., Remold, H.B., Welte, K., Rubin, B.Y., and Murray, H.W., Activation of human macrophages. Comparison of their cytokines with interferon-γ, J. Exp. Med., 160, 600, 1984.

75. Djeu, J.Y. and Blanchared, D.K., Gnereation of autoreactive killer cells against human monocytes/macrophages by IL-2, Lymphokine Res., 6 (Abstr.)., 1705, 1987.

76. Meltzer, M.S., Crawford, R.M., Finbloom, D.S., Phara, J., and Paul, W.E., BSF-1: a macrophage activation factor, Lymphokine Res., 6(Abstr.). 1719, 1987.

77. Esparza, I., Mannel, D., Ruppek, A., Falk, W., and Krammer, P.H., IFN-γ and lymphotoxin or TNF synergize to activate macrophages for tumoricidal and schistosomulicidal function, Lymphokine Res., 6(Abstr.), 1715, 1987.

78. Hoffman, M.K., The effects of TNF on the production of IL-1 by macrophages, Lymphokine Res., 5, 255, 1986.

79. Munker, R., Gasson, J., Ogawa, M., and Koeffer, H.P., Recombinant human TNF induces production of GM-CSF, Nature (London), 323, 79, 1986.

80. Medcalf, D., The granulocyte-macrophage colony stimulating factors, Science, 229, 16, 1986.

81. Weiser, W.Y., Van Niel, A., Clark, S.C., David, J.R., and Remold, H.G., Recombinant human GM-CSF activates intracellular killing of Leishmania donavani by human monocytes, Lymphokine Res., 6 (Abstr.), 1706, 1987.

82. Ralph, P., Warren, M.K., Nakoinz, I., Lee, M.T., Brindley, L., Sampson-Johannes, A., Kawaski, E.S., Ladner, M.B., Stickler, J.E., Boosman, A., Csejtey, J., and White, T.J., Biological properties and molecular biology of the human macriphage growth factor, CSF-1, Immunobiology, 172, 194, 1986.

83. Warren, M.K. and Ralph, P., CSF-1 stimulates human monocyte production of interferon, TNF and colony stimulating activity, J. Immunol., 137, 2281, 1986.

84. Beller, D.I. and Ho, K., Regulation of macrophage population. V. Evaluation of the control of macrophage Ia expression in vitro, J. Immunol., 129, 971, 1982.

85. Nacy, C.A., James, S.L., Osler, C.N., and Meltzer, M.S., Activation of macrophages to kill Rickettsiae and Leishmania: disassociation of intracellular microbicidal activities and extracellular diestruction of neoplastic cells and helminths, Contemp. Top . Immunobiol., 14, 147, 1984.

86. Pace, J.L. and Russell, S.W., Activation of mouse macrophages for tumor cell killing. I. Quantitative analysis of interactions between lymphokine and LPS, J. Immunol., 126, 1863, 1981.

87. Bonvini, E., Hoffman, T., Herberman, R.B., and Varesio, L., Selective augmentation by recombinant interferon-γ of the intracellular content of S-adenosylmethionine in murine macrophages, J. Immunol., 136, 2596, 1986.

88. Varesio, L., Imbalanced accumulation of ribosomal RNA in macrophages activated in vivo or in vitro to a cytolytic stage, J. Immunol., 134, 1262, 1985.

89. Hamilton, T.A., Becton, D.B., Somers, S.D., Gray, P.W., and Adams, D.O., Interferon gamma modulates protein kinase C activity in murine peritoneal macrophages, J. Biol. Chem., 260, 1378, 1985.

90. Strassmann, G., Springer, T.A., Somers, S.D., and Adams, D.O., Mechanisms of tumor cell capture by activated macrophages: evidence for involvement of lymphocyte function-associated (LFA)-1 antigen, J. Immunol., 136, 4328, 1986.

91. MacKay, R.J. and Russell, S.W., Protein changes associated with stages of activation of mouse macrophages for tumor cell killing, J. Immunol., 137, 1391, 1986.

MICROCARRIER-FACILITATED CULTURES FOR FIBROBLASTIC AND EPITHELIAL CELLS

Menemse Kiremitci*, Ismet Gurhan**, Erhan Piskin*

*Hacettepe University, Chemical Engineering Department
Beytepe, Ankara, Turkey
**Foot and Mouth Disease Institute, Ankara, Turkey

INTRODUCTION

Microcarrier-facilitated cell culture is an established technique for the growth of anchorage-dependent mammalian cells on an industrial scale[1,2]. In this approach, cells are propagated on the surface of small solid spheres, kept in suspension in the culture medium by slow stirring. Microcarriers must have the desired surface properties (wetability, surface charge, etc.) in order to support cell attachment and growth. They are non-toxic, non-rigid, transparent particles 100-220μ in diameter. Therefore, microcarrier cultures have the highest surface-to-volume ratio (S/V=150) of any other cell culture system. Furthermore, less labor, materials and media are required to produce a given quantity of cells in microcarrier cultures when compared to other systems. Microcarriers also have the best scale-up potential of any cell culture system3. Today microcarriers are used routinely on a large scale for the production of many important biologicals (viral vaccines, tissue type plasminogen activator and β-interferon). Institute Merieux produced inactivated polio and rabies vaccines in Vero cells using 1000 liter microcarrier cultures2.

Recently, progress in microcarrier culturing may lead to the development of several kinds of new supports, generally based on the natural or synthetic polymers, such as polystyrene, polyacrylamide, gelatin, cellulose and dextran4-6. Moreover, the use of synthetic peptides as a microcarrier surface instead of collagen or fibronectin has been studied7. Currently, a number of different microcarriers are commercially available. However, the properties most critical for a successful microcarrier are as yet not completely fulfilled.

In this study, the attachment and growth behavior of morphologically different anchorage-dependent cells, BHK-21 and IBRS-2, were investigated in microcarrier cultures. The commercially available microcarriers which are based on natural polymers, dextran and gelatin namely, (Cytodex-l and CultiSphere-G) were used comparatively with the synthetic polymeric based microcarrier, PHEMA, synthesized by Kiremitci, et.al[8,9]. Cytodex-l has been widely used in cell culture applications. Macroporous gelatin matrix, CultiSphere-G has been recently developed by Nilsson. Here, the relationship between morphological differences of cells and microcarrier properties are discussed.

MATERIALS AND METHODS

The Baby Hamster Kidney cells (BHK 21, passage 29) and the Swine Kidney cells (IBRS-2, passage 77) were used as model cell lines in order to investigate the effects of different morphology used on the cells characteristics (i.e., attachment, spreading and proliferation). BHK cells which have fibroblast-like morphology and IBRS-2 cells which have epithelial morphology grow only in monolayer culture and have a broad viral susceptibility.

The cross-linked PHEMA (polyhydroxyethylmethacrylate) microcarriers which have suitable surface and bulk characteristics were prepared by suspension copolymerization technique and were used in our previous cell culture studies successfully[7-9]. The commercial microcarriers, Cytodex-1 (DEAE-dextran beads) and CultiSphere-G (macroporous, gelatin beads) were supplied as a gift by Pharmacia (Sweden) and Percell Biolytica (Sweden), respectively.

In the cell culture studies the culture medium was Dulbecco's Modified of Eagle's MEM (DME, Gibco, USA). This medium was supplemented with 10% fetal calf serum (FCS, DIFC0, USA) and 100 μg/ml gentamycin (Gibco, USA).

Three major steps were involved in the preparation of microcarriers for cultivation: sterilization, swelling and elimination of buffer. The dry microcarriers were sterilized with gamma irradiation (^{60}Co. Gammacell 220, 2.5 MRads). They were swollen overnight at 37°C in PBS. All the microcarriers were rinsed with the fresh culture media (prior to cultivation).

The culture studies were performed in stationary and submerged culture conditions which are given in Table 1.

The cells were harvested from the microcarriers by incubation at 37°C for 15 min with trypsin solution; cells were centrifugted at 800 rpm for 5 min. The cells were stained with trypan blue and the cell concentration was determined by using hemacytometer. The

Table 1. Culture Conditions

PARAMETERS	Stationary Culture	Submerged Culture
Equipment	Bacteriological PS Petri Dishes(NUNC, 35x10 mm)	Spinner Flask (100ml. Bellco, Model 3002)
Media	DME + 10% FCS	DME + 10% FCS
Inoculation Density	$3x10^5$ cells/ml	6×10^5 cells/ml
Microcarrier Concentration	5g/l Cytodex-1 5g/l CultiSphere-G 10g/l PHEMA	5g/l Cytodex-1 5g/l CultiSphere-G 10g/l PHEMA
Stirring Speed	--	40 rpm
Incubation Media	CO_2 etuve	Etuve
Attachment Period	2 hrs4 hrs.	
Growth Period	4 days	4 days
pH	7.2-7.6	7.2-7.6
Temperature	37±0.5°C	37±0.5°C

attachment and spreading characteristics of the cells were followed microscopically by taking photographs (Olympus Light Microscope, Japan). In addition, the cell attachment and growth were followed by taking samples from the culture media at selected intervals and by counting the cells after trypsinization process. The possibility of bead to bead transfer and reuse of microcarriers was also investigated in stationary culture conditions.

RESULTS AND DISCUSSION

The attachment and growth curves obtained in this study were evaluated by surface saturation type mathematical expressions given below:

$$\frac{da}{dt} = kX \left(1 - \frac{a}{a_{max}}\right) \qquad (1)$$

$$\frac{dX}{dt} = \mu X \left(1 - \frac{X}{X_{max}}\right) \qquad (2)$$

Where X, t and k were free cell concentration, time and attachment rate constant, respectively, μ is the initial specific growth rate and X_{max} is the maximum attainable cell concentration.

$a=X-X_0 X_0$ is the inoculation density.

-4-

Equation (l) and (2) describe the cell attachment and growth kinetics on microcarriers in batch culture conditions. It is possible to calculate the k and μ values which show the rates of attachment and growth processes, respectively from these equations. The details of these mathematical models were given elsewhere[11].

Table 2. Attachment and Growth Rate Constants

Microcarrier	BHK-21 Cells k (min^{-1})	μ (day^{-1})	IBRS-2 Cells k (min^{-1})	μ (day^{1}
CYTODEX-1 0.045		1.48	0.038	0.72
CULTISPHERE-G	0.043	1.15	0.033	0.96
PHEMA	0.037	1.50	0.020	0.95

As seen from these values, for BHK cells, attachment and growth rates are higher than the IBRS-2 cells on all types of microcarriers. However, attachment to the Cytodex-l microcarriers is more rapid than the others. Although these initial values, the maximum growth yield was obtained in case of CultiSphere-G microcarriers because of their high surface area due to its macroporous structure.

According to the comparative results of this study, the main advantages and disadvantages of microcarriers can be summarized in Table 3.

Table 3. Comparative Properties of Microcarriers

	MICROCARRIER		
	PHEMA	CYTODEX-1	CULTISPHERE-G
Diameter (m)	131-210	170-270	120-200
Density (g/mil)	1.03	1.04	1.20
Swelling Factor	18	13-15	+
Mechanical Stability	--	--	+
Chemical Stability	--	--	+
Reusability	2,3 times	--	+
Bead to bead transfer	+	+	+
Efficiency of cell recovery	Low	High	Low
Microscopic Observations	Easy	Difficult	Easy
Growth Yield	High	Excellent	High

ACKNOWLEDGEMENT

The authors are grateful to the Percell Biolytica for their CultiSphere-G samples.

REFERENCES

1. A. L. van Wezel, Nature 216:64, 1967.
2. B. J. Montagnon, Vincent, J. C. Falquet, and B. Fangel, Develop. Biol. Stand. 55:37, 1984.
3. S. Reuveny, Adv. Biotecnol. Process. 2,1, 1983.
4. M. Hirtenstein, J. Clark, G. Lindgren, and P. Vretblad, Develop. Biol. Stand. 46, 109, 1980.
5. S. Reuveny, L. Silberstein, A. Shahar, E. Freeman, A. Mizrahi, In Vitro 18, 92, 1982.
6. V. Nielsen and A. Johansson, Develop. Biol. Stand. 46, 131, 1980.
7. M. D. Pierschbacker and E. Rouslahti, Nature 309:30, 1984.
8. M. Kiremitci, P.h.D. Thesis, Hacettepe University, Ankara, 1988.
9. M. Kiremitci, I. Gurhan, E. Piskin, Growth Characteristics and FMD Virus Susceptibilities of BHK-21 Cells Proliferated on PHEMA microcarriers, Proc. of 9th ESACT Meeting, 1980.
10. M. Kiremitci, M. Ozilgen, E. Piskin, Cell Culturing Characteristics of Newly Developed PHEMA Microcarriers, Biomat. Artif. Cells and ARtif. Organs (in press).
11. M. Kiremitci, M. Ozilgen, E. Piskin, Attachment and Growth Kinetics of Anchorage-Dependent BHK Cells on Microcarriers, Enzyme Microb. Technol. (in press)

STRATEGIES FOR IMMORTALISING HUMAN PRIMARY CELLS AND THE STEPS THAT LEAD TO IMMORTALITY

L.V. Mayne

Biological Sciences Division,
Sussex Centre for Medical Research, University of Sussex
Falmer, Brighton BN1 9QG, England

INTRODUCTION

Continuously growing human cell lines are a valuable resource for both industry and academia. Unfortunately, cell cultures obtained from human biopsies have only a limited *in vitro* lifespan and virtually never spontaneously give rise to continuously growing cell lines. This is in sharp contrast to the situation with rodent cells where immortal cells arise at a relatively high frequency (1). Not only are human cells highly refractory to spontaneous immortalisation, but the methods that have been successful in inducing immortality in rodent cells give rise to continuously growing human cell lines at a relatively low frequency (2). These results indicate that the mechanism leading to immortality in rodent cells is quite different from that in human cells and thus, strategies for immortalising human cells cannot simply be based upon those for rodents.

This paper will review the uses and properties of human cell lines and will discuss the various methods available for inducing immortality in human cells. The various approaches that have been used whether successful or unsuccessful, have given many insights into the cellular events that lead to immortality. This paper will also review the current level of understanding of these events.

APPLICATIONS OF HUMAN CELL LINES

The specifications we demand from our cell lines are very much determined by the way we intend to use them. Currently, the applications of cell lines can be broadly divided into two categories. The first of these is the use of cell lines as a substrate for the manufacture of biologically relevant compounds. The most common uses of cell lines for production are in the manufacture of:

1. Vaccines
2. Naturally occurring proteins or other cellular components
3. Recombinant proteins

One of the first commercial uses of cell lines for production of biologically relevant compounds was in the manufacture of vaccines, and this still remains a major application (3). More recently, cell lines have been used for the isolation of naturally occurring compounds such as cell surface receptors and monoclonal antibodies. The use of cell lines offers a number of advantages over using fresh tissues. In particular, they are

Pharmaceutical Applications of Cell and Tissue Culture to Drug Transport
Edited by G. Wilson *et al.*, Plenum Press, New York, 1991

readily accessible and the ability to culture a single cell type can simplify the isolation procedure.

The advent of recombinant DNA technology and the ability to express foreign genes in virtually any cell type, has allowed the large scale production of proteins from isolated and genetically engineered genes. Human growth hormone and tissue plasminogen activator (tPA) are examples of proteins that are produced from mammalian cells carrying genetically engineered genes for these proteins (3).

The other major application of cell lines is as a model system to study events at the cellular level and to study the interactions between cells and biologically relevant compounds. In many cases, cell lines are used to mimic *in vivo* interactions. Cell lines are easy to manipulate, conditions are very reproducible, external factors can be tightly controlled, and the material for study is readily accessible. Some of the uses of cell lines as a model system are listed below.

Cell lines are used to:

1. Study the specificity, uptake and transport of drugs
2. Screen drugs for potentially toxic effects
3. Study metabolic pathways
4. As a tool in the study of the biochemical basis of human disease.

Rodent cell lines, lower eukaryotes such as yeast, and bacteria have all played a major role both as substrates for the production of biologically interesting compounds and as model systems for studying various metabolic pathways. Why then should we pursue the use of human lines? Three main advantages are associated with the use of human cell lines. Products produced using human cell lines are more likely to be authentic human proteins, there is less likely to be contamination with non-human proteins and as a model system, human cells are more likely to mimic the human *in vivo* situation.

The biological activity of many proteins depends not only on their amino acid sequence but on a number of post-translational modifications. The enzymes required to mediate these modifications are often species specific. In some cases post-translational modifications are essential to produce a biologically active compound. While in others, post-translational modifications alter the biological activity and influence such factors as stability and half life, solubility and inter-or intra-cellular transport.

Another problem associated with the use of non-human cell lines for production of compounds for clinical use is the cross contamination of the product with non-human proteins. This is particularly relevant where there are repeated applications of the product as foreign proteins will elicit an immune response and limit its usefulness. The use of a human cell line can help to reduce the risk of eliciting an unwanted immune response.

SENESCENCE AND IMMORTALITY

Primary cell strains have a well defined and predictable life-span *in vitro*. After a fixed number of cell doublings the cells become quiescent and cell division ceases. This final phase of the cells life cycle is referred to as senescence and the cells are said to be in "crisis". This phenomenon was first described by Hayflick (4) for fibroblasts but has been observed now for all types of cells. Senescing cells eventually die, and as cell division has ceased, the culture itself disappears.

Given a research problem that requires the use of a continuously growing cell line, how does one go about making such a cell line? Ideally, the method used to generate immortal cell lines should be efficient and involve a minimum of work with a high success rate. It should be generally applicable to a wide variety of cell types, the resultant cell lines should retain the phenotype of the original cells, and this phenotype should be stable over many generations. The methods that have been used for generating immortal human cell lines are listed in Table 1.

Many attempts have been made to establish continuously growing cells from samples of *in vivo* tumours. In contrast to rodents, the majority of human tumours do not give rise to immortal *in vitro* cell lines (5). It is not clear whether these cells have the potential for continuous growth or whether their limited *in vitro* growth potential is a result of the failure to supply the necessary growth factors to maintain them. The use of tumour tissue has the added disadvantage that the number, type and general availability of tumours is limited.

TABLE 1. SOURCES OF CONTINUOUSLY GROWING HUMAN CELL LINES

1. Outgrowth of *in vivo* tumours
2. Treatment of primary cultures with mutagens
3. Introduction of oncogenes into primary cultures
    a. Viral encoded genes
    b. Oncogenes from cellular proto-oncogenes

A few cell lines have been generated by treating primary cell cultures with mutagens, such as radiation, in the form of uv-light or gamma-rays, or chemicals. This approach is designed to induce mutations in the genome in order to alter the pattern of gene expression. Unfortunately, this approach works only at a very low efficiency. Immortality appears to be a multi-step process and it is therefore necessary to collect a number of changes in the genome, all of which appear to occur at a low frequency (see below).

The third strategy available for generating continuously growing cell lines is to introduce oncogenes into the primary cell cultures. Two sources of oncogenes are available. Firstly, there are the isolated oncogenes that are derived from cellular proto-oncogenes (2) and secondly, the oncogenes that are an endogenous part of a number of mammalian viruses (11, 12, 14).

The use of oncogene 'cocktails' as a method for generating immortal cell lines was first developed in rodent cells (6). However, while a few human cell lines have been reported to arise from experiments with purified oncogenes, the method is inefficient and rarely produces a cell line. The inability of purified oncogenes when introduced into primary cells to overcome the growth limitations of the cells may simply indicate that we have failed to identify the right oncogene or combination of oncogenes. However, the problem is likely to be due, at least in part, to the observation that immortality is recessive to mortality (see below).

The most successful approach for generating immortal human cell lines has used viral encoded oncogenes. The most commonly used viral oncogenes are:

1. Adenovirus E1A and E1B
2. Epstein Barr Virus
3. SV40 large T antigen

The oncogenes can be introduced into the primary cells either as part of the intact virus or the oncogenes can be purified and handled as part of a DNA vector. When the whole virus is used, the oncogene is introduced during infection. Infection is the most efficient means of introducing DNA into cells. However, not all cell types carry the necessary receptors to allow infection.

When DNA vectors are used the DNA is introduced by artificial means such as electroporation or precipitation with calcium phosphate (9). Electroporation involves passing a high voltage electric current through a cell suspension in the presence of the exogenous DNA. The electric current temporarily opens channels in the membrane allowing free movement of DNA into the cells. This method can be extremely efficient for introducing DNA into cells and is particularly suitable for cells that grow in suspension culture.

DNA will form a precipitate when mixed with calcium chloride and phosphate buffers (9). When calcium phosphate precipitated DNA is added to the medium of monolayer cultures, the cells appear to take up the precipitate by pinocytosis. The frequency of transfer is much lower than with electroporation, but it is a relatively easy method to perform and does not require the purchase of any specialised apparatus.

Purified oncogenes can be genetically manipulated into human retroviruses to take advantage of the high frequency of gene transfer achieved by viral infection (10). However, human retroviruses have proved more difficult to work with than rodent retroviruses and in some cases, genetic changes have resulted in the oncogenes during the infection process. Work with retroviruses has the added disadvantage that it requires special laboratory conditions.

Each of the viral oncogenes listed above have been used to generate immortal human cell lines. However, by far the most successful strategies for immortalisation are based on SV40 virus. Adenovirus E1A and B (8, 11) and EBV (12, 13) have only a limited range of susceptible cells whereas SV40 has been used to immortalise a wide range of cell types. The frequency of immortalisation with E1A and B has been shown to be 50-fold less than with SV40 (8) and its use is largely restricted to embryonic cells. EBV is very specific for lymphocytes and the resultant cell lines are often genetically unstable (12).

SIMIAN VIRUS 40 (SV40)

SV40 is a member of the papovavirus family (reviewed in 14). Its genome size is about 3×10^6 daltons and it can be divided into an early and late region. The immortalising function of SV40 is entirely located within the early region. This region codes for two proteins: the large and small T antigens which are generated from overlapping RNA transcripts.

As only the early region of SV40 is required for immortalisation, many DNA vectors have been constructed that contain the early region alone. In the analyse of the role of the large and small T antigens in immortalisation, a number of modifications to the DNA have been used. Some SV40 based vectors contain a small deletion in the origin of replication (ORI-) (15-17), others contain temperature sensitive mutations or an inducible promoter (18, 19). Dominant selectable marker genes are a useful tool in gene transfer studies and are often incorporated into vectors containing SV40 DNA sequences (20, 21). These genes provide a new and selectable phenotype on the recipient cell. In most cases, the dominant selectable marker gene confers on the recipient cells the ability to grow in an otherwise toxic medium.

In our laboratory we have used a vector designated as pSV3*gpt* to immortalise human fibroblast cultures. This vector contains the entire early region of SV40 and the dominant selectable marker *gpt* (21, 22). This bacterial gene codes for the enzyme xanthine-guanine phosphoribosyltransferase and allows the cells to grow in the presence of the inhibitor mycophenolic acid (21).

Using this vector, we have successfully established immortal cell lines from a wide selection of normal individuals and patients with a number of inherited disorders (22, 23). The method we described (22) has proven very effective at systematically generating cell lines and has allowed us to establish lines from both affected individuals and other members of their families.

pSV3*gpt* was introduced into primary cell strains using the calcium phosphate precipitation method (9). The recipient cells were subsequently selected for the uptake of the vector by growing the cells in selective medium containing the inhibitor mycophenolic acid (21). Within six weeks, colonies of cells expressing the SV4O T antigen were present. Subcultivation of these foci gave rise to a robust vigorous culture. In all cases, the life span of these cultures exceeded those of the original primary culture. Despite displaying an increased lifespan, each culture eventually senesced, and entered crisis. However, from the background of senescing cells, rare foci of dividing cells began to appear after about 12 weeks. Though the average time for these foci to appear was 12 weeks, this period could be very short (the culture only appearing to briefly enter crisis) or be up to one year.

The post-crisis cells, when subcultured eventually gave rise to robust and vigorous cultures. The immortal cell lines established from skin biopsies of patients have in our experience, always retained the mutant phenotype of the original primary cells. These lines have been extremely useful for studying both cellular metabolism and for genetic manipulation (22, 23).

SV40 IMMORTALISATION OF DIFFERENTIATED CELLS

Fibroblasts are readily accessible cells that are easily cultured. Thus the majority of SV40-immortalised cell lines are derived from fibroblasts. In recent years, however, cell culture techniques have permitted *in vitro* growth of a wide range of differentiated cell types. The use of SV40 T antigen has proven to be the most widely applicable and successful approach for immortalising differentiated cell types. Table 2 lists some of the different cell types that have successfully been immortalised with SV40 T antigen. Included in this list are a number of references to cell lines that were reported to have significantly extended lifespan after introduction of SV40.

One of the problems sometimes associated with SV40-immortalisation of differentiated cell lines is the loss of differentiation markers. This phenomenon is also seen after *in*

TABLE 2

HUMAN CELL TYPES SHOWING AN EXTENDED LIFESPAN/CONTINUOUS GROWTH AFTER INTRODUCTION OF SV40

SKIN	Fibroblast (22)
	Keratinocyte (24,25)
EPITHELIAL CELLS FROM:	
	Mammary (26,27)
	Amnion (28)
	Colon (5)
	Kidney (16,29)
	Pharynx (5)
	Prostrate (5)
	Trachea (30)
OTHER:	
	Amniocytes (31)
	Choroid (32)
	Myoblasts/myocytes (33,34)
	Monocytes (35)
	Placenta (36)
	Endothelium (37)

vitro growth of differentiated primary cells. However, loss of differentiated characteristics does not necessarily accompany SV40-immortalisation and there are many reports of these characteristics being retained (26, 33, 34, 36, 37). Studies with SV40 T antigen are now being directed at genetically manipulating the T antigen in order to reduce the effects of T antigen on differentiation (e.g. 38).

SV40 AND THE STEPS THAT LEAD TO IMMORTALITY

The use of SV40 DNA that has been genetically manipulated has allowed us to ask fundamental questions on how SV40 induces immortality and has given us an insight into the cellular changes that lead to immortality. Such studies have identified the large T antigen as the sole requirement for immortalisation of human cells (19). While small t antigen is not essential, it may still play a role in maintaining immortality as it appears to do in rodent cells.

Senescing primary cells become blocked in G1 and are unable to continue synthesising DNA (39). Expression of SV40 T antigen in these cells appeared to be able to override this block and DNA synthesis was able to continue. Cells expressing T antigen have an extended life span when compared to the parallel culture of primary cells (5,19,22). In many cases, these cultures can provide enough material for research without having to pursue an immortal derivative. The extended life span is dependent on the continued expression of T antigen. This observation was made using both a temperature sensitive derivative of SV40 (40) (at the non-permissive temperature, T antigen is no longer expressed in the cells) and a derivative that used a promoter that allowed inducible expression of T antigen (T antigen expression was switched off by removing the inducing agent) (19). Once again the cells began to accumulate in G1 and DNA synthesis ceased.

The expression of T antigen is not sufficient to induce immortality (5,19,22). Though the cultures have an extended lifespan from the primary cultures, they still enter crisis and cease dividing despite the continued expression of T antigen. However, from the background of senescent cells, foci arise, which on subculturing produce actively dividing immortal cultures. Thus in a subpopulation of T antigen expressing cells, another event occurs that results in immortality.

Limited proliferative growth is a dominant feature over immortality (18, 41). This has been elegantly demonstrated by fusing SV40 immortalised cell lines with primary cell strains. The fusion hybrids that result have only a limited growth potential despite the continued expression of a functional T antigen. This implies that immortality is achieved by the loss of an endogenous gene function rather than by acquiring a new gene function. This may be why introducing 'oncogene cocktails' has failed to generate immortal lines.

Fusion studies between various immortal cell lines have indicated that there are at least two independent pathways to immortality (18,41). Fusion of SV40-immortalised cell lines with each other always gave rise to an immortal hybrid indicating that the secondary event leading to immortality was the same in all SV40 cell lines. However, fusion of SV40-immortalised lines with several other lines immortalised by other means gave rise to hybrids with only limited growth potential. This implies that the pathway to immortality in each parent cell line was different.

Continued expression of T antigen is required for proliferation of SV40-immortalised cells. This was elegantly demonstrated using cell lines that were immortalised with a temperature sensitive derivative of SV40 (40) or with T antigen under control of an inducible promoter (19). When expression of SV40 T antigen was switched off, the immortalised cell lines ceased to proliferate and the cells senesced.

Wright *et al* (19) have put forward a two step model for immortality in human cells to explain the above observations. They propose that there are two stages a cell must pass through in order to reach immortality. The first, Mortality Stage 1 is a block to DNA synthesis that leads to an accumulation in G1. It is this step that T antigen is able to bypass. Mortality Stage 2 is reached if cells are able to divide, but at this point cells attempt to divide but no longer remain viable. A cell destined to become immortal must be able to overcome this second step. T antigen appears to be necessary to overcome the first block to mortality and therefore to allow progression to the second block. However, having successfully surmounted the block at Mortality Stage 2, T antigen expression is still required to maintain DNA synthesis and immortality.

CONCLUSIONS AND SUMMARY

Immortal cell lines have a wide application in both production and research (3). Their use has many advantages over the use of fresh tissue and they can help to alleviate the need to sacrifice living animals. There is a growing need, however, to replace the use of non-human cell lines with human lines largely as the result of our developing awareness of the differences between prokaryotes and eukaryotes and between species.

Studies on the cellular mechanism leading to immortality have largely been done on rodents (1,2). This is a consequence of the easy availability of rodent cells and the fact that rodent cells will immortalise at a relatively high frequency in culture thus permitting the phenomenon to be studied. However, it appears that the cellular mechanisms leading to immortality in human cells are quite different from rodents and this unfortunately means that we are unable to simply extend the methods developed in rodents to humans.

The most efficient and successful strategies for developing human immortal cell lines have been based on the use of SV40 (22,27-37). The majority of cell lines established in this way have been derived from fibroblasts, however, the recent advances in cell culture techniques has opened the door to the use of SV40 based DNA vectors in the immortalisation of a wide range of differentiated cell types. The development and exploitation of immortalised differentiated cells is still in its infancy and much work is being focussed on retaining the differentiation characteristics of the cell lines (38).

The cellular mechanisms that lead to immortality have not been clearly defined. However, it is clear that it is a multi-step process and there is more than one way for a cell to become immortal (18). The changes that lead to immortality are genetically recessive and suggest that the loss of a dominant acting factor is required to achieve immortality.

A full understanding of the nature of cellular immortality in human cells will open up new methods for creating immortal cell lines, and will provide a greater range of immortal cell types for study.

ACKNOWLEDGEMENTS

LVM is a Wellcome Senior Research Fellow (Basic Biomedical Sciences). I am very grateful to Dr. C. MacDonald, University of Strathclyde for providing me with a pre-print of her publication (3) for use in preparation of this review.

REFERENCES

1. Meek, R.L., Bowman, P.D. and Daniel, C.W (1977) Exp.Cell Res. **107**, 277-284.
2. Bishop J.M. (1985), TIG Sept. 245-249.
3. MacDonald, C. (1989) CRC Reviews in Biotechnology, in press.
4. Hayflick, L. (1965), Exp. Cell Res.**37**:614-636.
5. Chang, S.E. (1986), Biochim Biophys Acta **823**, 181-194.
6. Land, H., Parada, L.F. and Weinberg, R.A. (1983), Nature **304**, 596-602.
7. Yoakum, G.H., Lechner, J.F., Gabrieson, E.W., Korba, B.E., Malan-Shibley, L., Willey, J.C., Valerio, M.G., Shamsuddin, A.M., Trump, B.F. and Harris, C.C. (1985), Science **227**, 1174-1179.
8. Gallimore, P.H., Grand, R.J.A. and Byrd, P.J. (1986), Anticancer Res. **6**, 499-508.
9. Gorman, C., (1985), in DNA Cloning Vol. II, A practical approach, Ed. D.M. Glover, IRL Press, Washington D.C. pg.143.
10. Jat, P.S., Cepko, C.L., Mulligan, R.C. and Sharp, P.A. (1986), Mol.Cell Biol **6**, 1204-1217.

11. Flint, S.J., (1980), in DNA Tumour Viruses Part II, Ed. J.Tooze, Cold Spring Harbour Laboratory pg. 443.
12. Sugden, B. (1989), Cell **57**, 5-7.
13. Griffin, B.E. and Karran, L. (1984), Nature **309**, 78-82.
14. Griffin, B.E., (1980) in DNA Tumour Viruses Part II, Ed. J.Tooze, Cold Spring Harbour Laboratory pg. 61.
15. Small, M.B., Gluzman, Y. and Ozer, H.I. (1982), Nature **296**, 671-672.
16. Major, E.O. and Matsumura, P. (1984), Mol.Cell Biol. **4**, 379.
17. Forest, C., Czerucka, D., Negrel, R. and Ailhaud, G. (1983), Cell Biol.Int.Reps. **7**, 73.
18. Pereira-Smith, O.M. and Smith, J.R. (1987), Mol.Cell Biol **7**, 1541-1544.
19. Wright, W.E., Pereira-Smith, O.M. and Shay, J.W. (1989), Mol.Cell Biol **9**, 3088-3092.
20. Su, R.T. and Chang, Y-C. (1989), Exp.Cell Res. **180**, 117.
21. Mulligan, R.C., and P.Berg, (1981), Proc.Natl.Acad. Sci. USA **78**, 2072-2076.
22. Mayne, L.V., Priestley, A., James, M.R. and Burke, J.F. (1986), Exp.Cell Res. **162**, 530.
23. Arlett, C. F., Green, M. H. L., Priestley, A., Harcourt, S. A., and Mayne, L.V., (1988), Int. J. Radiat. Biol.,**54**, 911-928.
24. Brown, K.W. and Parkinson, E.K. (1984), Int.J.Cancer **33**, 257.
25. Banks-Schlegel, S.P. and Howley, P.M. (1983), J.Cell Biol. **96**, 330-337.
26. Chang, S.E., Keen, J., Lane, E.B. and Taylor-Papadimitriou, J. (1982), Cancer Res. **42**, 2040.
27. Buehring, G.C. (1972), J.Nat.Cancer Inst. **49**, 1433-1434.
28. Gaffney, E.V., Fogh, F., Ramos, L., Loveless, J.P., Fogh, H. (1970), Cancer Res. **30**, 1668-1676.
29. Graham, F.L., Smiley, J., Russell, W.C. and Nairn, R. (1977), J.Gen.Virol. **36**, 59-72.
30. Gruenert, D.C., Basbaum, C.B., Welsh, M.J., Li, M., Finkbeiner, W.E. and Nadel, J.A. (1988), Proc.Natl.Acad.Sci. USA **85**, 5951.
31. Walen, K.H. and Arnstein, P. (1986), In Vitro **22**, 57-65.
32. Carruba, G., Dallapicolla, B., Brinchi, V. and DeGuili Morghen, C. (1983), In Vitro **19**, 443-452.
33. Hurko, O., McKee, L. and Zuurveld, J.G.E.M. (1986), Annls.Neurol. **20**, 573-582.
34. Miranda, A.F., Babbis, L.E. and Fischer, P.B. (1983), Proc.Natl.Acad.Sci.USA **80**, 6581-6585.
35. Nagata, Y., Diamond, B. and Bloom, B.R. (1983), Nature **306**, 597-598.
36. Chou, J.Y. (1978), Proc.Natl.Acad.Sci. USA **75**, 1854.
37. Gimbrone, M.A. and Fareed, G.C. (1976), Cell **9**, 685-693.
38. Cherington, V., Brown, M., Paucha, E., St. Louis, J., Spiegelman, B.M. and Roberts, T.M. (1988), Mol.Cell Biol **8**, 1380-1384.
39. Gorman, S.D. and Cristofalo, V.J. (1986) Exp.Cell Res. **167**, 87-94.
40. Radna, R.L., Caton, Y., Jha, K.K., Kaplan, P., Li, G., Traganos, F., and Ozer, H.L. (1989) Mol.Cell Biol. **9**, 3093-3096.
41. Harris, H. (1988), Cancer Res. **48**, 3302-3306.

CARCINOGEN METABOLISM IN IMMORTALISED HUMAN CELLS GROWN AS HYBRID CELLS IN CULTURE

Stuart Brown and Helen Ross

Biochemistry Department
University of Nottingham Medical School
Nottingham NG7 2UH, England

INTRODUCTION

Most cells from normal tissues have two characteristic properties. Their cell division is regulated in a particular way and they produce substances characteristic only of their tissue of origin. It was soon realised, however, that as cells from normal tissues were studied extensively, a limit to their long term cultivation was found. Normal cells died after a finite number of divisions. In contrast, tumour cells grew indefinitely in culture and usually did not express differentiated functions. In order to distinguish between these types of cells with finite or infinite lifespan in culture, Hayflick and Moorhead (1) used the term cell strain to denote normal cells with a finite lifespan and reserved the term cell line for cells which were established in culture and would divide indefinitely. They also noted that the property of infinite cell growth was usually associated with a change in the diploid nature of the cells and that a heteroploid karyotype was common in permanent cell lines.

More recently Pereira-Smith and Smith (2) have suggested that cellular immortality is a result of recessive alterations in the genetic program that limits the division of normal cells. Their studies further suggested that there was more than one way to alter this "normal program" in order that inifinite life was obtained.

Our knowledge of the changes which take place when a cell with finite life in culture becomes immortal is far from complete but has been greatly increased by studies on cell transformation and cancer. Agents which cause normal cells to become cancer cells, e.g. chemicals and viruses, have been studied extensively and the various stages involved in the transformation of a cell are being defined. From studies on both DNA and RNA tumour viruses it is clear that an early event in cell transformation involves the establishment or immortalisation of the cell. For example, adenovirus is able to transform primary cells in culture in stages. The first stage is connected with immortalisation of the cells while the second is required for full expression of the oncogenic phenotype. The EIA proteins are the first viral polypeptides to be synthesised after infection and they are responsible for the immortalisation. This is thought to occur through their interaction with certain cellular proteins. One of the

Pharmaceutical Applications of Cell and Tissue Culture to Drug Transport
Edited by G. Wilson *et al.*, Plenum Press, New York, 1991

most recent exciting discoveries in this field is the fact that the protein produced from the retinoblastoma susceptibility gene is one of the proteins the adenovirus protein EIA interacts with (7). Whatever the role of this retinoblastoma gene product, its loss by gene deletion or mutation or its effective removal by interaction with a viral protein produces the same end result - cell immortality. These studies are at an early stage and already other proteins as well as the 105KD retinoblastoma protein are being characteised as targets for the early adenovirus proteins (8,9). This work is now also being complemented by studies on SV40 and papilloma viruses and their transforming peptides. These, too, are able to bind to the cellular proteins identified in the adenovirus system (10,11,12).

The discovery of oncogenes in RNA tumour viruses has greatly increased our understanding of cell growth control and tumour initiation. Some oncogenes can encode proteins that are homologous to either growth factors or growth factor receptors, e.g. the oncogene v-sis (Simian sarcoma virus) is derived from the gene encoding one chain of the platelet derived growth factor (13,14). Also, the oncogene v-erbB (erythroblastosis virus) is derived from the normal epidermal growth factor gene (15). A further link between growth control and oncogene function is the common tyrosine kinase activity of many growth factor receptors and oncogene products (15,17,18). This relationship has motivated the research to explain non-virally induced cell transformation. Mutations which alter the expression of normal cellular genes, related to viral oncogenes, can have the same effects on cells as viral transformation. Thus, overexpression or mutation of particular cellular genes can lead to establishment or immortalisation of cells and cell transformation (19,20).

These studies on retroviral oncogenes have shown that the events involved in cell transformation are very complex. Initially it was suggested that the expression of a single oncogene, ras (rat sarcoma virus) was sufficient to immortalise and transform cells. Now it is apparent that oncogenes probably cooperate in the transformation process, i.e. expression of one oncogene may confer infinite life on a cell whereas a second oncogene product may be necessary for full transformation (20). This cooperation may involve DNA and RNA virus genes. Adenovirus EIA early gene products can immortalise cells which then become susceptible to transformation by the ras gene protein (3).

Certain RNA viral oncogene products appear to mimic the action of the adenoviral early gene products. V-myc (myelocytosis virus) oncogene codes for a protein found in the nucleus of infected cells. The RNA transcript of its cellular counterpart, c-myc, has been reported to increase 10-20 fold in a variety of mammalian cell types in response to mitogenic substances such as growth factors and hormones (21,22,23).

In addition, decreased c-myc RNA levels have been reported in cells withdrawing from the cell cycle and undergoing terminal differentiation (24). Thus aberrant expression of the c-myc gene leading to overproduction of its RNA and protein product has been linked to uncontrolled cellular proliferation. The complexity of this problem, however, is exemplified by recent studies by Nath et al. (25) who have shown an increase in c-myc RNA in chick lens cells undergoing terminal differentiation and also in studies on hepatocytes from young and old rats it has been found that their responsiveness to epidermal growth factor is the same in that the c-myc gene is activated but the old hepatocytes did not initiate cell division. They traversed from G_o to

G_1 in the cell cycle but did not enter S phase (26). A similar situation has been found in human senescent fibroblasts which retain the number and responsiveness of their platelet derived growth factor receptors as young cells in terms of c-myc expression, but again the cells did not enter S phase.

The idea that cancer is caused by a progression of events is not new but the individual steps are difficult to elucidate. The possibility that immortalisation is a preliminary event prior to full transformation is interesting in that it suggests cells can be immortalised without being transformed. This could be an important consideration if a commercially important gene product was to be produced possibly for human consumption. Parallel studies are being carried out on cell immortalisation and transformation using chemicals and a similar picture to that being developed by virus work is being constructed. Newbold et al. (27) have shown that cells treated with carcinogenic chemicals can produce rare immortal variants and these can then progress to cells with the transformed properties of anchorage independence and tumour formation in animals. They have found that they can produce cells which are either immortal or anchorage independent but not both and suggest that cell immortality is insufficient by itself for cell transformation.

The ability to produce cells which retain important differentiated properties and which are immortal is very valuable. Viruses and chemicals are being used to create immortal cells for cell metabolic studies (27,28). These agents, however, are not the only means by which normal cells can be given immortality in culture. In addition, the techniques of somatic cell hybridisation have been very successfully adapted to this problem especially with regard to immunoglobulin production. When non-secreting myeloma cells are fused with splenic lymphocytes, the resulting hybrid cells secrete a single immunoglobulin. This shows that the hybrid is expressing the differentiated function of the lymphocyte and yet will proliferate indefinitely in culture. This technique for producing antibodies was pioneered by Kohler and Milstein (29). Originally the technique was restricted to mouse x mouse hybridoma systems but now mouse x rat, mouse x human and even human x human systems are being developed (see ref. 30). Many problems still remain to be solved regarding the choice of ideal parental cells and optimal culture medium to maintain growth and maximum antibody secretion but the technique is clearly extremely valuable.

Our major interest is in the use of immortal cells which retain the differentiated functions required for metabolism of chemical carcinogens. We are interested in determining how the activity of the various enzymes involved in polycyclic aromatic hydrocarbon (PAH) metabolism function to either activate these chemicals to their mutagenic and carcinogenic forms or serve to detoxify these chemicals and process them for excretion. It is known that fresh tissue, primary cell cultures and continuous cell lines behave differently with respect to this metabolism of PAH and that the differences reflect the rapid loss of expression of many forms of the enzymes necssary as soon as the cells are cultured (31,32). We decided that if fresh tissue was hybridised with an established cell line then those hybrids which retained the differentiated functions from the fresh tissue and which possessed infinite life in culture would make a unique system for studying the metabolism of PAH and the role of the activation enzymes, aryl hydrocarbon hydroxylase (AHH) (33) and epoxide hydrolase (EH) (34) and of the detoxification enzyme UDP glucuronyltransferase (UDPGT) (35).

MATERIALS AND METHODS

Cell cultures

The continuous cell cultures used are listed in table 1. All cells were grown in Dulbecco's modified Eagles medium supplemented with 5% fetal calf serum (D5) except for human CEM cells which were grown in RPMl 1640 medium supplemented with 10% fetal calf serum. Cells were grown at 37°C in a humidified atmosphere supplemented with 10% CO_2 in 90 mm culture dishes. For each cell line used, cells were grown to 80% confluency and then fed with pre-warmed medium. Twenty-four hours later the cells were harvested. Cells were washed and removed from the dishes in ice-cold 0.9% saline and either used immediately or stored at -80°C before analysis.

Human cells

Nucleated cells from 10ml heparinised blood or 0.5ml of bone marrow aspirate were obtained from normal volunteers. The blood and bone marrow were centrifuged at room temperature and the buffy coat resuspended in plasma, washed by several centrifugations and finally diluted in D5 medium.

Cell fusion

Packed leukocytes (0.05-0.1ml) were added to 5 x 10^6 mouse RAG cells (BALB/cd 8-azaguanine resistant cells established from a renal adenocarcinoma (36)), the mixture centrifuged and resuspended in 1ml of D5 containing either 300 hemagglutinating units of inactivated Sendai virus or 50% (w/v) polyethylene glycol 1500 as fusing agents (37,38). The fusion mixtures were incubated for 15 min. at 4° and then 1 hour at 37° for Sendai virus fusion or 1 min at room temperature for the polyethylene glycol. The cells were then plated out in several 100mm petri dishes and hybrid clones selected in D5 medium containing HAT (hypoxanthine (10^{-4}M), thymidine (1.6 x 10^{-5}M) and aminopterin (4 x 10^{-7}M). The medium was changed every 3-4 days and colonies were picked 3-4 weeks later. The hybrids from the Sendai fusion were designated RBM and those from the polyethylene glycol were designated RKO. All lines were shown to be hybrid cells by isozyme analysis.

AHH assay

AHH activity was determined in parallel sets of cultures as described previously (30) except that 3ml of the organic phase was extracted with 1ml of 1M NaOH. Inducer medium was prepared by dissolving benz[a]anthracene in dimethylsulfoxide and adding this to complete growth medium to yield a final concentration of 1.0 ug/ml and 0.1% with respect to solvent. After preliminary experiments, aminophylline (0.7mM final concentration) was included in the inducer medium to enhance the level of inducible activity. AHH activity is expressed as fmol equivalents of 3-hydroxybenzo[a]pyrene formed per mg protein per min. Heat-inactivated homogenates were used to determine the "background" fluorescence values.

Epoxide hydrolase assay

EH was assayed in cell pellets by incubating solubilised extracts in 0.1M Tris pH 9.0 with ^{14}C-styrene oxide (200,000 dpm, 16nmol) for 5 and 10 minutes followed by differential extraction of the glycol product as originally described by Oesch (40). The substrate was extensively purified prior to use and blank values at each time point were always included.

UDP-glucuronyltransferase assay

Cell pellets were prepared by harvesting semi-confluent plates of cells and centrifuging the suspension for 1 min at 6000 r.p.m. The pellets were resuspended in 2 volumes of 0.05M tris buffer pH 7.6 and homogenised with 5 strokes of the homogeniser. The UDP-GT assay was performed essentially as described by Singh and Wiebel (41). A reaction mixture containing 100mM tris pH 7.6, 5mM $MgCl_2$, 2.5mM UDP-glucuronic acid and 20uM 3-hydroxybenzo[a]pyrene was incubated with the homogenates in a final volume of 0.4ml for 30 min. at 37^oC. The reaction was ended by the addition of 6ml chloroform: methanol (2:1, v/v) and 0.6ml water. The tubes were shaken vigorously and centrifuged to separate the phases. The aqueous methanol phase was treated with 0.1ml of 5M NaOH and the relative fluorescence measured using an Aminco-Bowman spectrofluorimeter set at 378nm excitation and 425nm emission. Quinine sulphate (10mg/ml) was used as standard.

Protein assay

Protein was determined by the method of Lowry *et al*. (42).

Isozyme analysis

Cells were harvested at the same time as cells grown for enzyme assays and stored at -80^oC until analysed. Homogenates were prepared in tris 0.05M pH 8.0 buffer to give a 20mg/ml suspension. After freezing and thawing 3 times homogenates were centrifuged for 30 min. at 20,000g. Supernatants were used for starch gel electrophoresis to determine the human chromosome content of the hybrids by analysing for enzymes assigned to all but chromosomes 3 and 22 using previously described methods (43,44).

RESULTS AND DISCUSSION

The cell lines examined in this study are listed in Table 1. Each cell line was grown up in large quantities and the cell pellets assayed for the three carcinogen metabolising enzymes aryl hydrocarbon hydroxylase, UDG glucuronyltransferase and epoxide hydrolase. As can be seen from the data the majority of the permanent cell lines examined did not express detectable levels of AHH activity and only expressed very low levels of the detoxification enzymes UDPGT and EH. This is as predicted by Pelkonen and Nebert (32). In fact, many other cell lines have been tested for UDPGT activity and of 29 human cell lines examined only one, NT29, a human renal adenocarcinoma line, expresses high levels of activity.

It has been claimed that the level of the AHH activating enzyme is directly related to the high incidence of lung cancer in cigarette smokers but these results have been questioned in subsequent investigations (reviewed in ref. 45). These arguments could indicate that the metabolism of carcinogenic hydrocarbons is not controlled solely at the level of activation. Clearly, if the activated intermediates are removed at different rates in different individuals they will represent different risks. Thus the fate of the carcinogen in a cell will be related to the balance between its rate of activation and its rate of detoxification. In order to study this process it is not applicable to only use isolated microsomes which do not express all the

Table 1. Enzyme activity in selected cell lines

Cell line	Species	AHH[a]	UDPGT[a]	EH[a]
HK	Human kidney	N.D	N.D	0.03
Chang	Human liver	N.D	0.12	N.T
CHL	Chimp liver	N.D	0.12	N.T
A549	Human lung	245	1.2	0.44
CEM	Human lymphocyte	N.D	N.D	N.D
BHK	Hamster kidney	20	1.1	0.04
E36	Hamster embryo	N.D	0.2	0.18
B82	Mouse L cell	N.D	0.6	0.07
RAG	Mouse renal cell	N.D	11.7	0.07
GM86	Mouse lymphocyte	N.D	N.D	N.D

[a] AHH, UDPGT and EH activities were determined using the assays described in the methods section. The values quoted represent the mean of at least two separate assays. AHH activity is quoted as fmol 3-hydroxy BP formed/min/mg protein; UDPGT activity as units (1 unit = 1 relative fluoresence unit/min/mg protein) and EH activity as nmol styrene glycol produced/min/mg protein. All assays were carried out at 37°C.

[b] N.D = not detectable

[c] N.T = not tested

necessary functions nor is it possible to use many cell lines because of the low rate of metabolism due to lack of enzyme activity as shown above. Hybrid cells, however, offer a solution to this problem. Not only will cells which express the required enzymes continue to express them in hybrids, but the hybrid cell will be immortal in culture. The level of enzyme expression should remain constant for long periods of time in culture. Human x rodent hybrid cells have one particular property which has to be accounted for. They segregate human chromosomes in a random manner as the cells are grown (43,44). This segregation is most rapid during the early passaging of the cells and after a short period in culture the chromosome composition becomes relatively stable. It is necessary to check the chromosome composition especially if the cells are being frozen and thawed. This property endows the hybrid cells with a unique character for studying the role of carcinogen metabolising enzymes since hybrids can be generated which express all the necessary enzymes or which have segregated particular functions. Thus the effect of altering the balance between rate of activation and rate of detoxification on whether the carcinogen is bound to cellular macromolecules or excreted can be studied.

In our initial studies we fused human leukocytes with mouse RAG cells and generated two sets of hybrid clones. These cells were grown up and analysed for the enzymes shown in Table 2 and for their human chromosome composition by isozyme analysis (43,44). As can be seen from Table 2, some of the hybrids did express high levels of AHH, UDPGT and EH activities. These activities have been measured at different times over several months in culture and very little change has been observed. Thus the hybrid cells produced are fulfilling the requirements of expressing the differentiated functions and yet achieving immortality in culture.

Table 2. Enzyme activities in hybrid cells

Cell line	AHH[a]	UDPGT[a]	EH[a]
RBM 3	538	11.7	0.34
RBM 5	297	12.4	0.21
RBM 20	558	11.8	0.31
RBM 2	N.D[b]	12.4	N.T[c]
RBM 8	N.D	13.1	N.T
RKO 3	562	9.3	0.11
RKO 12	640	12.3	0.19
RKO 1	N.D	11.4	0.41
RKO 22	N.D	13.0	0.29

[a] AHH, UDPGT and EH activities are expressed as described in the legend to table 1.
[b] N.D = not detectable
[c] N.T = not tested

The effect of human chromosome segregation can be seen with respect to AHH activity. The AHH activity detected in the hybrid cells is due to the human parent cells and thus when the chromosome which carries the gene(s) necessary for this function is lost, the AHH is no longer detectable. We have used this fact to study hybrid cells with regard to AHH expression and human chromosome content and have mapped the gene necessary for AHH expression in these cells to human chromosome 2 (43,44). Similar sets of hybrid cells with different parental cells have also been used to examine which chromosomes are necessary for EH and UDPGT activities. For EH we have found human chromosome 6 has a role (46) whereas for UDPGT, chromosome 17 is necessary (unpublished observations).

Some of the hybrid cells described in Table 2 have been used to study benzo[a]pyrene (BP) metabolism. BP is a common polycyclic aromatic hydrocarbon found in our environment and is highly carcinogenic (32). Hybrid cells expressing high levels of AHH activating enzyme and high levels of UDPGT detoxifying enzyme were incubated with radio-actively labelled BP for several days. The human lung cell line A549, which did express AHH activity but very low UDPGT activity, was used in similar incubations as a positive control whilst mouse RAG cells, which lacked AHH activity, were used as a negative control. The effect of the carcinogen on the cells was determined with regard to how much bound to the cells DNA and how cytoxic the carcinogen was in the cells.

The results of these tests are shown in Table 3 for a selection of the cell lines examined. It is very interesting to note that the human cell line, which could activate the carcinogen to its mutagenic and carcinogenic form, bound large quantities of BP to DNA and the BP was cytotoxic whereas the mouse RAG cells did not activate the BP, did not bind it to DNA and were not killed by it. The hybrids, whether AHH^+ or AHH^-, however, behaved in a similar manner to the mouse cells, i.e. they did not bind large quantities of BP to DNA and the BP was not cytotoxic. Thus, in those hybrid cells which could activate BP to its mutagenic and carcinogenic forms as efficiently as the human lung cell line, the presence of the high levels of UDPGT as detoxifying agent must be preventing the necessary accumulation of toxic intermediates. These studies are being extended to examine the role of all the enzymes in the pathway of BP metabolism but clearly show that AHH may not be the sole

Table 3. Metabolism of BP in human, mouse and hybrid cells

Cell line	AHH[a]	cytotoxicity[b]	BP-bound DNA[c]
A549	245	++++	250
RAG	N.D	none	0
RBM 2	N.D	+	30
RBM 3	538	none	0

[a] AHH was determined as quoted in tables 1 and 2.
[b] Cytotoxicity was determined by monitoring cell survival after treating the cultures with BP (50ug/ml) for up to 2 weeks. ++++ represents approximately 90% of the cells died after 2 weeks.
[c] ^{3}H-BP (1mC/ml; 40C/mmol) was incubated in cell cultures and DNA isolated after 16 hrs. The values represent dpm/ug DNA counted.

enzyme by which to predict the chance of lung cancer formation in people who may be heavy cigarette smokers (33,45).

CONCLUSION

Most normal cells when adapted for culture will only grow for a limited number of divisions before reaching a crisis point. They then will either senesce and die or become spontaneously transformed. The latter is usually associated with a change to heteroploidy and an altered differentiated state. In order to study complex metabolic processes using cultured cells it is necessary to develop methods of conferring 'immortality' on to the cell line preferably without full malignant transformation or loss of differentiated functions. The study of viral oncogenes is showing that immortality and transformation are not necessarily linked and it is possible to confer immortality on to a cell line using viral genes. We have shown in this paper that it is also possible to confer immortality and retain differentiated functions by fusing cells with cells previously adapted to infinite life in culture. This technique has been used to prepare monoclonal antibodies by Kohler and Milstein (29) and here we have used it to develop cells which metabolise chemical carcinogens. We have successfully examined which human chromosomes are necessary for carcinogen metabolising enzymes and have shown that the fate of the carcinogen can be dramatically influenced by the balance between its rate of formation and inactivation.

ACKNOWLEDGEMENTS

The studies carried out in the work described above have been generously supported by the SmithKline(1982)Foundation, the M.R.C., Nottingham University Medical School and Research Funds and the Dr. Hadwen Trust for Humane Research.

REFERENCES

1. L. Hayflick and P. S. Moorhead, The serial cultivation of human diploid cell strains, Exp. Cell Res. 25:585-621 (1961).
2. O. M. Pereira-Smith and J. R. Smith, Evidence for the recessive nature of cellular immortality, Science 221:964-966 (1983).
3. H. E. Ruley, Adenovirus early region 1A enables viral and cellular transforming genes to transform primary cells in culture, Nature 304:602-606 (1983).

4. P. van den Elsen, A. Houweling and A. J. van der Eb, Expression of region E1B of human adenovirus in the absence of region E1A is not sufficient for complete transformation, Virology 128:377-390 (1983).
5. B. Zerler, B. Moran, K. Maruyama, J. Moomaw, T. Grodzicker and H. E. Ruley, Adenovirus E1A coding sequences that enable ras and pmt oncogenes to transform cultured primary cells, Mol. Cell Biol. 6:887-899 (1986).
6. R. Cone, T. Grodzicker and M. Jaramillo, A retrovirus expressing the 12S adenoviral E1A gene products can immortalise epithelial cells from a broad range of rat tissues, Mol. Cell. Biol. 8:1036-1044 (1988).
7. P. Whyte, K. J. Buchkovich, J. M. Horowitz, S. H. Friend, M. Raybuck, R.A. Weinberg and E. Harlow, Association between an oncogene and an anti-oncogene: the adenovirus E1A proteins bind to the retinoblastoma gene product, Nature 334:124-129 (1988).
8. S-P. Yee and P. E. Branton, Detection of cellular proteins associated with human adenovirus type 5 early region E1A polypeptides, Virology 147:142-153 (1985).
9. E. Harlow, P. Whyte, B. R. Franza and C. Schley, Association of adenovirus early-region 1A proteins with cellular polypeptides, Mol. Cell Biol. 6:1579-1589 (1986).
10. N. Dyson, P. M. Howley, K. Munger and E. Harlow, The human papilloma virus-16 E7 oncoprotein is able to bind to the retinoblastoma gene product, Science 243:934-937 (1989).
11. P. Clertant and I. Seif, A common function for polyoma virus large T and papillomavirus E1 proteins, Nature 311:276-279 (1984).
12. M. E. Ewen, J. W. Ludlow, E. Marsillo, J. A. DeCaprio, R.C. Millikan, S. H. Cheng, E. Paucha and D. M. Livingston, An N-terminal transformation-governing sequence of SV40 large T antigen contributes to the binding of both $p110^{Rb}$ and a second cellular protein, p120, Cell 58:257-267 (1989).
13. M. D. Waterfield, G. T. Scrace, N. Whittle, P. Stroobant, A. Johnsson, A. Wasteson, B. Westermark, C-H. Heldin, J. S. Huang and T. F. Deuel, Platelet derived growth factor is structurally related to the putative transforming protein $p28^{sis}$ of simian sarcoma virus, Nature 304:35-39 (1983).
14. R. F. Doolittle, M. W. Hunkapiller, L. E. Hood, S. G. Devare, K. C. Robbins, S. A. Aaronson and H. N. Antoniades, Simian Sarcoma Virus onc. gene, v-sis, is derived from the gene (or genes) encoding a platelet derived growth factor, Science 221:275-277 (1983).
15. J. Downward, Y. Yarden, E. Mayes, G. Scrace, N. Totty, P. Stockwell, A. Ullrich, J. Schlessinger and M. D. Waterfield, Close similarity of epidermal growth factor receptor and v-erb B oncogene protein sequences, Nature 307:521-527 (1984).
16. M. S. Collett and R. L. Erikson, Protein kinase activity associated with the avian sarcoma virus src gene product, Proc. Natl. Acad. Sci. U.S.A. 75:2021-2024 (1978).
17. A. Ullrich, L. Coussens, J. S. Hayflick, T. J. Dull, A. Gray, A. W. Tam, J. Lee, Y. Yarden, T. A. Liebermann, J. Schlessinger, J. Downward, E. L. V. Mayes, N. Whittle, M. D. Waterfield and P. H. Seeburg, Human epidermal growth factor receptor cDNA sequence and aberrant expression of the amplified gene in A431 epidermoid carcinoma cells, Nature 309:418-425 (1984).
18. J. M. Bishop, Tricks with tyrosine kinases, Nature 319:722-723 (1986).
19. A. Balamin, Transforming ras oncogenes and multistage carcinogenesis, Brit. J. Cancer 51:1-7 (1985).

20. R. A. Weinberg, The action of oncogenes in the cytoplasm and nucleus, Science 230:770-776 (1985).
21. K. Kelly, B. H. Cochran, C. D. Stiles and P. Leder, Cell-specific regulation of the c-myc gene by lymphocyte mitogens and platelet derived growth factor, Cell 35:603-610 (1983).
22. J. M. Blanchard, M. Piechaczyk, C. Dani, J. C. Chambard, A. Franchi, J. Pouyssegur and P. Jeanteur, C-myc gene is transcribed at high rate in Go arrested fibroblasts and is post transcriptionally regulated in response to growth factors, Nature 317;443-445 (1985).
23. K. L. Fink, E. D. Wieben, G. E. Woloschak and T. C. Spelsberg, Rapid regulation of c-myc protooncogene expression by progesterone in the avian oviduct, Proc. Natl. Acad. Sci. U.S.A. 85:1796-1800 (1988).
24. D. L. Bentley and M. Groudine, A block to elongation is largely responsible for decreased transcription of c-myc in differentiated HL60 cells, Nature 321:702-706 (1986).
25. P. Nath, R. Getzenberg, D. Beebe, L. Pallansch and P. Zelenka, c-myc mRNA is elevated as differentiating lens cells withdraw from the cell cycle, Exp. Cell Res. 169:215-222 (1987).
26. N Sawada, Hepatocytes from old rats retain responsiveness of c-myc expression to EGF in primary culture but do not enter S phase, Exp. Cell Res. 181:584-588 (1989).
27. R. F. Newbold, R. W. Overell and J. R. Connell, Induction of immortality is an early event in malignant transformation of mammalian cells by carcinogens, Nature 299:633-635 (1982).
28. R. T. Su and Y-C. Chang, Transformation of human epidermal cells by transfection with plasmid containing Simian Virus 40 DNA linked to a neomycin gene in a defined medium, Exp. Cell Res. 180:117-133 (1989).
29. G. Kohler and C. Milstein, Continuous cultures of fused cells secreting antibody of predefined specificity, Nature 256:495-497 (1975).
30. L. Olsson and H. S. Kaplan, Human-human monoclonal antibody producing hybridomas: technical aspects, in Methods in Enzymology, vol. 92 pp. 3-16, eds. J. J. Langone and H. van Vunakis, Acad. Press, N.Y. (1983).
31. D. H. Bissell and P. A. Guzelian, Microsomal functions and phenotypic change in adult rat hepatocytes in primary monolayer cultures, in Gene expression and Carcinogenesis in Cultured Liver, ed. L. E. Gerschenson and E. B. Thompson, pp. 119-136, Acad. Press. N.Y. (1975).
32. O. Pelkonen and D. W. Nebert, Metabolism of Polycyclic Aromatic Hydrocarbons: Etiologic role in carcinogenesis, Pharmacol. Revs. 34:189-251 (1982).
33. B. Paigen, H. L. Gurtoo, J. Minowada, E. Ward, L. Houten, K. Paigen, A. Reilly and R. Vincent, Genetics of aryl hydrocarbon hydroxylase in the human population and its relationship to lung cancer, in Polycyclic Hydrocarbons and Cancer, ed. H. V. Gelboin and P. O. P. Ts'O, pp. 391-406, Acad. Press, N.Y. (1978).
34. P. Wang, J. Meijer and F. P. Guengerich, Purification of human liver cytosolic epoxide hydrolase and comparison to the microsomal enzyme, Biochemistry 21:5769-5776 (1982).
35. C. B. Kasper and D. Henton, Enzymatic basis of detoxification, vol. 2, pp. 3-36, ed. W. B. Jakoby, Acad. Press, N.Y. (1980).
36. B. Felluga, A. Claude and E. Mrena, Electron microscope observations on virus particles associated with a transplantable renal adenocarcinoma in BALB/cf/cd mice, J. Natl. Can. Inst. 43:319-333 (1969).

37. C. M. Croce, H. Koprowski and H. Eagle, Effect of environmental pH on the efficiency of cellular hybridisation, Proc. Natl. Acad. Sci. U.S.A. 69:1953-1956 (1972).
38. R. L. Davidson and P. S. Gerald, Improved techniques for the induction of mammalian cell hybridization by polyethylene glycol, Somat. Cell Genet. 2:165-170 (1976).
39. S. Brown, F. J. Wiebel, H. V. Gelboin and J. D. Minna, Evidence for linkage between aryl hydrocarbon hydroxylase expression and enzyme markers assigned to human chromosome 2 in human x mouse hybrid cells, in Polycyclic Hydrocarbons and Cancer, vol. 2, pp. 407-415, ed. H. V. Gelboin and P. O. P. Ts'O, Acad. Press, N.Y. (1978).
40. F. Oesch, D. M. Jerina and J. W. Daly, A radiometric assay for hepatic epoxide hydrase activity with 7-^{3}H styrene oxide, Biochim. Biophys. Acta. 227:685-691 (1971).
41. J. Singh and F. J. Wiebel, A highly sensitive and rapid fluorometric assay for UDP-glucuronyl transferase using 3-hydroxybenzo[a]-pyrene as substrate, Anal. Biochem. 98:394-401 (1979).
42. O. H. Lowry, N.J. Rosebrough, A. L. Farr and R. J. Randall, Protein measurement with the Folin phenol reagent, J. Biol. Chem. 193:265-275 (1951).
43. S. Brown, H. K. Oie, A. F. Gazdar, J. D. Minna and U. Francke, Requirement of human chromosomes 19, 6 and possibly 3 for infection of hamster x human hybrid cells with baboon M7 type C virus, Cell 81:135-143 (1979).
44. S. Brown, P. A. Lalley and J. D. Minna, Assignment of the gene for peptidase S to chromosome 4 in man and confirmation of peptidase D assignment to chromosome 19, Cytogenet. Cell Genet. 22:167-171 (1978).
45. G. Kellermann, M. Luyten-Kellermann, J. R. Jett, H. L. Moses and R. S. Fontana, Aryl hydrocarbon hydroxylase in man and lung cancer, Human Genet. Suppl., 1:161-168 (1978).
46. S. Brown and D. E. Chalmers, Microsomal epoxide hydrolase activity in human x mouse hybrid cells, Biochem. Biophys. Res. Comm. 137:775-780 (1986).

APPLICATION OF CELL AND TISSUE CULTURE IN THE PHARMACEUTICAL INDUSTRY?

D.C. Taylor

Formulation Research Section
ICI PLC
Mereside, Alderley Park, Macclesfield
Cheshire SK10 4TG, England

This chapter is an invited critique of cell and tissue culture methodologies presented during the meeting from an industrial researcher who had used predominantly *in vivo* methodologies.

Introduction

The discovery and development of the new chemical entities comprising modern medicines is a complex, technical and highly structured task which owes little to the serendipity and random screening processes of twenty years ago. It is now necessary to design drugs to achieve exacting standards of potency, specificity, biodistribution and lack of toxicity. The definition and understanding of drug transport processes is a key factor in this process both in terms of the discovery of new chemical entities with appropriate absorption, distribution and elimination characteristics and in the design of drug delivery systems which optimize and control the effects of these molecules. There is therefore a major requirement within the industry for experimental techniques which facilitate the investigation of drug transport and which will allow the building up of predictive correlations between molecular structure and transport characteristics.

Some of the potential applications of cell culture techniques in the study of drug transport have already been outlined by Professor Borchardt in the opening pages of this volume, i.e.:

i) For screening permeability properties of potential drug candidates
ii) For elucidating transcellular transport mechanisms
iii) For evaluating potential cellular sites of metabolism
iv) For testing strategies to prevent metabolism and/or enhance membrane permeability.

No one involved in research in drug and delivery system design could fail to be attracted by realistic *in vitro* models which offered such possibilities. This chapter therefore attempts to assess whether cell and tissue culture techniques live up to this promise or will do so in future. It is written in the light of the impressive research in cell culture presented in the foregoing chapters of this book and represents the viewpoint of an author who has spent several years in drug and dosage form design in the pharmaceutical industry. It attempts to set out the ways in which models of drug transport are used in the design of new drugs and delivery systems, to examine the strengths and shortcomings of existing models and to explore the potential value of cell culture techniques and the advantages they may offer.

The requirement for models of drug transport

The need to study drug transport as part of a process of rational design of drugs and delivery systems is referred to above. Specifically, this need arises in two broad areas of research in the pharmaceutical industry. Firstly, it arises in the relatively short-term integrated drug discovery programmes which are set up to find new chemical entities with defined pharmacological characteristics and which have potential for use in the treatment of specific clinical conditions. Secondly, it arises in longer-term basic research which is carried out to define new strategies for drug design and delivery.

1.Application in drug discovery

Integrated programmes of synthetic chemistry and pharmacological testing are at the core of the process of new drug discovery in the pharmaceutical industry. In a typical case the objective may be to find a new chemical entity which meets specific requirements of pharmacological specificity and potency, but which is also orally active. The process involved may be along the lines shown in Fig. 1. It would consist of a programme of synthetic medicinal chemistry based on exploring structural variations around a 'lead' molecule (or molecules) which shows some of the required pharmacological characteristics. It would typically involve a programme of intensive activity over a period of months or years in which hundreds or thousands of experimental compounds are synthesized and tested. Compounds would firstly be examined for intrinsic pharmacological activity in *in vitro* screens (for example, inhibition of a specific enzyme or effect on an isolated tissue) and then selected for *in vivo* testing by oral administration, usually in a rodent species. A common scenario is that compounds showing good intrinsic pharmacological characteristics are inactive after oral administration, due to poor absorption or to rapid clearance. The current understanding of these processes is not sufficient to allow the ab initio design of molecules with the appropriate characteristics. The finding of poor oral activity would therefore activate a 'feedback loop' which involved investigation of individual compounds to determine the pharmacokinetic/metabolic reasons for oral inactivity and to propose synthetic strategies for overcoming these difficulties. The quality and speed of such feedback, and the ultimate success of the whole programme, depends upon the experimental models available for this exercise. Ideally, it requires rapid and realistic models which can be used to evaluate specific compounds and to establish quantitative relationships between molecular structure and absorption/metabolism. Clearly, speed and accuracy of feedback is essential in this type of application.

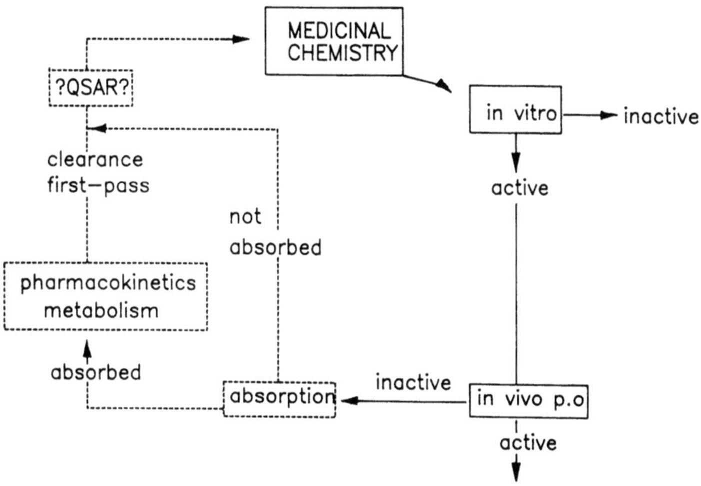

Figure 1. Search for an orally-active compound.

2. Application in basic drug delivery research

There is now a great deal of longer-term basic research aimed at identifying and exploiting new approaches to drug delivery. This is a response to the need to achieve greater efficiency and specificity of drug action and also to the problem of delivering some of the complex macromolecules now being made available through biotechnology.

For example, there is great interest throughout the pharmaceutical industry in new strategies for achieving the gastrointestinal absorption of peptides and other macromolecules. Clearly, effective methods for the systemic delivery of such molecules by oral administration would find widespread clinical usage in preference to most of the currently used methods of delivery. Some of the research approaches being taken include attempts to exploit the peptide transporter and bile acid transporter pathways in the intestinal membrane, and also to inhibit the active efflux pathways which may work against the systemic absorption of some molecules. However, this is a difficult area of research and one which requires good experimental models for the multiplicity of metabolic and facilitated transport processes which may be involved.

A major theme of current research in this field is the search for strategies to achieve tissue- or cell-selective delivery of drugs. This requires improved understanding of the mechanisms governing the transport of drug and carrier molecules across endothelia. For instance, there is much interest in the design of drugs which achieve access to the brain via the blood-brain barrier, either by passive diffusion or by utilizing one of the active transport processes present in the endothelial cells; this may require the design of specific prodrugs. In addition, the study of the extravasation of macromolecules across peripheral epithelia is a critical step in the design of drug-ligand complexes to achieve tissue- or cell-specific delivery.

There is also a great deal of current research on particulate/colloidal systems for targeting drugs to specific tissues and this requires detailed study of the interaction of these particles with the various cellular elements such as macrophages and Kupffer cells which are instrumental in the uptake and clearance of particles from the systemic circulation.

Shortcomings in existing models of drug transport

Most studies on drug transport are currently conducted using *in vivo* techniques in laboratory animals. For example, an *in situ* gut loop method in the anaesthetised rat, based on the technique originally described by Doluisio et al (1) is widely used to measure intestinal permeability to drugs. *In vivo/in situ* models have also been described for various other epithelia of interest in drug delivery (e.g., buccal, nasal membranes) and there has been a substantial amount of research on drug transport across the blood-brain barrier using *in vivo* preparations allowing access to the cerebral capillary endothelia (2-4).

Many of the *in vivo* and in situ models described above have been extensively studied over several years and have been developed into accurate and reliable methods which can give reasonably realistic assessment of the *in vivo* transport characteristics of the molecule of interest. However, they suffer a number of serious shortcomings which make the search for better models a priority. The models are invariably slow and inflexible and are difficult and expensive to run. Extensive interspecies differences in transport processes also pose a fundamental problem and often make it extremely difficult to use animal measurements to predict what will happen in the target species (man). Above all, there is the ethical question of whether it it acceptable to use live animals for research.

Alternatives to animal models have been extensively studied. *In vitro* models using isolated biological membranes have been widely discarded because of difficulties in isolation and keeping the membranes viable outside the animal, and because of the consequent poor correlation with drug transport *in vivo*. A notable exception to this is the widespread use of isolated human stratum corneum *in vitro* for studying the transdermal penetration of drugs. In this case, a more realistic *in vitro* situation is possible because of the unique nature of the stratum corneum.

Attempts have also been made to use physicochemical parameters to predict *in vivo* absorption, thus avoiding the need for animal measurements. This is clearly only theoretically feasible for passive transport processes and not for those which involve active or receptor-mediated mechanisms. However, even for passive processes, correlations have been shown in most cases to be poor and not capable for use for predictive purposes. For instance, the correlations between partition coefficient and drug transport in the small intestine (5,6) demonstrate only broad trends and are not precise and quantitative relationships. A similar pattern has been observed for drug transport across the blood-brain barrier (7,8). These observations show that the processes involved are more complex than the simple pH-partition mechanisms originally envisaged. Even when other factors such as convective absorption (solvent drug) and molecular weight effects on trans-membrane diffusion are taken into account, current evidence suggests that theoretical models based on simple physicochemical parameters may only allow the prediction of broad trends and that direct measurements on living tissue will still be required for detailed information on drug transport.*In vivo* models are therefore widely used because the *in vitro* alternatives have so far proved to be largely inadequate. Consequently, there is still a great deal of scope for the development and use of new model systems which overcome the disadvantages of existing models and which avoid the reliance on animal experimentation. There is an important future role for cell and tissue culture systems in the study of drug transport if they can be shown to meet the appropriate experimental criteria.

The potential offered by cell culture techniques

Cell and tissue culture techniques apparently offer immense potential for the study of drug transport. They allow the use of live human epithelial and endothelia cells in relatively simple *in vitro* systems and consequently may avoid many of the serious shortcomings outlined above for existing experimental models. In addition to avoiding the use of live animals, *in vitro* methods using cultured membranes are potentially much simpler, more rapid and less expensive to use. They are also experimentally more flexible systems, allowing access to both sides of the membrane, which is important in the study and definition of transport processes. The use of human cells (or cells from a non-human target species) also removes the problems of interspecies differences which continually occur with the use of animal models.

A further attraction of culture techniques is the range of tissues and membranes which are potentially accessible, many of which are difficult to study by other techniques. A compilation of the various cell lines discussed by previous authors in this book gives an impressive list of potential systems with immediate relevance to the study of drug transport:

Epithelia: There has clearly been a considerable amount of research on intestinal cell lines, particularly on the Caco-2 system, which may be of value in the study of intestinal transport processes. Work to characterize these systems has addressed passive drug transport, carrier-mediated systems for bile acids and amino acids, endocytic/vesicular pathways for vitamin B12 and other macromolecules, and intracellular metabolism by various routes. Buccal, nasal and skin epithelial systems have also been researched, all of which are of interest as potential sites of drug entry into the body.

Endothelia: Systems have been reported for the culture of the cerebral vascular membranes which constitute the blood-brain barrier and for the peripheral vasculature, both of which are of obvious interest in the study of drug distribution and extravasation.

Haematopoietic system: There has been work on the culture of cells of the haematopoietic system (i.e., monocytes and macrophages) which are of importance in the context of particulate delivery systems.

The amount of work which has been carried out to characterize these systems varies. Some of them are at a very early stage of development and there is still much to learn even about systems such as the intestinal Caco-2 cell line which has received a lot of attention. Nevertheless, these systems offer a great deal of potential for use as *in vitro* models of a whole range of transport processes which are important in the design of drugs and delivery systems. Clearly, this is the reason for the current intense interest and investment in these techniques in industry and acadaemia.

Do they meet the requirements for drug transport models?

The answers to two fundamental questions will determine whether cell culture systems realize their potential as models for the study of drug transport:

i) Are these systems practical in terms of their ease, speed, reliability and cost of operation?

ii) Is the correlation between *in vitro* and *in vivo* transport processes sufficiently reliable that cell culture systems can be used to predict *in vivo* events with an acceptable degree of accuracy?

The first of these is answered relatively easily. Clearly, the establishment of a new culture system or cell line is a significant research undertaking and the work involved in culturing and characterization is substantial. Setting up an established cell line is a different matter; it is often reasonably straightforward and can usually be accomplished successfully. Nevertheless, to run a cell line such as Caco-2 on a continual basis for use as a drug transport model does require considerable resources in terms of time, experimental facilities, staff and materials running costs. Overall, it probably demands an investment similar to that for running an *in vivo* animal model on a regular basis. This in itself would present few problems for most industrial and academic laboratories and these, and even more substantial, costs would be readily acceptable if they provided realistic models for drug transport which were an adequate alternative to existing *in vivo* techniques.

The second is the most important and fundamental question. It has been raised on numerous occasions in the literature and at various points in this book, but the number of cases in which quantitative *in vitro* - *in vivo* correlation has been attempted for transport characteristics is disappointingly small. This is no doubt due to the considerable time and effort required and, in many cases, to the paucity of good *in vivo* data for comparison. Nevertheless, it is clear from the foregoing chapters of this book that there may be significant biochemical and histological differences between cultured cells and the target cells *in vivo* and it is clearly not safe to assume that they will automatically show identical transport characteristics. This applies to both transformed cell lines such as Caco-2 and to primary cell cultures such as those described in previous chapters for the endothelial cells of the blood-brain barrier and for skin keratinocytes.

In fact, the small amount of data that is available indicates that the relationship between drug transport rates *in vitro* and *in vivo* is not straightforward. This is illustrated by the excellent work of Artursson et al. in this volume, which was aimed at establishing the validity of the Caco-2 cell culture system as a model for the passive transport of low molecular weight drugs in the intestinal tract, using a series of beta-blockers. Absorption rate constants measured in the *in vitro* system are compared with values obtained from a rat in situ intestinal model, which is known to be a good predictor of beta-blocker absorption in man. The results show, first of all, that *in vitro* rate constants are approximately an order of magnitude smaller than *in vivo* rates, which may be due to surface area effects. More importantly, there is a discontinuity across the series, with the hydrophilic beta-blockers atenolol and practolol being relatively much more poorly absorbed *in vitro*; in fact, they are virtually unabsorbed in this model.

This reason for this effect is clear from recent work on the mechanism of the intestinal absorption of these molecules (9). This shows that the more lipophilic beta-blockers are absorbed predominantly by a process of pH-partition via the mucosal cells.

However, this is an extremely slow process for the hydrophilic members of the group, and their absorption is almost exclusively by a process of convective absorption (or 'solvent drag') via aqueous channels in the intestinal membrane. This is a relatively inefficient process for molecules of this size and absorption is consequently slow compared to those molecules absorbed by pH-partition. It is also site-dependent, occurring in the small intestine and not in the large intestine. The Caco-2 system apparently shows beta-blocker absorption by the first of these mechanisms but not by the second, suggesting that it seriously underestimates the paracellular permeability of the rat and human small intestine and is, in fact, more characteristic of the large intestine.

Concerns about the paracellular permeability of cultured membranes have also arisen in work with brain capillary endothelium, where some workers have reported much higher permeabilities *in vitro* than *in vivo*. Borchardt in this volume, for instance, reports permeabilities for inulin and sucrose which are orders of magnitude higher than comparative *in vivo* data. Such effects are apparently due to the formation of membranes where the cells are less closely packed than *in vivo*. Conversely, the work of Van Bree in this volume, with cultured bovine brain endothelial cells suggest that such systems can give reasonable indication of blood-brain barrier permeability *in vivo*. Such observations emphasize that it is very important, at the outset, to characterize culture systems specifically in terms of the transport processes they are going to be used to study, to ensure that they are indeed valid models.

Conclusions

Cell culture systems clearly have a great deal of potential for use in the study of drug transport processes in the pharmaceutical industry and, in time, it is likely that these will allow the replacement of many of the existing *in vivo* models with good *in vitro* alternatives. However, there are legitimate concerns about the *in vivo* validity of many of the current systems which must be resolved if this potential is to be realized. Clearly, this emphasizes the need for further research to characterize cell lines in terms of their permeability characteristics and for the development of new culture systems which may have more appropriate characteristics.

KEY ISSUES IN THE USE OF CELL AND TISSUE CULTURE SYSTEMS

G. Wilson*, S.S. Davis+, L. Illum+, A. Zweibaum++

*SmithKline Beecham Pharmaceuticals, 709 Swedeland Road, King of Prussia, PA 19406, USA, +Pharmacy Department, University of Nottingham, University Park, Nottingham, England, ++ INSERM U178, 94807 Villejuif Cedex, France

This chapter is based on a panel discussion held at the end of the meeting to debate a number of key issues that had been raised in individual presentations.

CHOICE OF CULTURE SYSTEM

It was agreed that cell culture systems were already useful in providing essential information on aspects of drug metabolism and toxicity. The confluent monolayer system could also be used to probe transport mechanisms, especially those that involved vesicular trafficking and paracellular routes. The multiplicity of phases in drug delivery and the barriers to the transport of drugs was emphasized. Passage across epithelial or endothelial barrier, such as the systems discussed at this meeting are only one of a number of possible rate limiting events. In some situations, a confluent cell layer could be used to address questions related to barrier function but the choice of an appropriate cell line is essential. In all cell culture studies it is clear that the choice of systems should be dictated by the aim of the study. While the use of various systems had gained widespread acceptance for basic cellular and molecular biology studies, their relevance to specific questions relating to the transport of drug molecules had not yet received proper attention. Thus there was a lack of data to correlate the results from some systems with *in vivo* data, and to indicate their relevance and advantages over using physicochemical parameters to predict absorption. In addition there was also a danger of making cell culture systems too sophisticated in relation to their use. For example a number of simple confluent systems (e.g. MDCK cells) would probably be as useful for measuring the passive absorption of molecules. The conclusion was to choose the cell or tissue culture system for the correct purpose according to the question to be answered. The many technical limitations of monolayer systems were highlighted; however, advances in the use of three dimensional systems and co-culture techniques were starting to address some of these problems.

Pharmaceutical Applications of Cell and Tissue Culture to Drug Transport
Edited by G. Wilson *et al.*, Plenum Press, New York, 1991

Most discussion centered around the use of cultured gastrointestinal and skin cells.

The difference between intestinal lines in widespread use such as Caco-2, HT29 and T84 were discussed. Caco-2 cells were becoming the most widely used as they have properties similar to differentiated a number of small intestinal cells and are easy to culture. In contrast T84 cells do not have brush borders. HT29 cells can be induced to differentiate and produce different cell types, including goblet cells. Variabilities of the properties (e.g. electrical resistance) of Caco-2 cells grown by the different participants at the meeting were highlighted. Some participants were prepared to accept that Caco-2 cells had many characteristics of the distal ileum. Others pointed to the high electrical resistance properties as being more representative of the colon. Thus, Caco-2 cells appear to have hybrid properties of both the ileum and colon. Zweibaum emphasized that Caco-2 was a transformed cell line derived from a colon tumor, has a number of properties characteristics of transformed cell lines and thus should be treated as such. Other features of the small intestinal epithelium not represented in Caco-2 systems were the presence of M-cells, that could have an important role in the uptake of macromolecules and particles, and the presence of a mucin layer, even though attempts a co-culturing goblet cells and enterocytes were in progress. It was generally agreed that there were a number of deficiencies with cell culture systems and that they should not be generally considered as models of the normal intestinal mucosa. Their use for elucidating basic cell and molecular biological mechanisms was endorsed. The potential use for transport studies was more controversial. Although it was accepted that there were a number of clear differences between Caco-2 cells and normal ileal epithelial cells, many considered that this cell line was the best available for performing transport studies and may have some predictive value, particularly as the culture of normal intestinal epithelial cells, suitable for transport studies, has yet to be achieved.

Skin cultures were considered valuable as models to replace cadaver or surgical specimens. It was somewhat disappointing that keratinocyte systems with an intact stratum corneum were providing so difficult to culture but, nevertheless, available systems could be used as tools to follow the secretion of mediators that were markers of irritation or an immune response. The viability of skin models was believed to have special relevance to studies on environmental hazards where biotransformation (inactivation, activation) could be a critical factor. Results from the use of keratinocyte cultures for drug transport studies should be interpreted with care as such systems may not have the full compliment of cell types or skin lipids which may have important roles in transport processes.

While much progress has been made in the growth of bovine brain endothelial cells their permeability properties (e.g. low electrical resistance, high permeability to some molecular probes) still showed a number of fundamental differences to the intact barrier *in vivo*. There was considerable discussion on new approaches using co-cultures with astroglial cells or conditioned media to create a more physiological *in vitro* barrier. In addition the relevance of creating normal culture systems in relation to studies on transport of drugs into the brain during disease processes was questioned. The culturing of buccal and nasal cells (particularly those of human origin) was currently difficult but further progress was to be expected in the future.

CELL AND TISSUE CULTURES VS ANIMAL MODELS

A number of industrial participants were of the opinion that cell culture systems could be used to replace animals in the screening of drugs for their permeability properties. Others believed that presently available *in situ* models were just as predictive and were much easier to use and were less costly in terms of time and money. In addition isolated normal tissues that could be mounted in diffusion chambers also provided useful systems to study drug transport. These systems may have a number of significant advantages (normal phenotype, full metabolic capability) over transformed cell lines. All shared the view that it would be an advantage to use cell culture models where possible, especially for toxicity and irritancy testing. While it was agreed that an adequate database has not yet been developed, efforts to obtain good *in vitro-in vivo* correlations were viewed with some skepticism. However, from preliminary data, it was agreed that *in vitro* cell and tissue culture models might provide good predictive power. There was, however, a consensus that future studies on drug absorption will probably not rely solely on *in vitro* methodologies but will require a combination of *in vitro* and *in vivo* systems.

In conclusion, it was generally agreed that confluent layers of cells and tissues are already having an important role in fundamental aspects of drug transport studies and are being used by an increasing number of groups to answer specific questions regarding mechanisms and pathways. In some cases a well defined and easily grown and maintained system could be used to predict the permeability of a range of molecules in drug screening exercise. As such cell culture models might help to reduce the number of animal experiments presently being performed. As this whole area is in its infancy, close collaborations between the pharmaceutical industry and academic groups are highly desirable and already taking place. The hope that all useful data would be shared and published was endorsed. There was however, already evidence that the information on the use of confluent layers of cells to predict transport across epithelial and endothelial barriers, had proprietary value and was not being made generally available to the research community. In view of the rapid pace of progress in the area of cell and tissue culture systems an update meeting in 2-3 years time was considered desirable. In particular the need to examine specific transport pathways and mechanisms in more detail was expressed.

REFERENCES

1. Doluisio, J.T., Billups, N.F., Dittert, L.W., Sugita, E.T., Swintosky, J.V., J. Pharm. Sci. 1969, 58, 1196-1200.

2. Crone, C., 1963, The permeability of capillaries in various organs as determined by the use of the indicator dilution method. Acta Physiol. Scand. 58:292-305.

3. Fenstermacher, J.D., Blasberg, R.G., and Patlak, C.S., 1981, Methods for quantifying the transport of drugs across the blood-brain systems. Pharmac. Ther. 14: 217-248.

4. Oldendorf, W.H., 1974, Blood-brain barrier permeability to drugs. Ann.Rev.Pharmacol. 14:239-248.

5. Martin, Y.C., 1981, A practitioner's perspective of the role of quantitative structure-activity analysis in medicinal chemistry, J. Med. Chem. 24, 229-237.

6. Taylor, D.C., Pownall, R., and Burke, W.J., 1985, The absorption of β-adreno receptors antagonist in the rat *in situ* small intestine; the effect of lipophilicity, J. Pharm. Pharmacol. 37, 280-283.

7. Rapoport, S.I., Ohno, K. and Pettigrew, K.D., 1979, Drug entry into the brain. Brain Res. 172: 354-359.

8. Levin, V.A., 1989, Relationship of octanol/water partition coefficient and molecular weight of rat brain capillary permeability. J. Med. Chem. 23: 682-684.

9. Taylor, D.C., Lynch, J. and Leahy, D.E., 1990, Models for intestinal permeability to drugs, In Drug Delivery to the Gastrointestinal Tract (J.G. Hardy, S.S. Davis and C.G. Wilson, eds), pp. 133-145, Ellis Horwood Ltd., Chichester, U.K.

ADVANCED RESEARCH WORKSHOP ON
PHARMACEUTICAL APPLICATIONS OF CELL AND TISSUE CULTURE
September 4-9, 1989
Bandol, France

PARTICIPANTS

Dr. Veronique Andrie, Synthelabo Recherche, 34 Avenue L. Jouboux, 92160 Antony Z.I., France

Dr. P. Artursson, Department of Pharmaceutics, Uppsala University, Box 580, S-751, 23 Uppsala, Sweden

Dr. Ken Audus, Department of Pharmaceutical Chemistry, The University of Kansas, 3006 Malott Hall, Lawrence, Kansas 66045-2504 USA

Dr. Ronda Bartel, Marrow-Tech, 10933 North Torrey Pines Road, La Jolla, CA 92037 USA

Dr. Eugene Bell, Organogenesis, 83 Rogers Street, Cambridge, MA 02142 USA

Dr. Bert de Boer, Center for Bio-Pharmaceutical Sciences, P. O. Box 9503, 2300 RA Leiden The Netherlands

Professor Ronald T. Borchardt, Summerfield Professor and Chairman, Department of Pharmaceutical Chemistry, The University of Kansas, 3006 Malott Hall, Lawrence, Kansas 66045-2504 USA

Dr. Richard C. Boucher, University of North Carolina School of Medicine, 724 Burnett-Womack Building CB 7020, Chapel Hill, NC 27599-7020

Dr. Dirk Breitkrutz, German Cancer Research Centre, D-6900, Heidelberg, Federal Republic of Germany

Dr. S. Brown, Department of Biochemistry, Queen's Medical Centre, Nottingham, UK

Dr. R. Cecchelli, Institut Pasteur, Service de Recherche Sur Les Lipoproteines, et L'atherosclerose (Serlia), 1 rue du Professeur Calmette, B.P. 245, 59019 Lille Cedex, France

Ms. Guillemette Chevalier INSERM U178, Batiment INSERM, 16 Avenue Paul Vaillant Couturier, 94807 Villejuif Cedex, Paris, France

Professor S.S. Davis, Department of Pharmaceutical Sciences, University of Nottingham, University Park, Nottingham, UK

Dr. George Delaconstantinos, University of Athens Medical School, Department of Experimental Physiology, GR-11527 Athens, Greece

Dr. A. Dendorfer, Department of Physiology, University of Munich, Pettenkoferstr. 12, D-8000 Muchen 2, West Germany, Dr. Gerhard Scriba, Institut fur Pharmaceutische Chemie, Universitat Munster, Hittorfstrasse 58-62, 4400 Munster, West Germany

Dr. G. Fabre, Sanofi Recherche, Sanofi Recherche Rue du Pr. J. Blayac, 34082 Montpellier Cede, France

Dr. Jacques Frances, Pharmaceutical Forms Manager, SIPSY, Locataire-Gerant de Jouveinal S.A., 7 Allee des Jacheres, Cidex 423-94263 Fresnes Cedex, France

Dr. G. Fricke, Sandoz Ltd., Pharma Division, CH-4002 Basle, Switzerland

Dr. J.F. Ghersi-Egea, Centre du Medicament, Faculte des Sciences Pharmaceutiques et Biologiques, Universite de Nancy - I, 30 Rue Lionnois, B.P. 403, F-54001 Nancy Cedex, France

Dr. I. Hassan, Ciba Geigy Pharmaceuticals, Wimblehurst Road, Horsham, West Sussex, England

Dr. Jerome Hochman, InterX Corporation, 2201 West 21st Street Lawrence, KS 66044 USA

Professor C. Hopkins, Department of Biochemistry, Imperial College of Science and Technology, Imperial College Road, London SW7 2AZ, England

Dr. Christian Huet, Unite de Biologie des Membranes, Department de Biologie Moleculaire, Institut Pasteur, 25 Rue de Roux, 75724 Paris Cedex 15, France

Dr. L. Illum, Department of Pharmaceutical Sciences, University of Nottingham, University Park, Nottingham, UK

Dr. Arne Jepsen, Tissue Culture Laboratory, Department of Oral Pathology, Royal Dental School, Aarhus 8000, Denmark

Dr. M. Kiremitci, Hacetepe University, Chemical Engineering Department, Brytepe, Ankara, Turkey

Dr. Richard Kirsh, SmithKline Beecham Pharmaceuticals, P. O. Box 1539, King of Prussia, PA 19406-0939 USA

Dr. C. Laboisse, Laboratorie de Biologie et de Physiologie des Cellules, Digestives, U239 INSERM, Faculte Xavier Bichat, 16 Rue Huchard, 75018 Paris, France

Ms. Elizabeth Launay, INSERM U178, Batiment INSERM, 16 Avenue Paul Vaillant Couturier, 94807 Villejuif Cedex, Paris, France

Dr. Hans Leuenberger, Institute of Pharmacy, CH-4051, Basel, Switzerland

Dr. Daniel Louvard, Unite de Biologie des Membranes, Department de Biologie Moleculaire, Institut Pasteur, 25 Rue de Dr. Roux, 75724 Paris Cedex 15, France

Professor Chris Marriott, Pharmaceutical Sciences Research Group, Department of Pharmacy, Brighton Polytechnic, Crockcroft Building, Moulescomb, Brighton, West Sussex BN2 4GJ, England

Dr. L. V. Mayne, School of Biological Sciences, University of Sussex, Falmer, Brighton, Sussex BN1 9QG, England

Ms. Joanne Miller, SmithKline Beecham Pharmaceuticals, P.O. Box 1539, King of Prussia, PA 19406-0939, USA

Dr. K. Miller, Sandoz Ltd., Pharma Division, CH-4002 Basle, Switzerland

Ms. Corinne Moureau, Millipore S.A., BP 307, 78054 St. Quentin Yvelines Cedex, France

Dr. Gail K. Naughton, Marrow-Tech, 10933 North Torrey Pines Road, La Jolla, CA 92037, USA

Dr. Thomas Raub, Drug Delivery Systems Research Unit, The Upjohn Company, Kalamazoo, MI 49001, USA

Dr. Jean Francois Renaud de la Faverie, Laboratories Fondax (Servier), 7 Rue Ampere, 92800 Puteaux, France

Dr. A. Rogerson, Reckitt & Coleman Pharmaceutical Division, Dansom Lane, HULL HU8 5DS, England

Dr. Michele Rommain, Laboratories Cassenne, 17 Rue de Pontoise, Osny 95520, Paris, France

Dr. Gerrit Scherphof, Laboratory of Physical Chemistry, State University, 9712 KZ Groningen, The Netherlands

Dr. Birgit Sehested, Department of In Vitro Biology, Novo-Nordisk, 1-Niels-Steensensvej, 2820 Gentofte, Denmark

Dr. Chick Shasteen, Monsanto Company, 700 Chesterfield Village Parkway, St. Louis, MO 63198 USA

Dr. Nuria Suesa, R&D Departments of Biochemistry, Laboratorios Menarini S.A., Alfonso XII, 587, 08912 Barcelona, Spain

Dr. David Taylor, Formulation Research Section, Pharmaceutical Department, ICI PLC, Mereside, Alderley Park, Macclesfield, Cheshire SK10 4TG, England

Dr. N. Testa, Paterson Institut for Cancer Research, Christie Hospital & Holt Tadium Institute, Wilmslow Road, Manchester M20 9BX, England

Dr. Joost van Bree, Sandoz AG, CH-002 Basel, Switzerland

Dr. Andrew Walker, Alcan International, Southam Road, Banbury, Oxon OX16 7SP, England

Dr. Fiona Watt, Imperial Cancer Research Fund, P. O. Box 123, Lincoln's Inn Fields, London WC2A 3PX, England

Dr. Glynn Wilson, SmithKline Beecham Pharmaceuticals, P. O. Box 1539, King of Prussia, PA 19406-0939, USA

Professor Alain Zweibaum, INSERM U178, Batiment INSERM, 16 Avenue Paul Vaillant Couturier, 94807 Villejuif Cedex, Paris, France

INDEX

MIX
Papier aus verantwortungsvollen Quellen
Paper from responsible sources
FSC® C105338

If you have any concerns about our products,
you can contact us on
ProductSafety@springernature.com

In case Publisher is established outside the EU,
the EU authorized representative is:
Springer Nature Customer Service Center GmbH
Europaplatz 3, 69115 Heidelberg, Germany

Printed by Libri Plureos GmbH
in Hamburg, Germany